Gianfranco Capriz • Guido Stampacchia (Eds.)

New Variational Techniques in Mathematical Physics

Lectures given at a Summer School of the
Centro Internazionale Matematico Estivo (C.I.M.E.),
held in Bressanone (Bolzano), Italy,
June 17-26, 1973

Springer

FONDAZIONE
CIME
ROBERTO CONTI

C.I.M.E. Foundation
c/o Dipartimento di Matematica "U. Dini"
Viale Morgagni n. 67/a
50134 Firenze
Italy
cime@math.unifi.it

ISBN 978-3-642-10958-4 e-ISBN: 978-3-642-10960-7
DOI:10.1007/978-3-642-10960-7
Springer Heidelberg Dordrecht London New York

Reprint of the 1st ed. C.I.M.E., Ed. Cremonese, Roma 1974
With kind permission of C.I.M.E.

Printed on acid-free paper

Springer.com

CENTRO INTERNAZIONALE MATEMATICO ESTIVO
(C.I.M.E.)

II Ciclo - Bressanone - dal 17 al 26 giugno 1973

« NEW VARIATIONAL TECHNIQUES IN MATHEMATICAL PHYSICS »

Coordinatori: Proff. G. Capriz, G. Stampacchia

CENTRO INTERNAZIONALE MATEMATICO ESTIVO

(C. I. M. E.)

PROBLÈMES A FRONTIÈRE LIBRE LIÉS A' QUESTIONS D'HYDRAULIQUE

CLAUDIO BAIOCCHI

Corso tenuto a Bressanone dal 17 al 26 giugno 1973

PROBLÈMES A' FRONTIÈRE LIBRE LIÉS A' QUESTIONS D'HYDRAULIQUE.

par C. Baiocchi - Université de Pavie et Laboratoire d'Analise Numérique du C.N.R. de Pavie.

N.1.- L'étude du mouvement des liquides à travers des matériaux poreux conduit en général à des "problèmes à frontière libre". Un cas typique peut être schématisé sous la forme suivante: sur une base imperméable deux bassins d'eau, de niveaux différents, sont en communication à travers une digue en matérieau poreux. L'eau filtre du niveau le plus élevé au niveau le moins élevé; et on veut déterminer la "partie mouillée" de la digue, ainsi que les grandeurs physiques (telles que la pression, la vitesse, le débit...) associées au mouvement.

On se bornera au cas plus simple (pour une description générale, ainsi que pour plus de détails sur le plan physique, ou consultera par exemple les textes 6 , 13 , 16 , 18); précisément on envisagera le cas correspondant à un fluxe stationnaire, irrotationnel, incompressible; le matérieau composant la digue est supposé isotrope, homogène et ne donnant pas lieu à des phénomènes de capillarité. On considèrera comme "Problème modèle" le cas où la digue est à base horizontale et à parois verticales planes et parallèles (la fig. 1 est

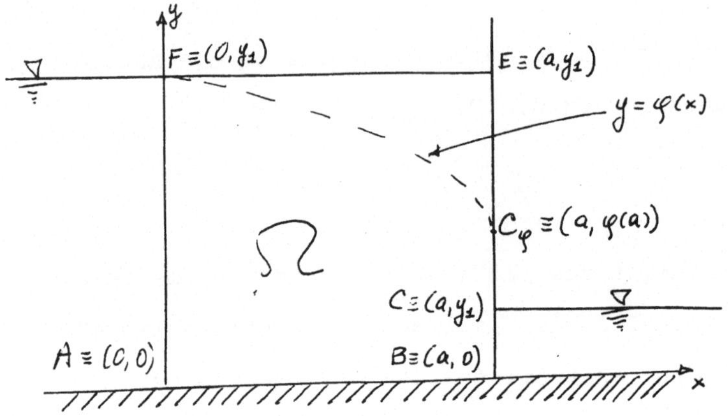

FIGURE N. 1

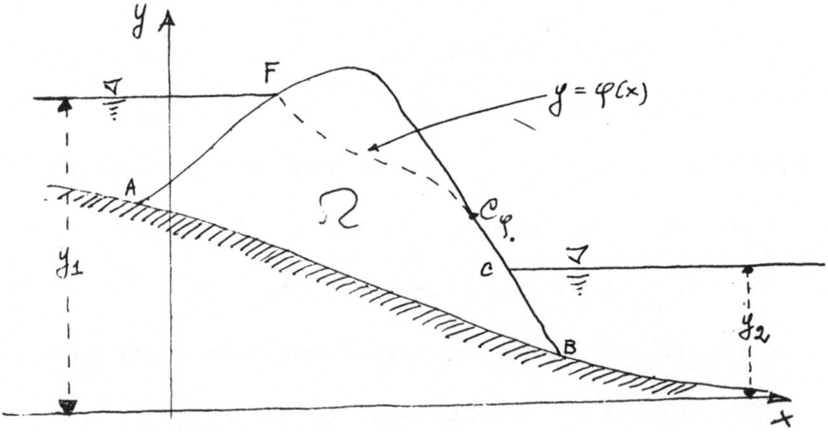

FIGURE N. 2

une section orthogonale aux parois: a est l'épaisseur de la digue, y_1 et y_2 sont les hauteurs des deux bassins).

Plus en général on désignera par D une section verticale de la digue (cf. fig. 2; dans la fig. 1 on a $D =]0,a[\times]0,y_1[$) et on supposera que, dans la direction orthogonale à la figure la digue est infiniment étendue et à section constante (de façon à étudier un problème bidimensionnel).

On désignera par Ω la "partie mouillée" de D; par $y=\varphi(x)$ l'équation du "bord supérieur" de Ω ; par $p(x,y)$, $\overrightarrow{V(x,y)}$ respectivement la pression et la vitesse de l'eau dans le point (x,y) de Ω (x axe horizontal, y axe vertical); par $u(x,y)$ la "hauteur piézométrique", à savoir:

$$u(x,y) = y + \frac{p(x,y)}{\gamma} \tag{1.1}$$

γ étant le poids spécifique du liquide. La loi de DARCY (cf. toujours les textes cités plus haut) assure que u est un "potentiel de vitesse", à savoir que l'on a:

$$\overrightarrow{V(x,y)} = -\tilde{k} \operatorname{grad} u \tag{1.2}$$

où $\tilde{k} = \frac{\gamma}{\mu} k$, μ étant la viscosité du liquide et k étant le coefficient de perméabilité ([1]).

([1]) Sous les hypothèses faites k (et donc $\tilde{k}$) est constant; plus en général k est une fonction de (x,y) si la digue n'est pas homogène, et un tenseur symétrique si la digue n'est pas isotrope.

L'incompressibilité du liquide et la loi de continuité donnent alors

(1.3) $\quad \text{div}\ \tilde{k}\ \text{grad}\ u = 0 \qquad \text{dans}\ \Omega$

(en particulier u est harmonique dans Ω si le matérieau est homogène isotrope).

À la relation (1.3) on doit ajouter des conditions aux limites. D'abord, le long des parties de $\partial\Omega$ qui sont des lignes de courant doit s'annuller la dérivée normale de u:

(1.4) $\quad \frac{\partial u}{\partial n} = 0 \qquad \text{sur}\ AB\ ;$

(1.5) $\quad \frac{\partial u}{\partial n} = 0 \qquad \text{sur}\ FC\ ;$

ensuite, le long des parties de $\partial\Omega$ à contact avec l'athmosphère on doit avoir $p(x,y)=0$ donc (cf. (1.1)) $u=y$:

(1.6) $\quad u(x,y) = y \qquad \text{sur}\ FC_\varphi\ ;$

(1.7) $\quad u(x,y) = y \qquad \text{sur}\ C_\varphi C\ ;$

finalement, le long des parois à contact avec les bassins, la pression est donnée par la pression de l'eau qui est en haut; (1.1) donne:

(1.8) $\quad u(x,y) = y_1 \qquad \text{sur}\ AF;$

(1.9) $\quad u(x,y) = y_2 \qquad \text{sur}\ BC.$

Il s'agit d'un classique problème à frontière libre; sur un domaine inconnu Ω on doit résoudre le problème aux limites (1.4), (1.5), (1.6), (1.7), (1.8), (1.9) pour l'équation (1.3): on a donc des conditions surhaboundantes (cf. (1.5) et (1.6)) sur la partie incomme $\partial\Omega \cap D$ de la frontière de Ω.

N.2.- Une des premières méthodes proposées dans la litérature speécialisée pour la resolution du problème (1.3),....(1.9) est basée sur la théorie des fonctions de variable complexe et s'appuie sur la transformation:

(2.1) $x+iy \rightarrow p+iq \; ; \quad p = -\frac{\partial u}{\partial x}, \quad q = \frac{\partial u}{\partial y}$

(transformation qui est conforme si le coefficient $\tilde{k}$ dans (1.3) est constant; (p,q) est le "plan de l'odographe"). Par exemple (pour plus de détails et de généralité cf. 16 , 18) dans le cas du Problème modèle illustré en fig. 1 le domaine Ω se transforme en un domaine Ω' du plan (p,q), dont le bord $\partial\Omega'$ est parfaitement connu et, sauf pour ce qui concerne la position sur $\partial\Omega'$ des points A',B' transformés de A,B,Ω' est indépendent de a, y_1, y_2; cela fornit au problème deux dégrés de liberté (en accord avec ce qui se passe sur le plan (x,y), où l'on peut choisir comme paramètres $\frac{y_1}{a}$, $\frac{y_2}{a}$, tout étant invariant par homotéties). A partir de cette famille à deux paramètres de domains Ω' la transformation $p+iq \rightarrow x+iy$ inverse de (2.1) fornit une famille de solutions Ω . Toutefois les paramètres que l'on peut se donner à priori sont ceux du plan de l'odographe (et non $\frac{y_1}{a}$, $\frac{y_2}{a}$); on ne sait pas démontrer la biunivocité de la corrèspondence entre les paramètres physiques et ceux de l'odographe; et d'ailleurs on aurait bésoin de "beaucoup de régularité" pour justifier les passages du plan physique à celui de l'odographe et viceversa!

Une méthode plus récente ([2]) est basée sur les considérations suivantes (on se borne toujours, pour simplifier, au cas du Problème modèle). A toute courbe "régulière" $y=\varphi_o(x)$ on associe le "sous-graphe" $\Omega_o=\{(x,y)\,|\,0<x<a;\ 0<y<\varphi_o(x)\}$; et sur Ω_o on résoud dans l'inconnue u_o le problème mêlé correspondant à (1.3), (1.4), (1.7), (1.8), (1.9) et <u>une seulement</u> entre (1.5) et (1.6); puis on modifie φ_o de façon à remplire l'autre entre (1.5) et (1.6); et on itère le procédé. Par exemple si pour évaluer u_o on impose (1.5), on posera $\varphi_1(x)=u_o(x,\varphi_o(x))$; le problème sera résolu (à savoir (1.6) aussi sera vérifiée) si l'on a $\varphi_1\equiv\varphi_o$; donc le problème à frontière libre correspond à trouver les points fixes de la transformation $\varphi_o \to \varphi_1$. Si, au contraire, on avait choisi (1.6) pour la détermination de u_o, on cherchera à minimiser, par rapport à φ_o, une convenable norme de la trace sur $y=\varphi_o(x)$ de la dérivée normale de u_o; plus en général on pourrait minimiser, par rapport au triplet Ω_o, φ_o, u_o une

---

([2]) que l'on peut d'ailleurs appliquer à la resolution numérique d'une vaste classe de problèmes à frontière libre; cf. |11| pour une vue d'ensemble sur ces procédés.

fonctionelle du type:

$$\int_{\Omega_0} \frac{1}{2}\tilde{k}\ |\text{grad}\ u_0|^2 dx\, dy + \int_0^{y_1} [u_0(0,y)-y_1](u_0)_x(0,y)dy -$$
$$- \int_0^{y_2} |u_0(0,y)-y_2|(u_0)_x(a,y)dy - \int_{y_2}^{\varphi_0(a)} [u_0(a,y)-y](u_0)_x(a,y)dy$$
$$- \int_0^a [u_0(x,\varphi_0(x))-\varphi_0(x)]\ \frac{\partial u_0}{\partial \nu_{\varphi_0}}\ (x,\varphi_0(x))dx \qquad (^3).$$

Il s'agit de procédés qui peuvent être adaptés à la re solution numérique des problèmes envisagés ([4]) et qui, de ce point de vue, ont donné des résultats satisfaisants; toutefois, du point de vue théorique, on ne sait pas justifier ces procédés (par exemple on ne connait ni existence ni unicité de points fixes pour la transofrmation $\varphi_0 \to \varphi_1$; on ne connait pas l'unicité du point de minimum pour les fonctionnelles considérées, et on nesait pas si le minimum vaut zéro....).

N.3.- Par moyen d'un convénable changement de fonction inconnue j'ai donné en 1971 un théorème d'existence et unicité de la solution du Problème modèle, en ramenant ce problème à une inéquation variationnelle (cf. 1). Pour décrire ce résultat

---

([3]) Pour un traitement numérique basé sur ces idées cf. 17 .

([4]) Tout en rencontrant des difficultés de programmation non indifférentes: on doit resoudre une famille de problèmes mêlés sur des domaines qui, à chaque étape, varient en fonction de l'étape précédente.

il faut d'abord préciser le problème, en particulier pour ce qui concerne la régularité de la courbe $y=\varphi(x)$ et de la fonction $u(x,y)$ (de façon à donner un sens précis aux relations (1.3)...(1.9)). Pour faire ça il est commode ([5]) de faire usage, outre que du potentiel de vitesse u, de la "fonction de courant" $v(x,y)$ liée à u par:

(3.1) $$\frac{\partial u}{\partial x} = \frac{\partial v}{\partial y}\ ; \qquad \frac{\partial u}{\partial y} = -\frac{\partial v}{\partial x} \quad \text{dans} \quad \Omega\ (^6);$$

et de remplacer les condition de type Neumann sur u (cf. (1.4), (1.5)) par des conditions de constance sur v; v étant déterminée à une constante additive près, ou traduira (1.5) par:

(3.2) $$v=0 \qquad \text{sur} \qquad FC_\varphi$$

et (1.4) en imposant l'existence d'une constante q ([7]) telle que:

(3.3) $$v=q \qquad \text{sur} \qquad AB$$

Ceci étant, on appellera <u>solution faible du Problème modèle</u> une quintuple $\{\varphi, \Omega, u, v, q\}$ telle que:

(3.4) $\varphi : x \to \varphi(x)$ est continue de $[0,a]$ dans $]y_2, y_1]$, décroissante et telle que $\varphi(0)=y_1$.

([5]) Mais non indispensable; dans |1| on a travaillé en termes de u, sans introduire v; la présentation donnée ici suit l'exposé |2|.

([6]) A' savoir $x+iy \to u+iv$ est holomorphe dans Ω; l'existence d'un telle v équivaut à (1.3) lorsque $\hat{k}$ est constant.

([7]) à un coefficient dimensionnel près le paramètre q fournit le débit de la digue.

(3.5) $\Omega = \{(x,y) \mid 0<x<a;\ 0<y<\varphi(x)\}$

(3.6) $u,v \in C^0(\bar{\Omega}) \cap H^1(\Omega)$ (8); $q \in \mathbb{R}$

(3.7) $\left\{\begin{array}{l} u,v \text{ satisfont (3.1) au sens de } \mathcal{D}'(\Omega), \text{ et (1.6), (1.7),} \\ \text{(1.8), (1.9), (3.2), (3.3) au sens de } C^0(\bar{\Omega}) \end{array}\right.$

Remarque 3.1.- On va appeller faible une telle solution car, du point de vue physique, une solution sera acceptable seulement si elle satisfait d'autres relations, soit

qualitatives, soit quantitatives; par exemple la pression doit être positive dans Ω donc (cf. (1.1)):

(3.8) $u(x,y)>y$ dans Ω;

et d'ailleurs φ,u,v devraient être "plus régulières"; on obtiendra des propriétés de ce type comme conséquence de la définition de solution faible.

Remarque 3.2.- Pour ce qui concerne la valeur de q on obtiendra la formule explicite:

(3.9) $$q = \frac{y_1^2-y_2^2}{2a}$$

connue sous le nom de "formule de DUPUIT" et usuellement obtenue comme formule approchée (et déduite en supposant que la courbe y= (x) est une parabole).

Un'idée naturelle pour étudier le problème consiste à prolonger u,v à $\bar{D}$ tout entier ($\bar{D}$, fermeture de D, est $[0,a]\times[0,y_1]$) en posant:

(8) Notations usuelles: u,v sont des fonctions continues sur la fermeture $\bar{\Omega}$ de Ω et dont les dérivées distributionnelles sont de carré sommable sur Ω.

(3.10) $\tilde{u}(x,y)=\begin{cases} u(x,y) & \text{pour } (x,y)\in\overline{\Omega} \\ y & \text{pour } (x,y)\in\overline{D}\setminus\overline{\Omega}\end{cases}$; $\tilde{v}(x,y)=\begin{cases} v(x,y) & \text{pour } (x,y)\in\overline{\Omega} \\ 0 & \text{pour } (x,y)\in\overline{D}\setminus\overline{\Omega}\end{cases}$

et à voir si $\tilde{u},\tilde{v}$ satisfont un "problème bien posé" dans D; ça serait suffisant car, grâce au principe du maximum, on démontre aisément la validité de (3.8) et donc, connaissant $\tilde{u}$, on obtiendrait Ω en posant:

(3.11) $\Omega=\{(x,y)\mid (x,y)\in D\ ;\ u(x,y)>y\}.$

Toutefois ce n'est pas le cas: aucun problème au limites sur $\tilde{u},\tilde{v}$ ne semble être bien posé; il faut donc encore transformer le problème.

Remarquons maintenant que l'on a:

(3.12) $\tilde{u},\tilde{v}\in C^{0}(\overline{D})\cap H^{1}(D)$

et que de (3.1) on déduit:

(3.13) $(-\tilde{v})_{y}=(y-\tilde{u})_{x};\quad (-\tilde{v})_{x}+(y-\tilde{u})_{y}=\chi_{\Omega}$

où $(\ )_{y}$ et $(\ )_{x}$ désignent les dérivées partielles et χ_{Ω} est la fonction caractéristique de Ω . La première de (3.13) assure qu'il a un sens de considérer des intégrales curvilignes du type:

(3.14) $w(P)=\int_{F}^{P}-\tilde{v}\,dx+(y-\tilde{u})dy \qquad \forall P\in\overline{D}$

et l'on aura:

(3.15) $w_{x}=-\tilde{v}\ ;\qquad w_{y}=y-\tilde{u};$

donc, grace à (3.12) et à la deuxième de (3.13):

(3.16) $w\in C^{1}(\overline{D})\cap H^{2}(D)$

(3.17) $\Delta w = \chi_\Omega$ $\quad (\Delta = \frac{\partial^2}{\partial x^2} + \frac{\partial^2}{\partial y^2})$

Remarquons tout de suite que le valeurs de w sur ∂D sont connues: en effet on a (cf. (3.10), (3.15), (1.7), (1.8), (1.9)):

(3.18) $$\begin{cases} w = 0 \quad \text{dans } \overline{D}\setminus\overline{\Omega} \quad \text{(donc sur FE)} \\ w_y = y-\tilde{u} = y-y_1 \quad \text{sur FA} \; ; \quad w_y = y-\tilde{u} = 0 \quad \text{sur EC} \\ w_y = y-\tilde{u} = y-y_2 \quad \text{sur CB} \end{cases}$$

et d'après (3.15), (3.3):

(3.19) $w_x = -\tilde{v} = -q$ sur AB; donc $w_{xx} = 0$ sur AB;

donc si l'on définit g sur ∂D par la formule:

(3.20) $$\begin{cases} g = 0 \quad \text{sur} \quad FE \cup EC \; ; \quad g = \dfrac{(y-y_2)^2}{2} \quad \text{sur CB;} \\ g = \dfrac{(y-y_1)^2}{2} \quad \text{sur AF;} \quad g \text{ linéaire sur AB} \end{cases}$$

on aura nécessairement

(3.21) $w_{|\partial D} = g$

<u>Remarque</u> 3.3.- De (3.20) on tire que la pente de g sur AB est donnée par $\frac{g(B)-g(A)}{a} = \frac{y_2^2-y_1^2}{2a}$; donc de (3.19), (3.21), on tire la validité de (3.9).

La connaissance de w fournirait automatiquement $\{\varphi, \Omega, u, v, q\}$; en effet on a vu que q est donnée par (3.9); d'après (3.11), (3.15) on tire

(3.22) $\Omega = \{(x,y) \mid (x,y) \in D; \; w_y(x,y) < 0\}$

et, encore de (3.15), on aura aussi:

(3.23) $u = y - w_y|_\Omega \; ; \quad v = -w_x|_\Omega \; ;$

finalement de (3.22) on évaluera φ par la formule:

(3.24) $\varphi(x) = \max\{y \mid (x,y) \in \bar{\Omega}\}$ $\quad 0 \leq x \leq a$

La caractérisation (3.22) de Ω n'est toute_fois pas encore suffisante: en effet, si l'on cherche à combiner (3.17), (3.22) de façon à faire disparaître l'inconnue Ω on tombe sur l'équation non linéaire:

(3.25) $\Delta w \in H(-w_y)$

où $t \to H(t)$ est le graphe maximal monotone associé à la fonction de Heaveside; et le problème (3.25), (3.21) n'est pas bien posé ([9]).

En effet on peut faire mieux: de (3.18), (3.14) on déduit que l'on a identiquement:

(3.26) $w(x,y) = \int_y^{y_1} [\tilde{u}(x,t)-t]\,dt \qquad \forall (x,y) \in \bar{D}$

et alors, grâce à (3.8), on aura:

(3.27) $w(x,y) \geq 0$ dans $\bar{D}$

(3.28) $\Omega = \{(x,y) \mid (x,y) \in D;\ w(x,y) > 0\}$.

Maintenant la combinaison de (3.17), (3.28) donne:

(3.29) $\Delta w \in H(w)$

et le problème (3.29), (3.21) est bien posé: il admet une et une seule solution dans $H^1(D)$ (cf. par ex. le cours de M. MOREAU dans ce meme volume) donc (on a déjà vu que à partire

---

([9]) Par exemple il admet comme solution la solution w du problème $\Delta w=1$; $w_{|\partial D}=g$; ce qui donnerait (cf. (3.22)) $\Omega=D$.

de w on évaluait $\{\varphi,\Omega,u,v,q\}$):

(3.30) $\left\{\right.$ le problème modèle admet au plus une solution faible

Remarque 3.4.- On peut présenter le problème (3.29), (3.21) sous forme de problème de minimum en posant:

$$(3.31)\begin{cases} \mathcal{K}^+=\{z\,|\,z\in H^1(D);\ z_{|\partial D}=g\} & \text{(g donnée par (3.20))} \\ J^+(z) = \frac{1}{2}\int_D |\text{grad } z|^2 dx\,dy + \int_D z^+\,dx\,dz & (t^+=\frac{|t|+t}{2}) \end{cases}$$

et alors w est l'unique solution de:

$$(3.32)\quad w\in\mathcal{K}^+;\ J^+(w)\leq J^+(z) \qquad \forall z\in\mathcal{K}^+ ;$$

d'ailleurs (3.22) n'est pas la seule formulation variationnelle que l'on peut tirer des reinseignements que l'on a sur w; par exemple si l'on pose:

$$(3.33)\quad \mathcal{K}=\{z\,|\,z\in\mathcal{K}^+;\ z\geq 0\};\quad J(z)=\int_D\{\frac{1}{2}\,|\text{grad } z|^2+z\}dx\,dy$$

grâce à (3.27) on a aussi:

$$(3.34)\quad w\in\mathcal{K};\quad J(w)\leq J(z) \qquad \forall z\in\mathcal{K}$$

ou bien l'inéquation variationnelle équivalente:

$$(3.35)\quad w\in\mathcal{K};\ a(w,\ z-w)\geq L(z-w)\quad \forall z\in\mathcal{K}$$

avec $a(\xi,\mu)=\int_D \text{grad } \xi .\ \text{grad } \mu\, dx\,dy$ et $L(\xi)=-\int_D \xi dx\, dy$

Du point de vue numérique c'est la présentation (3.34) qui semble être la meilleure; d'ailleurs on a intérêt à exploiter beucoup de formulations car, voulant généraliser la méthode à des problèmes plus compliqués, on devra choisir, suivant le cas, l'une ou l'autre voie.

Pour ce qui concerne l'existence d'une solution faible

du problème modèle on doit maintenant démontrer que, partant de l'unique solution w du problème (3.35) (par exemple; ou équivalentement de (3.34); ou de (3.32)) les formules (3.28), (3.24), (3.23), (3.9) fournissent une solution. Je n'entrerai pas dans les détails (pour lesquels je renvoye à 1), en me bornant ici à souligner que les phases essentielles de la démonstration sont

a) grâce aux théorèmes de régularité des solutions des inéquations variationnelles (cf. 14) la solution w de (3.35) satisfait:

(3.36) $\quad w \in W^{2,p}(D) \quad$ pour tout p fini;

en particulier on a (3.16); Ω défini par (3.28) est ouvert; et on a (3.17).

b) grâce à (3.36) on peut appliquer le principe du maximum à w_x, w_y et démontrer qu'il s'agit de fonctions non positives; d'ici on tire que Ω est borné supérieurement par une fonction φ qui satisfait (3.4).

Une fois obtenu le théorème d'existence et unicité des solutions faibles, se pose le problème de la régularité; toujours sans entrer dans les détails je me bornerai à remarquer que, en adaptant un discours de Caccioppoli (cf. |15|) on peut démontrer la relation:

$$\varphi : x \to \varphi(x) \text{ est analytique sur }]0,a[\qquad (10)$$

et que, pour ce qui concerne la régularité de u,v, de (3.36) et (3.23) ou a $u,v \in W^{1,p}(\Omega)$ pour tout p fini; il s'agit d'une régularité optimale car on peut démontrer (cf. toujours |1|) que les relations $u,v \in W^{1,\infty}(\Omega)$ sout fausses.

N.4.- On termine cet exposé par quelques considérations à caractère numérique et par une vue d'ensemble sur les généralisation de la méthode.

Pour ce qui concerne la résolution numérique des inéquations variationnelles on connaît des nombreux procédés à la fois mathématiquement rigoureux et pratiquement efficaces (cf. par ex. |12|). Dans |10| on a étudié l'inéquation variationnelle (3.35) par discrétisation en différences finies, et resolvant le problème discret par la méthode de S.O.R. et projection; la comparaison avec les méthodes "traditionnelles" indiquées au N.2 a montré un gain sensible à la fois du point de vue simplicité de programmation et du point de vue rapidité d'exé-

(10) Pour un traitement systématique du problème de la regularité de la "ligne de détachement" pour les solutions d'inéquations avec obstacle on consultera la conférence de D. KINDERLEHRER sur ce meme cours CIME.

cution (cf. toujours |10|). Il est toutefois à remarquer que la méthode est "correcte" mathématiquement dans le sens l'on obtient une suite $\{w_h\}$ de solutions approchées qui converge, dans une topologie convénable, vers la solution w de (3.35); mais si l'on prend, comme "approximation" de Ω, l'ensemble de positivité Ω_h de w_h (analoguement à (3.28)) la convergence de w_h à w n'assure pas, à priori, la convergence de Ω_h à Ω. Cette difficulté a été surmontée en |5| où l'on a montré que l'on a:

(4.1) $\quad \Omega$ = intérieur de $\liminf_{h\to 0^+} \Omega_h$.

Passons maintenant à quelques généralisations. Le cas de digues à perméabilité variable (à savoir dans lequel le coefficient $\tilde{k}$ figurant dans (1.3) n'est pas constant) pose de nombreuses difficultés. Dans |3|,|4| on a traité le cas où $\tilde{k}(x,y)$ est constant par morceaux par rapport à une des variables et constant par rapport à l'autre ([11]); dans |7| on traite le cas de $\tilde{k}(x,y)$ de la forme $k_1(x).k_2(y)$ mais sous des hypothèses restrictives sur la régularité de k_1,k_2.

Le cas où l'on a plusieurs liquides immiscibles de densités différentes peut aussi être traité par la meme méthode; dans |3|,|4| on étudie le problème de la débouchée à la

---

([11]) Ce qui correspond à digues en plusieures couches, horizontales ou verticales, de matérieaux différents.

mer d'une lame phréatique (on a donc, outre que l'usuelle surface libre, une surface de séparation entre la terre mouillée par l'eau douce et la terre mouillée par l'eau de mer); (3.25) est remplacée par une inéquation à "double obstacle", les deux frontières libres étant obtenues comme frontières des zones de contact avec les deux obstacles.

Dans |3|,|4| on a aussi étudié le cas où la parois à contact avec le premier bassin est imperméable le long du morceau $[c,y_1]$ avec $0<c<y_1$. Dans ce cas la valeur de q n'est plus une fonction explicite des données (à savoir on n'a plus une formule du type (3.9)) et la condition aux limites sur w pour ce qui concerne le morceau $\{(0,y)|c<y<y_1\}$, au lieu d'être de type dérivée tangentielle (cf. (3.18)) est de type dérivée normale. Sur la partie restante de ∂D, supposant connue la valeur de q, on peut évaluer les valeurs $g_q(x,y)$ de la trace de w; on peut alors construire, pour tout $q \in \mathbb{R}$, un convexe $\mathcal{K}_q^+$ du type (3.31) et la solution w_q du problème de minimum correspondant (analogue à (3.32)); les inéquations associées resolvent des problèmes aux limites de type mêlé (au lieu que de Dirichlet) pour lesquels, en général, la validité de (3.36) est fausse; dans |4| on a montré l'existence et unicité de une valeur q^* de q en correspondence à laquelle la solution w_{q^*} satisfait (3.36); ce qui a permis encore

de conclure avec un théorème de existence et unicité. Un algorithme numérique (à caractère physique-euristique) entroduit dans |3| pour l'approximation de q^*, à été complètement justifié dans |5|.

Pour ce qui concerne la possibilité de adapter la méthode à des digues de géométrie plus compliquée on peut remarquer que, si la parois adjacente au bassin de droite n'est pas verticale, devient fausse l'une des relations fondamentales de la méthode, à savoir la validité de (3.27), (3.28); donc on va supposer que la parois de droite est verticale ([12]).

Sous ces restrictions les relations obtenues au N.3 restant encore valables jusqu'à (3.18) inclue; toutefois (3.18) (et l'analogue de (3.19) qui donne $w_x=-q$ sur AB) fournissent pour w, au lieu que des données de Dirichlet, des données du type "dérivée oblique". L'étude théorique des inéquations correspondentes, dans le cas général, n'à pas encore été abordé; dans 3 , 4 on s'est borné aux cas particuliers correspondents à: base horizontale et parois inclinée; ou base inclinée et

---

([12]) Des essais numériques faits avec parois de droite inclinée suggèrent que la méthode devrait marcher aussi dans ce cas, quitte à introduire des solutions "a plusieures paramètres"; par exemple, outre à la valeur du débit q, l'abscisse s du point C_φ (on a s=a si la parois est verticale; cf., pour plus de details, |5|).

parois verticale ([13]). Il s'agit encore d'étudier une famille d'inéquations ([14]) dépendante du paramètre q, la bonne valeur du paramètre étant à individuer par moyen d'une "condition de régularité" de type (3.36). Dans |4| on est arrivé jusqu'au théorème d'unicité; le théorème d'existence a été donné dans |8| par une méthode "numérique", en passant à la limite sur des "solutions approchées".

D'autres problèmes analogues, tels que le problème modèle en présence de évaporation et le problème de l'eau qui filtre à travers les parois perméables d'un canal, ont été récemment étudiés par la méthode ici proposée (cf. respectivement 19 et 20).

Je voudrais finalment conclure en rappellant que la méthode décrite dans le N.3 pour transformer un problème à frontière libre dans une inéquation variationnelle (ou éventuellement dans une famille, à un ou plusieurs paramètres, d'inéquations) semble avoir un domaine d'applicabilité plus ample que

---

([13]) Plus récemment, dans |9| , on est arrivé à traiter le cas où le parois et la base sont toutes les deux inclinées; il s'agit d'un problème de "dérivée oblique qui saute", donc de type non variationnel.

([14]) Qui traduisent un problème de dérivée oblique, donc la forme a (ξ,μ) qui intervient dans (3.35) n'est plus symétrique et le problème n'est plus équivalent à un problème de minimum.

celui relatif aux mouvements de filtration; elle a en effet été adaptée à la résolution de problèmes à frontière libre qui sourgissent dans l'étude de problèmes de fluxe de fluides compressibles autour d'un obstacle (sans ou avec sillage) et, dans un contexte un peu différent (problème d'évolution au lieu que stationnair) à l'étude d'un problème de type Stéfan; mais pour ces problèmes je renvoye au cours de M. DUVAUT dans ce meme volume.

C. Baiocchi

BIBLIOGRAPHIE

1 C. BAIOCCHI- Su un problema di frontiera libera connesso a questioni di idraulica. Annali di Matematica XCII (1972) p.107-127; note préliminaire aux C. R. Acad. Sc. Paris, 273 (1971), p.1215-1217.

2 C. BAIOCCHI - Sur quelques problèmes à frontière libre. Astérique, 2 et 3 (1973), p. 69-85.

3 C. BAIOCCHI, V. COMINCIOLI, L. GUERRI, G. VOLPI - Free boundary problems in the theory of fluid flow through porous media: numerical approach. Calcolo, X (1973), p.1-86.

4 C. BAIOCCHI, V. COMINCIOLI, E. MAGENES, G.A. POZZI - Free boundary problems in the theory of fluid flow through porous media: existence and uniqueness theorems. Annali di Matematica XCII (1973), 1-82.

5 C. BAIOCCHI, E. MAGENES - Problemi di frontiera libera in idraulica. Atti del Convegno intern. "Metodi valutativi nella Fisica Matematica", Acc. dei Lincei, Roma, 1972.

6 J. BEAR - Dynamics of fluids in porous media. Amer. Els. Publ., New York, 1972.

7 V. BENCI - Su un problema di filtrazione attraverso un mezzo poroso. A paraitre aux Annalid di Matematica.

8 V. COMINCIOLI - A theoretical and numerical approach to some free boundary problems. A paraitre aux Annali di Matematica.

9 V. COMINCIOLI - Travail en préparation.

10 V. COMINCIOLI - L. GUERRI - G. VOLPI - Analisi numerica di un problema di frontiera libera connesso col moto di un fluido attraverso un mezzo poroso. Pubbl. n.17 du L.A.N., Pavia, 1971.

11 C.W. CRYER - On the approximate solution of free boundary problems using finite differences. J.A.C.M. 17 (1970) p.379-411.

12 R. GLOWINSKI - J.L. LIONS- R. TREMOLIERES - Résolution numérique des inéquations de la Mécanique et de la Physique. Livre à paraitre près DUNOD.

13 M.E. HARR - Groundwater and seepage. Mc Graw Hill, New York 1962.

14 H. LEWY, G. STAMPACCHIA - On the regumarity of the solution of a variational inequality. Comm. P.A.M. 22 (1969), p.153-188.

15 C. MIRANDA - Su un problema di frontiera libera. Symp. Math. 2 (1968), p.71-83.

16 U. MUSKAT - The flow of homogeneous fluid through porous media. Mc Graw Hill, New York 1937.

17 S.T. NEUMANN, P.A. WITHERSPOON - Finite element method of analysing steady seepage with a free surface, Wather Resources Research , 6 (1970), 889-897.

18 P. Ya. POLUBARINOVA-KOCHINA - The theory of groundwater movement (translaté du russe). Princeton University Press, Princeton 1962.

19 G.A. POZZI - On a free boundary problem arising from fluid through a porous medium, in the presence of evaporation. Travail à paraître.

20. S. TORELLI - Su un problema di filtrazione da un canale. Travail à paraitre.

CENTRO INTERNAZIONALE MATEMATICO ESTIVO

(C. I. M. E.)

INTEGRALES CONVEXES DUALES

CHARLES CASTAING

Corso tenuto a Bressanone dal 17 al 26 giugno 1973

INTEGRALES CONVEXES DUALES

par

CHARLES CASTAING

(Univ. de Montpellier)

Introduction. Dans la première partie de ce papier on démontre deux théorèmes de dualité des intégrales convexes. Dans la deuxième partie, on donne deux théorèmes de fermeture directement liés au problème de râfle d'un convexe étudié par Moreau ([7], prop. 8.i).

En ce qui concerne l'étude systématique des intégrales convexes et leurs applications, on renvoie aux travaux de Rockafellar ([8], [9], [10], [11]), Castaing ([3]) et Valadier ([12]).

I - Théorèmes de dualité des intégrales convexes

Notations. Soient T un espace localement compact polonais muni d'une mesure de Radon positive μ, E un espace de Banach réflexif, E' son dual fort. Une application v de T dans un espace topologique est dite μ-mesurable si elle est Lusin μ-mesurable ([1]). Si f est une fonction convexe semi-continue inférieurement sur E à valeurs dans $]-\infty, +\infty]$ non partout étale à $+\infty$, sa duale g est définie par

$$g(x') = \sup_{x \in E} [\langle x', x \rangle - f(x)] \quad (x' \in E')$$

Si K est un convexe fermé non vide de E, on désigne par $\delta(., K)$

la fonction indicatrice de K :

$$\delta(x, K) = \begin{cases} 0 & \text{si } x \in K \\ +\infty & \text{si } x \notin K \end{cases}$$

et par $\delta^*(., K)$ la fonction d'appui de K :

$$\delta^*(x', K) = \sup \{\langle x', x\rangle \mid x \in K\}$$

Il est évident que δ et δ^* sont duales l'une de l'autre.

Voici un résultat de mesurabilité qui intervient directement dans la démonstration des théorèmes de dualité que nous avons en vue.

Proposition . Soient f une fonction convexe semi-continue inférieurement sur E à valeurs dans $]-\infty, +\infty]$ non partout égale à $+\infty$, v une application μ-mesurable de T dans E' et α une fonction réelle μ-mesurable sur T telle que

$$\alpha(t) > \inf \{f(x) - \langle v(t), x\rangle \mid x \in E\} , \quad \forall t \in T$$

Alors, il existe une application μ-mesurable, u, de T dans E telle que

$$f(u(t)) - \langle v(t), u(t)\rangle \leq \alpha(t) \qquad \mu\text{-p.p.}$$

Démonstration. On se ramène aussitôt au cas où T est compact. Posons

$$\Gamma(t) = \{x \in E \mid f(x) - \langle v(t), x\rangle \leq \alpha(t)\} , \quad \forall t \in T$$

Alors $\Gamma(t)$ est convexe fermé non vide, $\forall t \in T$. Il existe par ailleurs une partition de T en une suite de compacts (T_n) et un μ-négligeable N telle que les restrictions de v et α à chacun des T_n soient continues. Par suite la restriction de la multi-application Γ à chacun des K_n est de graphe séquentiellement fermé dans $K_n \times E_\sigma$ (E_σ désignant l'espace vectoriel E muni de la topologie affaiblie $\sigma(E, E')$). D'après un résultat de ([2], Cor 4 du théor. 1), il existe une application μ-mesurable, u_n, de K_n dans E telle que $u_n(t) \in \Gamma(t)$, $\forall t \in K_n$. Alors, l'application, u, obtenue en recollant les u_n :

$$u(t) = \begin{cases} u_n(t) & \text{si} \quad t \in K_n \\ p & \text{arbitraire si} \quad t \in N \end{cases}$$

vérifie les conditions de l'énoncé.

Théorème 1.1. On suppose T compact métrisable. Soient f une fonction convexe semi-continue inférieurement sur E à valeurs dans $]-\infty, +\infty]$, non partout égale à $+\infty$ et g sa duale. On pose

$$I_f(u) = \int_T f(u(t))\,\mu(dt), \quad \forall u \in L^1_E(T, \mu)$$

$$I_g(v) = \int_T g(v(t))\,\mu(dt), \quad v \in L^\infty_{E'}(T, \mu)$$

Alors I_f et I_g sont duales l'une de l'autre, c'est à dire,

$$\forall u \in L^1_E, I_f(u) = \sup \{\langle u, v\rangle - I_g(v) \mid v \in L^\infty_{E'}(T, \mu)\}$$

$$\forall v \in L^\infty_{E'}, I_g(v) = \sup \{\langle u, v\rangle - I_f(u) \mid u \in L^1_E(T, \mu)\}$$

Démonstration. Il suffit de vérifier la formule

$$I_f(u) = \sup \{\langle u, v\rangle - I_g(v) \mid v \in L^{\infty}_{E'}(T, \mu)\}$$

Quelque soient $u \in L^1_E$ et $v \in L^{\infty}_{E'}$, on a

$$I_f(u) + I_g(v) = \int_T f(u(t))\, \mu(dt) + \int_T g(v(t))\, \mu(dt)$$
$$\geq \int_T \langle u(t), v(t)\rangle\, \mu(dt) = \langle u, v\rangle$$

D'où

$$I_f(u) \geq \sup \{\langle u, v\rangle - I_g(v) \mid v \in L^{\infty}_{E'}\}$$
$$= -\inf \{I_g(v) - \langle u, v\rangle \mid v \in L^{\infty}_{E'}\} = (I_g)^*(u)$$

Soient u un élément de $L^1_E(T, \mu)$ et r un nombre réel tels que $r < I_f(u)$. Soit α une fonction réelle intégrable telle que

$$\begin{cases} \alpha(t) < f(u(t)), & \forall t \in T \\ \int_T \alpha(t)\, \mu(dt) > r \end{cases}$$

Alors on a

$$-\alpha(t) > \inf \{g(x') - \langle x', u(t)\rangle \mid x' \in E'\}$$

Pour tout $t \in T$, posons

$$\Gamma(t) = \{x' \in E'_s \mid g(x') - \langle x', u(t)\rangle \leq -\alpha(t)\}$$

D'après la proposition précédente appliquée à l'espace E'_s muni de la topologie $\sigma(E', E)$, il existe une application μ-mesurable, v, de T dans E' telle que $v(t) \in \Gamma(t)$ presque partout. Il existe alors une suite croissante d'ensembles compacts (K_n) et un négligeable

N tel que la restriction de v à chacun des K_n soit bornée et telle que $v(t) \in \Gamma(t)$ pour $t \in T \backslash N$. Soit $\tilde{v} \in L^{\infty}_{E'}$, telle que $I_g(\tilde{v}) < +\infty$

Posons

$$v_n(t) = \begin{cases} v(t) & \text{si } t \in K_n \\ \tilde{v}(t) & \text{si } t \in T\backslash K_n \end{cases}$$

Alors les v_n appartiennent à $L^{\infty}_{E'}$ de sorte que les $\langle u, v_n\rangle$ sont intégrables. Pour tout n, on a

$$\int_{K_n} \alpha(t)\,\mu(dt) \leq \int_{K_n} [\langle u(t), v(t)\rangle - g(v(t))]\cdot \mu(dt)$$

$$= \langle u, v_n\rangle - I_g(v_n) - \int_{T\backslash K_n} [\langle u(t), \tilde{v}(t)\rangle - g(\tilde{v}(t))]\,\mu(dt)$$

avec $\langle u, v_n\rangle - I_g(v_n) = \int_T [\langle u(t), v_n(t)\rangle - g(v_n(t))]\cdot \mu(dt)$

Or l'intégrale

$$\int_{T\backslash K_n} [\langle u(t), \tilde{v}(t)\rangle - g(\tilde{v}(t))]\cdot \mu(dt)$$

peut être rendue arbitrairement petite dès que n est suffisamment grand et comme on a

$$\lim_{n\to\infty} \int_{K_n} \alpha(t)\,\mu(dt) = \int_T \alpha(t)\,\mu(dt) > r$$

On en déduit que, pour n suffisamment grand,

$$\langle u, v_n\rangle - I_g(v_n) > r$$

ce qui implique $(I_g)^*(u) > r$. Comme r est un nombre arbitraire vérifiant $I_f(u) > r$, on a $I_g(u) = (I_g)^*(u)$.

<u>Remarque.</u> Ce théorème reste valable si l'on remplace le couple $(L^1_E, L^\infty_{E'})$ par un couple (L, M) d'espaces décomposables au sens de Rockafellar.

Nous allons donner maintenant une variante du théorème 1.1., variante qui s'applique directement au problème de râfle d'un convexe développé très récemment par Moreau ([5], [6], [7]) dans une série d'exposés du Séminaire d'Analyse Convexe Montpellier 1971-72-73 .

Dans tout le reste de cette partie, I est l'intervalle [0, 1] muni de la mesure de Lebesgue dt . Une multi-application Γ de I à valeurs dans les convexes fermés non vides de E est dite à <u>variation continue</u> s'il existe une fonction réelle continue r définie sur I telle que

$$\forall t \in I \ , \quad \forall \tau \in I \ , \quad h(\Gamma(t), \Gamma(\tau)) \leq |r(t) - r(\tau)|$$

où h désigne la distance de Hausdorff des ensembles fermés non vides de E.

<u>Théorème 1.2.</u> <u>Soient</u> Γ <u>une multi-application à variation continue de</u> I <u>à valeurs dans les convexes fermés non vides de</u> E, $g(t, x')$ <u>et</u> $f(t, x)$ <u>les fonctions d'appui et indicatrice de l'ensemble</u> $\Gamma(t)$:

$$g(t, x') = \sup \{\langle x', x\rangle \mid x \in \Gamma(t)\}$$

$$f(t, x) = \begin{cases} 0 & \text{si } x \in \Gamma(t) \\ +\infty & \text{si } x \notin \Gamma(t) \end{cases}$$

<u>et soit</u>

$$S_\Gamma = \{u \in L^1_E(I) \mid u(t) \in \Gamma(t) \quad p.p.\}$$

<u>Alors,</u> S_Γ <u>est un convexe fermé non vide de</u> $L^1_E(I)$ <u>et les fonctions convexes</u> I_g <u>et</u> I_f <u>définies sur</u> $L^\infty_{E'}(I)$ <u>et</u> $L^1_E(I)$ <u>respectivement</u> :

$$I_g(v) = \int_I g(t, v(t))\, dt\ , \quad v \in L^\infty_{E'}(I)$$

$$I_f(u) = \int_I f(t, u(t))\, dt\ , \quad u \in L^1_E(I)$$

<u>sont à valeurs dans</u> $]-\infty, +\infty]$, <u>non partout égales à</u> $+\infty$ <u>et duales l'une de l'autre.</u>

<u>Démonstration.</u> Le principe de démonstration est le même que dans la démonstration du théorème 1.1.

Les points essentiels à vérifier sont :

a) la non vacuité de S_Γ

b) la mesurabilité des fonctions $t \longmapsto g(t, v(t))$ et $t \longmapsto f(t, u(t))$.

a) S_Γ <u>est non vide</u> car Γ est à variation continue, donc semi-continue inférieurement pour la topologie forte de E. Par suite Γ admet une section continue d'après un résultat de Michael ([4], théor.3.2)

b) la mesurabilité des fonctions $t \longmapsto f(t, u(t))$ est automatiquement assurée car Γ est de graphe fermé dans $I \times E_\sigma$ et, celle des fonctions $t \longmapsto g(t, v(t))$ est assurée grâce au lemme suivant :

Lemme. *Soit φ une fonction réelle définie sur $I \times E_\sigma$ telle que pour tout réel λ, l'ensemble*

$$A_\lambda = \{(t,x) \in I \times E_\sigma \mid \varphi(t,x) > \lambda\}$$

soit réunion dénombrable de compacts, soit Δ une multiapplication de I à valeurs dans les fermés non vides de E_σ. Si le graphe $G(\Delta)$ de Δ est séquentiellement fermé dans $I \times E_\sigma$, alors la fonction

$$\rho : t \longmapsto \sup \{\varphi(t,x) \mid x \in \Delta(t)\} \quad (t \in I)$$

est universellement mesurable sur I.

Soit λ un nombre réel. On a

$$\{t \in I \mid \rho(t) > \lambda\} = \mathrm{proj}_I \, [G(\Delta) \cap A_\lambda]$$

Comme A_λ est réunion dénombrable de compacts de $I \times E_\sigma$, $G(\Delta) \cap A_\lambda$ est aussi réunion dénombrable de compacts de $I \times E_\sigma$, donc $\mathrm{proj}_I \, [G(\Delta) \cap A_\lambda]$ est réunion dénombrable de compacts de I, donc universellement mesurable.

Corollaire. *Si v est une application mesurable de I dans E', alors la fonction*

$$t \longmapsto g(t, v(t)) = \sup \{\langle v(t), x\rangle \mid x \in \Delta(t)\}$$

est mesurable sur I.

En effet, on se ramène immédiatement au cas où v est continue, de sorte que la fonction $(t, x) \longmapsto \langle v(t), x\rangle$ vérifie les condition d'applications du lemme précédent:

II - Théorèmes de fermeture

Théorème 2.1. *Soient* T *un espace compact métrisable muni d'une mesure de Radon positive* μ, E *un espace de Banach réflexif,* f *une fonction convexe semi-continue inférieurement sur* E *à valeurs dans* $]-\infty, +\infty]$ *non partout égale à* $+\infty$, $\partial f(x)$ *le sous différentiel de* f *au point* x. *Soient* (u_n) (resp. v_n) *une suite dans* $L^{\infty}_{E}(T, \mu)$ (resp. $L^{1}_{E'}(T, \mu)$) *telle que* :

(i) (u_n) *est uniformément bornée et converge uniformément vers* $\tilde{u} \in L^{\infty}_{E}(T, \mu)$

(ii) (v_n) *converge pour* $\sigma(L^{1}_{E'}, L^{\infty}_{E})$ *vers* $\tilde{v} \in L^{1}_{E'}(T, \mu)$

(iii) $v_n(t) \in \partial f(u_n(t))$ μ-*presque partout*

Alors on a

$$\tilde{v}(t) \in \partial f(\tilde{u}(t)) \quad \mu\text{-presque partout}$$

Démonstration. D'après la condition (iii), on a pour presque tout t,

$$f(u_n(t)) + g(v_n(t)) + \langle u_n(t), v_n(t)\rangle = 0$$

Les conditions (i) et (ii) impliquent

(1) $$\lim_{n \to \infty} \int_T \langle u_n(t), v_n(t)\rangle \mu(dt) = \int_T \langle \tilde{u}(t), \tilde{v}(t)\rangle \mu(dt)$$

D'après le théorème 1.1. , les fonctions convexes

$$I_f(u) = \int_T f(u(t)) \mu(dt)$$

$$I_g(v) = \int_T g(v(t)) \mu(dt)$$

sont semi-continue inférieurement sur $L^\infty_E(T, \mu)$ et $L^1_{E'}(T, \mu)$ pour toute topologie localement convexe compatible avec la dualité mise sur ces espaces. Par suite, on a

(2) $$\liminf_{n \to \infty} (I_f(u_n) + I_g(v_n)) \geq I_f(\tilde{u}) + I_g(\tilde{v})$$

car d'après (i), (u_n) converge uniformément vers $\tilde{u}$, donc converge vers $\tilde{u}$ pour la topologie forte de $L^1_E(T, \mu)$ et d'après (ii), (v_n) converge pour $\sigma(L^1_{E'}, L^\infty_E)$ vers $\tilde{v}$. Tenant compte de (1) et (2), on obtient finalement :

(3) $$I_f(\tilde{u}) + I_g(\tilde{v}) - \langle \tilde{u}, \tilde{v}\rangle \leq 0$$

puisque I_f et I_g sont duales l'une de l'autre.

Or, (3) équivaut à

$$\tilde{v} \in \partial I_f(\tilde{u}) \iff \tilde{v}(t) \in \partial f(\tilde{u}(t)) \quad \mu\text{-presque partout.}$$

Théorème 2.2. Les hypothèses et notations étant celles du théorème 1.2, soit D le domaine de définition de $g(t, x')$, B' la boule unité de E', (v_n) une suite d'application mesurables de I à valeurs dans $B' \cap D$ convergeant vers $\tilde{v}$ pour $\sigma(L^\infty_{E'}, L^1_E)$, (u_n) une suite d'applications continues de I dans E convergeant uniformément vers une section continue $\tilde{u}$ de Γ, (θ_n) une suite d'applications mesurables de I dans I. On suppose :

(i) $\lim_{n\to\infty} \theta_n(t) = t \ , \ \forall t \in I$

(ii) $v_n(t) + \partial f\,(\theta_n(t),\, u_n(\theta_n(t))) \ni 0 \quad , \quad \forall t \in I$.

Alors on a

$$\tilde{v}(t) + \partial f(t,\, \tilde{u}(t)) \ni 0 \qquad \text{presque partout.}$$

Démonstration. Pour tout t et tout n, on a, d'après (ii)

$$\begin{cases} u_n(\theta_n(t)) \in \Gamma(\theta_n(t)) \\ g(\theta_n(t),\, -v_n(t)) + \langle u_n\,(\theta_n(t)),\, v_n(t)\rangle = 0 \end{cases}$$

La condition (i) implique

(1) $$\lim_{n\to\infty} \int_I \langle u_n\,(\theta_n(t)),\, v_n(t)\rangle\, dt = \int_I \langle \tilde{u}(t),\, \tilde{v}(t)\rangle\, dt$$

Comme les v_n prennent leurs valeurs dans $(-D) \cap B'$, on a

$$|g(\theta_n\,(t),- v_n(t)) - g(t,- v_n(t))| \le |r\,(\theta_n(t)) - r(t)|$$

D'où

(2) $$\lim_{n\to\infty} |g(\theta_n(t),\, - v_n(t)) - g(t,\, - v_n(t))| = 0$$

D'où

$$\lim_{n\to\infty} \int_I [g(\theta_n(t),\, - v_n(t)) - g(t,\, - v_n(t))]\, dt = 0$$

Comme la fonction convexe

$$L^{\infty}_{E'}(I) \ni v \mapsto I_g(v) = \int_T g(t,\, v(t))\, dt$$

est semi-continue inférieurement sur $L^{\infty}_{E'}$ pour $\sigma(L^{\infty}_{E'}, L^{1}_{E})$ d'après le théorème 1.2 , on a

(3) $$\liminf_{n \to \infty} I_g(-v_n) \geq I_g(-\tilde{v})$$

tenant compte de la condition (ii) et des relations (1), (2) et (3), on obtient

$$0 \geq \langle\tilde{u}, \tilde{v}\rangle + I_g(-\tilde{v}) \iff \langle\tilde{u}(t), \tilde{v}(t)\rangle + g(t, -\tilde{v}(t)) \leq 0$$

presque partout

$$\iff -\tilde{v}(t) \in \partial f(t, \tilde{u}(t))$$

presque partout.

C. Castaing

[1] N. BOURBAKI, Intégration, chapitre 1,2,3,4, Hermann, Paris 1965

[2] CH. CASTAING, Proximité et Mesurabilité - Un théorème de compacité faible. Colloque sur la théorie mathématique du controle optimal, Bruxelles 1969.

[3] CH. CASTAING, Intégrales convexes duales. Séminaire d'Analyse convexe Montpellier 1973, Exposé n° 6

[4] E. MICHAEL, Continuous selections I, Ann. of Math. 63, 1956 p. 361-382.

[5] J.J. MOREAU, Râfle par un convexe (Première partie), Séminaire d'Analyse convexe Montpellier 1971, Exposé n° 15.

[6] J.J. MOREAU, Râfle par un convexe variable (Deuxième partie), Séminaire d'Analyse convexe Montpellier 1972, Exposé n° 3.

[7] J.J. MOREAU, Retraction d'une multi-application, Séminaire d'Analyse convexe Montpellier 1972, Exposé n° 13.

[8] R.T. ROCKAFELLAR, Integrals which are convex functionals, pacific Journal of Math.,24, 1968, p.525-539.

[9] R.T. ROCKAFELLAR, Integrals which are convex functionals II, Pacific Journal of Math., 39, 1971, p.439-469.

[10] R.T. ROCKAFELLAR, Weak compactness of level sets of integrals functionals, Broceedings of the Liege Symposium of Functionals Analysis, Septembre 1970, H.G. Garnir, Editeur.

[11] R.T. ROCKAFELLAR, Convex Integrals functionals on duality, Contribution to Non linear functional Analysis, Ac. Pres, 1971, p. 215-236.

[12] M. VALADIER, Convex integrands on Souslin locally convex spaces, Séminaire d'Analyse convexe Montpellier 1973, Exposé n° 2.

CENTRO INTERNAZIONALE MATEMATICO ESTIVO

(C.I.M.E.)

ETUDE DE PROBLEMES UNILATERAUX EN MACHANIQUE PAR DES METHODES VARIATIONNELLES

GEORGES DUVAUT

Corso tenuto a Bressanone dal 17 al 26 giugno 1973

PREMIERE CONFERENCE

GENERALITES - ELASTICITE AVEC FROTTEMENT

par G. Duvaut (Univ., Paris, XIII)

1) Généralités.

Les méthodes variationnelles ont été appliquées en mécanique depuis quelques années à des problèmes divers :

i) Elasticité classique (P. Germain [1]), unilatérale (problème de Signorini) (G. Fichera [1], J.L. Lions et G. Stampacchia [1], Boucher [1]), avec frottement (G. Duvaut et J.L. Lions [1][2]), puis généralisation à la viscoélasticité (G. Duvaut [1], Boucher [1]).

ii) Plasticité avec la loi de Hencky (H. Lanchon [1][2]), avec la loi de Prandtl-Reuss (B. Nayrolles [1][2], J. Moreau [1][2]), élastoviscoplasticité (G. Duvaut et J.L. Lions [2], G. Duvaut [2]), matériaux à blocage (W. Prager [1], F. Léné [1]).

iii) Parois semi-perméables (en thermique, mécanique des fluides en milieux poreux, physique des solutions) (G. Duvaut et J.L. Lions [2], H. Brézis [1]).

iv) Théorie des plaques linéarisées (G. Duvaut et J.L. Lions [2]) ou en forte flexion (G. Duvaut et J.L. Lions [3].

v) Ecoulements des fluides de Bingham.

vi) Electromagnétisme (théorie du claquage d'antenne) et magnétohydraudynamique (G. Duvaut et J.L. Lions [4])

vii) Ecoulement à frontière libre à travers une digue poreuse

(C. Baiocchi [1]).

viii) Ecoulement d'un fluide parfait avec ou sans sillage (H. Brézis et G. Stampacchia [1], H. Brézis et G. Duvaut [1]).

ix) Problèmes de Stéfan : phénomène de fusion ou cristallisation (G. Duvaut [3]).

2) Un problème d'élasticité avec frottement.

2.1. Enoncé mécanique du problème. Soit une région Ω formée de R^3 de frontière régulière Γ composée de 3 parties Γ_1, Γ_2, Γ_3 avec

(1) $\qquad \text{mes } \Gamma_1 > 0.$

La partie Γ_1 est supposée fixée, Γ_2 est soumise à une densité surfacique de forces donnée, Γ_3 est le siège de forces de frottement. Le matériau qui, avant déformation, occupe la région Ω est élastique linéaire. Les équations et conditions aux limites sont les suivantes, où $\{u,\sigma\}$ sont les champs de déplacements et contraintes recherchés :

(2) $\qquad \frac{\partial}{\partial x_j} \sigma_{ij} + f_i = 0 \qquad$ dans Ω (1)

(3) $\qquad \sigma_{ij} = a_{ijkh}\varepsilon_{kh}(u)$

(4) $\qquad u = 0 \qquad$ sur Γ_1

(5) $\qquad \sigma_{ij}n_j = F_i \qquad$ sur $\Gamma_2 \qquad$ (F_i donné)

(6) $\qquad \sigma_N = F_N \qquad$ sur $\Gamma_3 \qquad$ ($F_N < 0$ donné)

---

(1) On applique la convention de sommation sur les indices répétés, soit

$$x_i y_i = \sum_{i=1}^{3} x_i y_i \quad .$$

(7) $$\begin{cases} |\sigma_T| \leq g & (g > 0 \text{ donné}) \\ |\sigma_T| < g \implies & u_T = 0 \\ |\sigma_T| = g \implies & \exists k \geq 0, \text{ tel que } u_T = -k\sigma_T. \end{cases}$$

Les quantités f_i sont les composantes d'une densité volumique de forces données sur Ω . Les coefficients a_{ijkh} sont les coefficients d'élasticité du matériau et satisfont aux relations suivantes

(8) $$\begin{cases} a_{ijkh} = a_{ijkh} = a_{khij} \\ a_{ijkh}\mu_{ij}\mu_{kh} \geq \alpha\mu_{ij}\mu_{ij}, \quad \alpha > 0, \quad \forall \mu_{ij} = \mu_{ji}. \end{cases}$$

De plus (σ_N,σ_T) et (u_N,u_T) sont les composantes normales et tangentielles de $\{\sigma_{ij}\eta_j\}$ et u.

Remarque 1.

Si $\Gamma_3 = \emptyset$ on a un problème d'élasticité sans frottement. La méthode décrite ici en donne la solution.

2.2. Formulation variationnelle.

Soit (u,σ) une solution du problème (2) - (7). On suppose (u,σ) assez régulière pour que les calculs suivants aient un sens. Soit v un champ de vecteurs sur Ω tels que $v_{|\Gamma_1} = 0$. Multiplions (2) par $v_i - u_i$ et intégrons sur Ω ; il vient après intégration par parties

(9) $$\int_\Omega \sigma_{ij}\varepsilon_{ij}(v-u)dx = \int_\Omega f_i(v_i-u_i)dx + \int_\Gamma \sigma_{ij}\eta_j(v_i-u_i)d\Gamma \ .$$

Le dernier terme de l'égalité (9) se décompose en

(10) $$\int_\Gamma \sigma_{ij}\eta_j(v_i-u_i)dx = \int_{\Gamma_1} \sigma_{ij}\eta_j(v_i-u_i)dx + \int_{\Gamma_2} F_i(v_i-u_i)d\Gamma + \int_{\Gamma_3} F_n(v_N-u_N)d\Gamma + \int_{\Gamma_3} \sigma_T(v_T-u_T)d\Gamma \ .$$

Compte tenu de (4) et de $v_{|\Gamma_1} = 0$ on a

$$\int_{\Gamma_1} \sigma_{ij}\eta_j(v_i-u_i)d\Gamma = 0 .$$

Les conditions de frottement (7) donnent l'inégalité

$$(11) \qquad \int_{\Gamma_3} \sigma_T(v_T-u_T)d\Gamma \geq \int_{\Gamma_3} (-g|v_T| + g|u_T|)d\Gamma .$$

Introduisons les notations

$$(12) \qquad a(u,v) = \int_\Omega a_{ijkh}\varepsilon_{kh}(u)\ \varepsilon_{ij}(v)dx$$

$$(13) \qquad L(v) = \int_\Omega f_i v_i dx + \int_{\Gamma_2} F_i v_i d\Gamma + \int_{\Gamma_3} F_N v_N d\Gamma$$

$$(14) \qquad j(v) = \int_{\Gamma_3} g|v_T|d\Gamma .$$

L'égalité (9) devient alors, en tenant compte de (3) et (11),

$$(15) \qquad a(u,v-u) + j(v) - j(u) \geq L(v-u), \quad \forall v, \quad v_{|\Gamma_1} = 0.$$

L'inégalité (15) avec la propriété (4) constitue la propriété variationnelle satisfaite par un champ de déplacement régulier solution du problème mécanique posé par (2) - (7). Cette propriété va nous permettre de poser un problème mathématique précis qui, nous l'espérons et le vérifierons ensuite, va résoudre en un certain sens le problème mécanique initial.

2.3. Cadre variationnel.

Nous introduisons l'espace V

$$(16) \qquad V = \{v \mid v \in (H^1(\Omega))^3, \quad v_{|\Gamma_1} = 0\}$$

qui est un espace de Hilbert pour la structure engendrée par $(H^1(\Omega))^3$, l'espace $H^1(\Omega)$ étant l'espace de Sobolev des (classes de) fonctions de carrés sommables sur Ω et dont les dérivées (au sens des distributions sur Ω) sont des fonctions de carrés sommables sur Ω. Les théorèmes de trace sur $H^1(\Omega)$ (J.L. Lions et E. Magenes [1]) montrent que V est un sous-espace fermé de $(H^1(\Omega))^3$.

L'introduction de V et les hypothèses suivantes sur les données

$$(17)\quad \begin{cases} f_i \in L^2(\Omega),\ F_i \in L^2(\Gamma_2),\ F_N \in L^2(\Gamma_3) \\ a_{ijkh} \in L^\infty(\Omega),\ g_1 \geq g \geq g_o > 0,\ g_o,\ g_1 \text{ constantes,} \end{cases}$$

permettent de reposer le problème en termes précis. On cherche u tel que

$$(18)\quad \begin{cases} u \in V \\ a(u,v-u) + j(v) - j(u) \geq L(v-u), \quad \forall v \in V. \end{cases}$$

Mais le problème (18) est équivalent (J.L. Lions [1]) au problème

$$(19)\quad \begin{cases} \text{Trouver } u \in V \text{ tel que} \\ I(v) \geq I(u), \quad \forall v \in V \\ \text{où } I(v) = \frac{1}{2}\, a(v,v) + j(v) - L(v). \end{cases}$$

2.4. Existence d'une solution unique de (18) ou (19).

Nous avons les résultats suivants :

Lemme 1. (Inégalité de Korn). Sur $(H^1(\Omega))^3$ une norme équivalente à la norme classique est

$$(20)\quad ||v|| = \left[|v|^2_{L^2(\Omega))^3} + \sum_{i,j} |\varepsilon_{ij}(v)|_{L^2(\Omega)} \right]^{1/2}.$$

Démonstration : Nous renvoyons le lecteur à G. Duvaut et J.L. Lions [2] (Théorème 3.1) ou aux références de cet ouvrage.

Lemme 2. Sur V avec l'hypothèse (1) la quantité $[a(v,v)]^{1/2}$ est une norme équivalente à la norme induite par $(H^1(\Omega))^3$.

Démonstration : Nous renvoyons le lecteur à G. Duvaut et J.L. Lions [2] (Théorème 3.3).

On peut alors montrer le

Théorème.

Sous les hypothèses (1), (8), (17) il existe u unique solution de (18) ou (19).

Démonstration.

i) Unicité : Supposons que u et u^* soient solutions de (18). Choisissons $v = u^*$ (resp. $v = u$) dans l'inégalité (18) relative à u (resp. u^*) et ajoutons membre à membre les inégalités obtenues ; il vient

(21) $$- a(u-u^*, u-u^*) \geq 0$$

d'où $u = u^*$ d'après le lemme 2.

ii) Existence : La fonctionnelle $v \to I(v)$ est convexe s. c. i. sur V faible. En effet $v \to L(v)$ est une forme linéaire continue sur V faible. L'application $v \to j(v)$ est convexe et $v \to v|_{\Gamma_1}$ est compacte de V dans $L^2(\Gamma_1)$, donc l'application $v \to j(v)$ est continue sur V faible. Par ailleurs a(v,v) est une forme quadratique positive sur V donc est convexe s. c. i. sur V faible (vérification par un calcul élémentaire).

De plus $I(v) \to +\infty$ si $||v|| \to \infty$. En effet d'après le lemme 2 on a

(22) $$I(v) \geq ||v||^2 - C||v||$$

où C est une constante.

Il en résulte (cf J.L. Lions [1]) qu'il existe u solution de (19).

2.5. Retour au problème mécanique.

La solution u est caractérisée par (18). Choisissons dans cette inégalité $v = u \pm \phi$ où $\phi = \{\phi_i\}$, $\phi_i \in \mathcal{D}(\Omega)$ (*), ce qui est loisible car alors $v \in V$. Il vient

(23) $$\int_\Omega a_{ijkh}\varepsilon_{kh}(u)\varepsilon_{ij}(\phi)dx = \int_\Omega f_i\phi_i dx\ , \quad \forall\ \phi_i \in \mathcal{D}(\Omega),$$

et donc, en posant $\sigma_{ij} = a_{ijkh}\varepsilon_{kh}(u)$

(24) $$\frac{\partial}{\partial xj}\sigma_{ij} + f_i = 0$$

au sens des distributions dans Ω.

Il en résulte que

(25) $$\frac{\partial}{\partial xj}\ \sigma_{ij} \in L^2(\Omega)\ ,$$

ce qui permet de définir $\sigma_{ij}n_j|_\Gamma$ comme élément de $H^{-1/2}(\Gamma)$ (**) par ,

(26) $$<\sigma_{ij}n_j,\ \psi> = \int_\Omega \sigma_{ij}\frac{\partial\Psi}{\partial xj}\ dx + \int_\Omega \frac{\partial\sigma_{ij}}{\partial xj}\Psi\ dx, \quad \forall\ \Psi \in H^{1/2}(\Gamma)$$

(*) On désigne par $\mathcal{D}(\Omega)$ l'ensemble des fonctions ∞ - différentiables à support compact dans Ω.

(**) $H^{-1/2}(\Gamma)$ est le dual de $H^{1/2}(\Gamma)$. cf J.L. Lions et E. Magénès [1]

où Ψ est un relèvement linéaire continu de ψ de $H^{1/2}(\Gamma)$ dans $H^1(\Omega)$ (Théorème de trace dans $H^1(\Omega)$. cf J.L. Lions et E.Magénès [1]). Le premier membre de (26) est alors une forme linéaire continue sur $H^{1/2}(\Gamma)$. On vérifie qu'elle est indépendante du relèvement : en effet le deuxième membre de (26) est nul pour $\Psi \in \mathcal{D}(\Omega)$, donc pour $\Psi \in H^1_o(\Omega)$ à cause de la densité de $\mathcal{D}'(\Omega)$ dans $H^1_o(\Omega)$. On notera désormais le crochet du 1er membre de (26) comme une intégrale sur Γ, soit

$$< \sigma_{ij}n_j, \psi > = \int_\Gamma \sigma_{ij}n_j\psi d\Gamma \ .$$

Multipliant alors (24) par $v_i - u_i$, où $v = \{u_i\} \in V$, et utilisant (26) on obtient

$$(27) \qquad \int_\Omega \sigma_{ij}\varepsilon_{ij}(v-u)dx = \int_\Omega f(v-u)dx + \int_\Gamma \sigma_{ij}n_j(v_i-u_i)d\Gamma$$

et en comparant avec (18)

$$\int_\Gamma \sigma_{ij}n_j(v_i-u_i)d\Gamma + j(v) - j(u) \geq \int_{\Gamma_2} F_i(v_i-u_i)d\Gamma + \int_{\Gamma_3} F_N(u_N-u_N)d\Gamma$$

Choisissant alors $v \in V$, $v|_{\Gamma_3} = u|_{\Gamma_3}$ on obtient

$$(28) \qquad \int_{\Gamma_2} \sigma_{ij}n_j(v_i-u_i)d\Gamma = \int_{\Gamma_2} F_i(v_i-u_i)d\Gamma \ ,$$

ce qui implique

$$(29 \qquad \sigma_{ij}n_j = F_i \qquad \text{sur } \Gamma_2$$

au sens de $H^{-1/2}(\Gamma_2)$. (*)

(+) Les éléments de $H^{-1/2}(\Gamma_2)$ sont les restrictions à Γ_2 des éléments de $H^{-1/2}(\Gamma)$ ce qui a un sens car $H^{-1/2}(\Gamma)$ est un espace de distributions.

On montre de manière analogue que

(30) $$F_N = \sigma_N \qquad \text{sur } \Gamma_3$$

au sens de $H^{-1/2}(\Gamma_3)$.

Il subsiste alors de (28),

(31) $$\int_{\Gamma_3} \left[\sigma_T(v_T - u_T) + g(|v_T| - |u_T|)\right] d\Gamma \geq 0, \qquad \forall v \in V$$

L'espace V étant vectoriel on peut remplacer v par λv dans (31), $\lambda > 0$. Faisant alors tendre λ vers $+\infty$ et vers zéro on obtient

(32) $$\int_{\Gamma_3} (\sigma_T v_T + g|v_T|) d\Gamma \geq 0 \qquad \forall v \in V$$

(33) $$\int_{\Gamma_3} (\sigma_T u_T + g|u_T|) d\Gamma = 0.$$

On déduit de (32) que

(34) $$\left| \int_{\Gamma_3} \sigma_T v_T d\Gamma \right| \leq \int_{\Gamma_3} g|v_T| d\Gamma .$$

Il en résulte que l'application linéaire $v_T \rightarrow \int_{\Gamma_3} \sigma_T v_T d\Gamma$ de $(H^{1/2}(\Gamma_3))^2$ dans $\mathbb{R}$ est linéaire continue de $(L^1(\Gamma_3))^2$ dans $\mathbb{R}$. Il s'en suit que

(35) $$\sigma_T \in (L^\infty(\Gamma_3))^3$$

et que

(36) $$|\sigma_T| \leq g \qquad \text{sur } \Gamma_3 .$$

Il en résulte alors que

(37) $$\sigma_T u_T + g|u_T| \geq 0$$

et d'après (33)

(38) $$\sigma_T u_T + g|u_T| = 0 \qquad \text{pp. sur } \Gamma_3$$

ce qui implique

$$(39) \qquad \begin{cases} |\sigma_T| < g \longrightarrow u_T = 0 \\ |\sigma_T| = g \Longrightarrow \exists k \geq 0 \text{ tel que } u_T = - k\, \sigma_T . \end{cases}$$

et donc (7) est satisfait par la solution u.

REFERENCES

C. BAIOCCHI [1]. Sur un problème à frontière libre traduisant le filtrage de liquides à travers des milieux poreux. Note aux C.R.A.S. Dec. 1971.

BOUCHER [1]. Thèse de 3e cycle. Université Paris VI. A paraître.

H. BREZIS [1]. Thèse de doctorat d'état. Université Paris VI (1971).

H. BREZIS et G. DUVAUT [1]. Ecoulement avec sillage autour d'un profil symétrique sans incidence. Note aux C.R.A.S. Paris t. 276 (mars 1973) série A, page 875.

H. BREZIS et G. STAMPACCHIA [1]. Note aux C.R.A.S. Paris t. 276, série A 1973, page 129.

G. DUVAUT [1]. Le problème de Signorini en viscoélasticité linéaire C.R.A.S. Paris, tome 268 (1969), page 1044.

[2]. Problème dynamique en élastoviscoplasticité et plasticité parfaite avec frottement. Séminaire de plasticité Ecole Polytechnique (1972).

[3]. Résolution d'un problème de Stéfan. A paraître aux C.R.A.S. Paris 1973.

G. DUVAUT et J.L. LIONS [1]. Elasticité avec frottement. Journal de Mécanique, 1971, vol. 10, n° 3.

[2]. Les inéquations en mécanique et en physique. Dunod 1972

[3]. Problèmes unilatéraux dans la théorie des plaques en

forte flexion. A paraître au Journal de Mécanique 1973.

[4]. Inéquations en Thermoélasticité et magnétohydrodynamique. Arch. f. Rat. Mech. Anal. Vol 46, n°4, 1972, p. 241.

G. FICHERA [1]. Mem. Acad. Naz. Lincei, 8(7), 1964, p.91

P. GERMAIN [1]. Mécanique des milieux continus. Masson Paris (19).

H. LANCHON [1]. Problème d'élastoplasticité avec la loi de Hencky. C.R.A.S. Paris, tome 271 (1970), p. 888.

[2]. Thèse de doctorat d'état. Université Paris VI (1972).

F. LENE [1]. Sur les matériaux à énergie de déformation non quadratique. Thèse de 3e cycle, Université Paris VI, 1973.

J.L. LIONS [1]. Contrôle optimal. Dunod 1968.

J.L. LIONS et E. MAGENES [1]. Problèmes aux limites non linéaires et applications. Tome 1, Dunod (19).

J.L. LIONS et G. STAMPACCHIA [1]. Variational inequalities. Com. Pure Applied math. XX, 1967, p. 493-519.

J. MOREAU [1]. Sur l'évolution d'un système élasto-visco-plastique. C.R.A.S. t. 273, 1971, p. 118-121.

[2]. Rafle par un convexe variable. Séminaire d'analyse unilatérale Montpellier. Exposés n° 15 (1971) et 3 (1972).

B. NAYROLES [1]. Essai de théorie fonctionnelle des structures rigides plastiques parfaites. Journal de Mécanique. Vol 9, n°2 (1970) p. 491.

[2]. Opérations algébriques en Mécanique des structures. C.R.A.S. Paris, t. 273 (1971), p. 1075-1078.

G. Duvaut

W. PRAGER [1]. On ideal lecking materials. Transactions of the society of rheology. Vol 1, 1957, p. 169-175.

G. Duvaut

DEUXIEME ET TROISIEME CONFERENCES

PROBLEME DYNAMIQUE EN ELASTO-VISCOPLASTICITE ET PLASTICITE PARFAITE AVEC CONDITIONS DE FROTTEMENT A LA FRONTIERE

1) Position du problème.

On désigne par Ω un ouvert borné de $\mathbb{R}^3$ de frontière Γ régulière, Ω étant situé localement d'un même côté de Γ. On suppose que Ω représente la région occupée par un corps plastique dans son état non déformé. Nous nous intéressons ici aux déformations subies par ce corps lorsqu'il est maintenu par des forces de frottement sur une partie Γ_1, de mesure strictement positive, de Γ, et soumis de plus à une densité variable $\{f_i(t,x)\}$ de forces volumiques.(1)

Nous désignons par $u = \{u_i\}$, $\sigma = \{\sigma_{ij}\}$, i, j prenant les valeurs 1, 2, 3 les champs de déplacements et contraintes dans Ω. Le matériau est supposé obéir à une loi élasto-viscoplastique dans le premier temps, puis élasto-plastique (loi de Prandtl-Reuss). La loi de frottement retenue est la loi de Coulomb, mais dans une première étape, nous utiliserons une loi régularisée par rapport à cette dernière. Au n°2 nous donnerons les équations et conditions du problème.

---

(1) On pourra trouver une étude des problèmes de plasticité sans frottement dans G. Duvaut et J.L. Lions [1].

Au n°3 les diverses formulations variationnelles seront établies. Le n°4 est destiné à l'énoncé des résultats sous forme de trois théorèmes d'existence et unicité des solutions des problèmes. Les démonstrations détaillées des théorèmes seront données dans un article de l'A. à paraître au Journal de Mécanique.

2) Mise en équations.

Avec les notations introduites précédemment nous avons :

$$(1) \qquad u''_i = \sigma_{ij,j} + f_i \quad \text{dans } \Omega \quad (1)$$

où la densité du matériau a été prise égale à l'unité, ce qui ne restreint pas la généralité. Nous introduisons un ensemble convexe fermé K de l'espace vectoriel des tenseurs des contraintes, l'intérieur de K correspondant au comportement élastique, et nous désignons par P_K l'opérateur de projection orthogonale sur cet ensemble convexe K. Une loi de comportement élasto-viscoplastique s'énonce alors :

$$(2) \quad \begin{cases} \varepsilon_{ij}(u') = A_{ijkh}\sigma'_{kh} + \lambda_{ij} \\ \lambda_{ij} = \dfrac{1}{2\mu} \quad \beta_{ij}(\sigma) \end{cases}$$

où $\beta(\sigma) = \sigma - P_K(\sigma)$, et où le scalaire positif μ joue le rôle d'une

---

(1) On utilise les notations suivantes : $X' = \dfrac{\partial X}{\partial t}$, $X_{,j} = \dfrac{\partial X}{\partial x_j}$

et la convention de sommation sur les indices répétés, ainsi :

$$\sigma_{ij,j} = \sum_{j=1}^{3} \sigma_{ij,j} .$$

viscosité [1].

Les coefficients A_{ijkh} sont des coefficients d'élasticité satisfaisants à :

$$A_{ijkh} = A_{jikh} = A_{khij}$$

$$A_{ijkh}\tau_{ij}\tau_{kh} \geq C_o\tau_{ij}\tau_{ij} \quad , \quad \forall \tau_{ij} = \tau_{ji}$$

où C_o est une constante strictement positive.

De plus, on a posé classiquement :

$$\varepsilon_{ij}(v) = \frac{1}{2}\left(\frac{\partial v_i}{\partial x_j} + \frac{\partial v_j}{\partial x_i}\right) .$$

Les conditions aux limites s'écrivent :

(3) $\sigma_{ij}n_j = 0$ sur $\Gamma - \Gamma_1 = \Gamma_2$

où $\vec{n} = \{n_1, n_2, n_3\}$ est la normale extérieure unitaire à Γ.

Désignons par σ_N et σ_T les composantes normales et tangentielles du vecteur $\sigma_{ij}n_j$ en un point de Γ. Nous imposons

(4) $\sigma_N = F_N$ sur Γ_1

où F_N est un scalaire donné négatif indépendant de t. La loi de frottement de Coulomb conduit alors à :

(5)
$$\begin{cases} |\sigma_T| \leq g & \\ |\sigma_T| < g & \Longrightarrow \quad u'_T = 0 \\ |\sigma_T| = g & \Longrightarrow \quad \exists k \geq 0 \text{ tel que } u'_T = -k\sigma_T \ , \end{cases}$$

où g est un scalaire donné positif indépendant de t.

---

(1) Les résultats de ce travail peuvent être généralisés au cas d'une loi avec potentiel viscoplastique (G. Mandel [1]).

Enfin on impose les conditions initiales :

(6) $$u(0,x) = u'(0,x) = \sigma(0,x) = 0.$$

Nous allons immédiatement faire quelques

Remarques :

i) L'inconnue u(t,x) n'intervenant que dans (6) nous poserons comme inconnue à la place de u

(7) $$v = u'$$

de sorte que grâce à (6)

(8) $$u(t,x) = \int_0^t v(t',x)dt' .$$

ii) Pour les besoins des démonstrations la loi de frottement de Coulomb devra dans un premier temps être remplacée par la loi de frottement régularisée

(9) $$\begin{cases} v_T = -\frac{1}{\varepsilon} \lambda(\sigma) \\ \lambda(\sigma) = 0 \qquad \text{si} \quad |\sigma_T| < g \\ \lambda(\sigma) = (1 - \frac{g}{|\sigma_T|}) \sigma_T \quad \text{si} \quad |\sigma_T| \geq g, \end{cases}$$

où ε est un paramètre positif. On retrouve, à partir de (9) formellement, la loi de Coulomb lorsque $\varepsilon \to 0$.

iii) Lorsque $\mu \to 0$ dans la loi élasto-viscoplastique (2) on trouve la loi de Prandtl-Reuss (P. Germain [1]),

(10) $$\begin{cases} \varepsilon_{ij}(v) = A_{ijkh}\sigma'_{kh} + \lambda_{ij} \\ \lambda_{ij} \in \{\partial\chi_K(\sigma)\}_{ij} \end{cases}$$

où $\partial\chi_K(\sigma)$ désigne le sous-gradient de la fonction indicatrice χ_K du convexe K (Moreau [1], Roccafellar [ ]).

iv) Physiquement le problème posé est celui de la déformation d'une pièce retenue sur la surface Γ_1 par une machoire et soumise par ailleurs à des forces volumiques f(t,x) variables. Il est clair que ce problème n'a de sens que si, en l'absence des forces f(t,x), il existe un équilibre possible, c'est-à-dire, un tenseur des contraintes convenable soit :

$$(11)\quad \begin{cases} \exists \text{un champ de tenseur des contraintes } \tau_o \text{ tel que :} \\ \tau_{oij,j} = 0 \ , \quad \beta(\tau_o) = 0 \quad \text{dans } \Omega \\ \tau_{oij}n_j = 0 \quad \text{sur } \Gamma_2 \ , \\ \tau_{oN} = F_N \ , \quad \lambda(\tau_o) = 0 \qquad \text{sur } \Gamma_1 \ . \end{cases}$$

Nous retiendrons l'hypothèse (11) dans toute la suite de ce travail.

3) Formulations variationnelles.

On introduit les espaces :

$$H = (L^2(\Omega))^3 \quad , \qquad V = (H^1(\Omega))^3 \ ,$$

munies des structures hilbertiennes respectives

$$(12)\qquad (v,w) = \int_\Omega v_i w_i dx \ , \quad \forall v,\ w \in H \ ,$$

$$(13)\qquad ((v,w)) = (v,w) + \int_\Omega \varepsilon_{ij}(v)\varepsilon_{ij}(w)dx \ , \quad \forall v,\ w \in V,$$

$$(14)\qquad \tilde{H} = \{\tau \mid \tau = \{\tau_{ij}\},\ \tau_{ij} = \tau_{ji} \ ,\ \tau_{ij} \in L^2(\Omega)\ \}$$

$$(15)\qquad \tilde{V} = \{\tau \mid \tau \in \tilde{H} \ ,\ \tau_{ij,j} \in L^2(\Omega)\ \}$$

munis des structures hilbertiennes respectives

$$(16)\qquad (\tau,\sigma) = \int_\Omega \tau_{ij}\sigma_{ij}dx \qquad \forall \tau,\ \sigma \in \tilde{H}$$

$$(17)\qquad ((\tau,\sigma)) = \int_\Omega \tau_{ij,j}\sigma_{ik,k}dx + (\tau,\sigma) \ , \ \forall \tau,\ \sigma \in \tilde{V}.$$

Enfin nous posons :

(18) $\Sigma ad = \{\tau \mid \tau \in \tilde{V}, \tau_{ij}n_j = 0 \text{ sur } \Gamma_2, \tau_N = F_N \text{ sur } \Gamma_1\}$,

ce qui a un sens. De plus, on vérifie immédiatement que Σad est un sous-espace affine fermé de $\tilde{V}$.

Nous allons maintenant obtenir les formulations variationnelles des divers problèmes envisagés suivant que le matériau est élasto-viscoplastique ou obéit à la loi de Prandtl-Reuss et que le frottement obéit à la loi de Coulomb ou à la loi de frottement régularisée (9).

a) Loi élasto-viscoplastique - Frottement régularisé.

Supposons que $\{v,\sigma\}$ satisfasse aux équations et conditions

(19) $v \in V, \sigma \in \tilde{V} \cap \Sigma ad, \quad \forall t \in]0,T[, \quad T > 0$

(20) $v_i' = \sigma_{ij,j} + f_i \text{ dans } \Omega, \quad \forall t \in]0,t[$

(21) $v(o)x) = \sigma(o)x) = 0 \text{ dans } \Omega$

ainsi qu'aux lois (2), où u' est remplacé par v, et (9).

Soit alors $\tau \in \Sigma ad$.

Multiplions (2) par $\tau_{ij} - \sigma_{ij}$ et intégrons sur Ω ; il vient après quelques transformations :

(22) $$\mathcal{A}(\sigma',\tau-\sigma) + \frac{1}{\mu}(\beta(\sigma),\tau-\sigma) + \int_\Omega v_i(\tau_{ij,j} - \sigma_{ij,j})dx = \\ = \int_{\Gamma_1} V_T(\tau_T - \sigma_T)d\Gamma ,$$

où on a posé :

(23) $$\mathcal{A}(\tau,\sigma) = \int_\Omega A_{ijkh}\sigma_{kh}\tau_{ij}dx .$$

Utilisant (9) et posant :

(24) $$(\lambda(\sigma),\tau)_{\Gamma_1} = \int_{\Gamma_1} \lambda(\sigma)\, \tau_T d\Gamma$$

il vient,

$$\mathcal{A}(\sigma',\tau-\sigma) + \frac{1}{\mu}(\beta(\sigma),\tau-\sigma) + \frac{1}{\varepsilon}(\lambda(\sigma),\tau-\sigma)_{\Gamma_1} + \int_\Omega v_i(\tau_{ij,j}-\sigma_{ij,j})dx = 0,$$

$$\forall \tau \in \Sigma ad .$$

Par ailleurs, si $w \in V$, on obtient à partir de (20),

$$(v',w) - \int_\Omega \sigma_{ij,j} w_i dx = (f,w), \qquad \forall w \in V.$$

<u>En résumé, si</u> $\{v,\sigma\}$ <u>est une solution du problème</u> (19)(20)(21)(2)(9) <u>on a</u> :

(25) $\{v,\sigma\} \in V \times \Sigma ad , \qquad \forall t \in]0,T[$

(26) $\mathcal{A}(\sigma',\tau-\sigma) + \frac{1}{\mu}(\beta(\sigma),\tau-\sigma) + \frac{1}{\varepsilon}(\lambda(\sigma),\tau-\sigma)_{\Gamma_1} +$

$+ \int_\Omega v_i(\tau_{ij,j}-\sigma_{ij,j})dx = 0 , \qquad \forall \tau \in \Sigma ad$

$\forall t \in]0,T[$

(27) $(v',w) - \int_\Omega \sigma_{ij,j} w_i dx = (f,w), \quad \forall w \in V , \quad \forall t \in]0,T[$

(28) $v(0,x) = \sigma(0,x) = 0, \qquad \forall x \in \Omega .$

On dira que (25)-(28) constitue la <u>formulation variationnelle</u> du problème posé.

<u>Inversement</u>, on peut montrer (G. Duvaut [1]) que si le couple $\{v,\sigma\}$ satisfait (25)-(28), il satisfait aussi (19)(20)(21)(2)(9) au moins en un certain sens.

b) <u>Loi élasto-viscoplastique. Loi de frottement de Coulomb.</u>

Supposons que $\{v,\sigma\}$ satisfasse (19)-(21)(2)(5), où n'a été remplacé par v. Soit alors $\tau \in \Sigma ad$, $|\tau_T| \leq g$ sur Γ_1. Multiplions (2) par $\tau_{ij} - \sigma_{ij}$ et intégrons sur Ω. Il vient :

(29) $\mathcal{A}(\sigma',\tau-\sigma) + \frac{1}{\mu}(\beta(\sigma),\tau-\sigma) + \int_\Omega v_i(\tau_{ij,j} - \sigma_{ij,j})dx =$

$$= \int_{\Gamma_1} v_T(\tau_T - \sigma_T)d\Gamma \ .$$

On vérifie alors facilement, grâce à la loi de Coulomb, que le dernier terme est positif ou nul. Il en résulte que :

Si $\{v,\sigma\}$ est une solution de (19)-(21)(2)(5) on a :

(30) $\{v,\sigma\} \in V \times \Sigma ad, \quad |\sigma_T| \leq g \quad \text{sur } \Gamma_1$,

(31) $\mathcal{A}(\sigma',\tau-\sigma) + \frac{1}{\mu}(\beta(\sigma),\tau-\sigma) + \int_\Omega v_i(\tau_{ij,j}-\sigma_{ij,j})dx \geq 0$

$$\forall \tau \in \Sigma ad, \ |\tau_T| \leq g \ \text{sur } \Gamma_1 ,$$

(32) $(v',w) - \int_\Omega \sigma_{ij,j} w_i dx = (f,w) , \qquad \forall w \in V$

(33) $v(o) = \sigma(o) = 0$.

ces propriétés étant satisfaites quel que soit $t \in]0,T[$.

Ceci constitue la formulation variationnelle du problème (19)-(21)(2)(5).

Inversement on peut montrer que si $\{v,\sigma\}$ satisfait (30)-(33), il satisfait aussi (19)-(21)(2)(5) en un certain sens.

c) Loi de Prandtl-Reuss. Loi de frottement de Coulomb.

Supposons que le couple $\{v,\sigma\}$ satisfasse (19)-(21)(5)(10). Les équations (10) sont équivalentes à :

(34) $\sigma \in K$

(35) $A_{ijkh}\sigma'_{kh}(\tau_{ij}-\sigma_{ij}) \geq \varepsilon_{ij}(v)(\tau_{ij}-\sigma_{ij}), \qquad \forall \tau \in K$.

ce qui va nous permettre d'obtenir une formulation variationnelle. Introduisons à cet effet l'ensemble convexe :

(36) $\mathcal{K} = \{\tau \mid \tau \in \Sigma ad \ , \{\tau_{ij}\} \in K \ \text{p.p. dans } \Omega, |\tau_T| \leq g \text{ sur } \Gamma_1\}$.

Choisissons alors $\tau \in K$ dans (35) et intégrons sur Ω. Il vient :

(37) $$\mathcal{A}(\sigma',\tau-\sigma) \geq - \int_\Omega v_i(\tau_{ij,j} - \sigma_{ij,j})dx + \int_{\Gamma_1} v_T(\tau_T - \sigma_T)d$$
$$\geq - \int_\Omega v_i(\tau_{ij,j} - \sigma_{ij,j})dx,$$

en utilisant la loi de Coulomb.

On peut alors énoncer,

si $\{v,\sigma\}$ est une solution de (19)-(21)(2)(5) on a :

(38) $\{v,\sigma\} \in V \times \mathcal{K} \ , \ \forall t \in]0,T[$

(39) $\mathcal{A}(\sigma',\tau-\sigma) + \int_\Omega v_i(\tau_{ij,j} - \sigma_{ij,j})dx \geq 0, \qquad \forall \tau \in \mathcal{K}$

(40) $(v',w) - \int_\Omega \sigma_{ij,j} w_i dx = (f,w), \qquad \forall w \in V$

(41) $v(o) = \sigma(o) = 0,$

ces propriétés ayant lieu $\forall t \in]0,T[$.

Inversement on peut montrer (38)-(41) impliquent (19)-(21)(2)(5) en un certain sens.

4) Résultats.

Dans le cas d'un matériau élasto-viscoplastique avec la loi de frottement régularisée, on obtient le :

Théorème 1

Sous les hypothèses données au n° 1 sur Ω, l'hypothèse (11) et

(42) $f, f' \in L^\infty(0,T \ ; H), \ F_N \in H^{-1/2}(\Gamma_1)$ (1)

---

(1) L'hypothèse $F_N \in L^2(\Gamma_1)$ serait en fait suffisante pour les applications.

<u>il existe une unique solution</u> $\{v_{\varepsilon\mu}\sigma_{\varepsilon\mu}\}$ <u>de (25)-(28) dans la classe fonctionnelle</u>

(43) $\{v_{\varepsilon\mu},\sigma_{\varepsilon\mu}\} \in L^{\infty}(0,T\ ;\ H \times \tilde{H}) \cap L^{2}(0,T\ ;\ V \times \tilde{V})$

(44) $\{v'_{\varepsilon\mu},\sigma'_{\varepsilon\mu}\} \in L^{\infty}(0,T\ ;\ H \times \tilde{H})$.

Dans le cas d'un matériau élasto-viscoplastique avec la loi de frottement de Coulomb, on obtient le :

<u>Théorème 2</u>

<u>Sous les hypothèses du théorème 1, il existe une unique solution</u> $\{v_{\mu}\sigma_{\mu}\}$ <u>de (30)-(33) dans la classe fonctionnelle</u>

(45) $\{v_{\mu}\sigma_{\mu}\} \in L^{\infty}(0,T\ ;\ H \times \tilde{H}) \cap L^{2}(0,T\ ;\ V \times \tilde{V})$

(46) $\{v'_{\mu}\sigma'_{\mu}\} \in L^{\infty}(0,T\ ;\ H \times \tilde{H})$.

<u>De plus cette solution</u> $\{v_{\mu}\sigma_{\mu}\}$ <u>est limite de</u> $\{v_{\varepsilon\mu}\sigma_{\varepsilon\mu}\}$, <u>lorsque</u> ε <u>tend vers zéro, dans l'espace</u> $L^{\infty}(0,T\ ;\ H \times \tilde{H})$ <u>faible étoile</u> [(1)].

Dans le cas d'un matériau obéissant à la loi de Prandtl-Reuss (élastique parfaitement plastique) et la loi de frottement de Coulomb, on obtient le :

---

(1) Sur le dual X' d'un espace de Banach X non réflexif, on appelle topologie faible étoile celle liée à la dualité entre X' et X, soit,

si $f_n \in X'$, $f \in X'$ on dira que :

$f_n \to f$, quand $n \to \infty$, dans X' faible étoile

si $f_n(x) \to f(x)$, quand $n \to \infty$, $\forall x \in X$.

Théorème 3

Sous les hypothèses du théorème 1, il existe une unique solution $\{v,\sigma\}$ de (38)-(41) dans la classe,

$$(47) \qquad \{v,\sigma\} \in L^\infty(0,T\ ;\ H \times \tilde{H}) \cap L^2(0,T\ ;\ V \times \tilde{V})$$

$$(48) \qquad \{v',\sigma'\} \in L^\infty(0,T\ ;\ H \times \tilde{H}).$$

De plus cette solution $\{v,\sigma\}$ est limite de $\{v_\mu,\sigma_\mu\}$, lorsque tend vers zéro, dans l'espace $L^\infty(0,T\ ;\ H \times \tilde{H})$ faible étoile.

5) Démonstration du théorème 1.

5.1. Unicité. Supposons qu'on ait deux solutions $\{v,\sigma\}$ et $\{v^*,\sigma^*\}$ du problème élasto-visco-plastique avec frottement régularisé (1).

Choisissons dans (26) et (27) écrites pour la solution $\{v,\sigma\}$ (resp v^*,σ^*), $\tau = \sigma^*$ et $w = v^* - v$ (resp $\tau = \sigma$ et $w = v = v^*$) et ajoutons membre à membre. Il vient en tenant compte de la monotonie des applications $\tau \to \beta(\tau)$ et $\tau \to \lambda\ (\tau)$

$$(49) \qquad -\frac{d}{dt}\left[\mathcal{A}(\sigma-\sigma^*) + |v - v^*|^2\right] \leq 0$$

où l'on a posé $\mathcal{A}(\tau) = \mathcal{A}(\tau,\tau)$. Il résulte de (49) et des conditions initiales que $\sigma = \sigma^*$ et $v = v^*$

5.2. Existence.

Nous introduisons les notations

$$(50) \qquad [v,w] = \int_\Omega v_{i,j}w_{i,j}dx,\ [\tau,\sigma] = \int_\Omega \tau_{ij,j}\sigma_{ik,k}dx\ ,$$

et un scalaire η positif.

---

(1) Dans ce paragraphe nous négligeons, pour alléger l'écriture, les indices ε,μ.

Si on remplace les équations (26) et (27) par les équations régularisées elliptiquement

(51) $$\mathcal{A}(\sigma',\tau-\sigma) + \eta[\sigma,\tau-\sigma] + \frac{1}{\mu}(\beta(\sigma),\tau-\sigma) + \frac{1}{\varepsilon}(\lambda(\sigma),\tau-\sigma)_{\Gamma_1} + \int_\Omega v_i(\tau_{ij,j}-\sigma_{ij,j})dx \geq 0 \quad , \quad \forall \tau \in \Sigma ad$$

(52) $$(v',w-v) + \eta[v,w-v] - \int_\Omega \sigma_{ij,j}(w_i-v_i)dx = (f,w-v), \forall w \in V,$$

on obtient un système d'équations paraboliques monotones qui possède (J.L. Lions [1]) une unique solution $\{v_\eta,\sigma_\eta\}$ dans la classe

(53) $$\{v_\eta,\sigma_\eta\} \in L^2(0,T ; V \times \tilde{V}), \{v'_\eta,\sigma'_\eta\} \in L^2(0,T ; V' \times \tilde{V}')$$

où V' (resp $\tilde{V}'$) est le dual de V (resp $\tilde{V}$) quand on identifie H (resp $\tilde{H}$) à son dual.

La méthode de démonstration consiste à obtenir des estimations sur $\{v_\eta,\sigma_\eta\}$ de manière à pouvoir passer à la limite lorsque η tend vers zéro dans (51)(52).

Estimations I.

Choisissons $\tau = \tau_o$ introduit par (11) dans (51) et $w = 0$ dans (52) et ajoutons membre à membre. On obtient, en négligeant d'écrire l'indice η,

$$\mathcal{A}(\sigma',\tau_o-\sigma) + \eta[\sigma,-\sigma] + \frac{1}{\mu}(\beta(\sigma),\tau_o-\sigma) + \frac{1}{\varepsilon}(\lambda(\sigma),\tau_o-\sigma)_{\Gamma_1} - (v',v) - \eta[v,v] = -(f,v).$$

Les opérateurs β et λ étant monotone, on a, grâce à (11),

$$(\beta(\sigma),\tau_o-\sigma) = (\beta(\sigma)-\beta(\tau_o),\tau_o-\sigma) \leq 0$$

$$(\lambda(\sigma),\tau_o-\sigma)_{\Gamma_1} = (\lambda(\sigma)-\lambda(\tau_o),\tau_o-\sigma)_{\Gamma_1} \leq 0.$$

En intégrant sur (0,t) il vient alors

$$(54)\qquad \frac{1}{2}\left[\mathcal{A}(\sigma-\tau_o) + |v|^2\right] + \eta\int_o^t([v,v] + [\sigma,\sigma])dt' \leq \int_o^t (f,v)dt' \leq$$

$$\leq \frac{1}{2}\int_o^t |f(t')|^2 dt' + \frac{1}{2}\int_o^t |v(t')|^2 dt'$$

d'où il résulte que

$$(55)\qquad \begin{cases} \{v_\eta,\sigma_\eta\} \in \text{borné de } L^\infty(0,T\ ;\ H\times\tilde{H}) \\ \eta^{1/2}\{v_\eta,\sigma_\eta\} \in \text{borné de } L^2(0,T\ ;\ V\times\tilde{V}), \end{cases}$$

ces majorations étant indépendantes de ε, μ et η.

Estimations II.

Faisant t = 0 dans (51) (52) on obtient

$$(56)\qquad \begin{cases} \mathcal{A}(\sigma_\eta'(o),\tau) = 0 & \forall\tau\in\Sigma ad\ , \\ (v_\eta'(o),w) = (f(0),w), & \forall w\in V, \end{cases}$$

ce qui entraîne que $\sigma_\eta'(o)$ et $v_\eta'(o)$ sont bornés dans $\tilde{H}$ et H respectivement.

Soit alors un scalaire h > 0 destiné à tendre vers zéro. Ecrivons (51)(52) pour l'instant (resp t + h) avec $\{w,\tau\} = \{v(t+h),\ \sigma(t+h)\}$, (resp $\{w,\tau\} = \{v(t),\sigma(t)\}$) et ajoutons membre à membre les égalités obtenues

$$(57)\qquad \mathcal{A}(\sigma'(t+h)-\sigma'(t),\ (t+h)-\sigma(t)) + (v'(t+h)-v'(t),v(t+h)-v(t)) +$$

$$+\ \eta\{[\sigma(t+h)-\sigma(t),\sigma(t+h)-\sigma(t)] + [v(t+h)-v(t),\ v(t+h)-v(t)]\} \leq$$

$$\leq (f(t+h)-f(t),\ v(t+h)-v(t)).$$

On divise cette inégalité par h^2, puis fait tendre h vers zéro. On obtient alors que,

$$(58)\quad \begin{cases} \{v'_\eta(t),\sigma'_\eta(t)\} \in \text{borné de } L^\infty(0,T\ ;\ H\times \tilde{H}) \\ \eta^{1/2}\{v'_\eta,\ \sigma'_\eta\} \in \text{borné de } L^2(0,T\ ;\ V\times \tilde{V})\ , \end{cases}$$

ces estimations étant indépendantes de ε, μ, η.

Quel que soit $w \in V$ nous avons

$$(59)\qquad (v'_\eta,w) + \int_\Omega \sigma_{ij}\varepsilon_{ij}(w) + \eta[v_\eta,w] - \int_{\Gamma_1} F_N w_N d\Gamma - \\ - \int_{\Gamma_1} \dot{\sigma}_{\eta T} w_T d\Gamma = (f,w).$$

Il en résulte, compte tenu des estimations I et II que l'application

$$w \in V \to \int_{\Gamma_1} \sigma_{\eta T} w_T d\Gamma = \langle \xi(\sigma),w\rangle_{\Gamma_1}$$

est bornée dans $L^2(0,T\ ;\ V')$.

Introduisons alors

$$(60)\qquad \tilde{V}_o = \{\tau \mid \tau \in \tilde{V},\quad \tau_N = 0 \quad \text{sur } \Gamma_1\}$$

et l'élément $\Lambda(\sigma)$ de $\tilde{V}'_o$, dual de $\tilde{V}_o$ quand on identifie $\tilde{H}$ et son dual, par

$$(61)\qquad \langle\Lambda(\sigma),\tau\rangle = \frac{1}{\mu}(\beta(\sigma),\tau) + \frac{1}{\varepsilon}(\lambda(\sigma),\ \tau_T)_{\Gamma_1}\ .$$

Soit alors $\nu \in \tilde{V}_o$; choisissons $\tau = \sigma_\eta + \nu$, ce qui est loisible.

Alors

$$(62)\quad \mathcal{A}(\sigma'_\eta,\nu) + \eta[\sigma_\eta,\nu] + \langle\Lambda(\sigma_\eta),\nu\rangle + \int_\Omega v_{\eta i}\nu_{ij,j}dx = 0,\quad \forall \nu \in \tilde{V}_o.$$

Compte tenu des estimations I et II il en résulte que

$$(63)\qquad \Lambda(\sigma_\eta) \in \text{borné de } L^2(0,T\ ;\ \tilde{V}'_o)\ .$$

<u>Passage à la limite</u>.

Il résulte des estimations précédentes que l'on peut extraire de v_η, σ_η une sous-suite, encore notée v_η, σ_η telle que

$$v_\eta, v'_\eta \to v, v' \quad \text{dans} \quad L^\infty(0,T\ ;\ H) \quad \text{faible étoile}$$

$$\sigma_\eta, \sigma'_\eta \to \sigma, \sigma' \quad \text{dans} \quad L^\infty(0,T\ ;\ \tilde{H}) \quad \text{faible étoile}$$

$$\Lambda(\sigma_\eta) \to \chi \quad \text{dans} \quad L^2(0,T\ ;\ \tilde{V}'_o) \quad \text{faible}$$

$$\xi(\sigma_\eta) \to \xi \quad \text{dans} \quad L^2(0,T\ ;\ V') \quad \text{faible.}$$

Nous pouvons alors, en utilisant les méthodes exposées dans J.L. Lions [1], passer à la limite dans

$$(64)\begin{cases} \mathcal{A}(\sigma'_\eta, \tau-\tau_o) + \eta[\tau_\eta, \tau-\tau_o] + \langle\Lambda(\sigma_\eta), \tau-\tau_o\rangle + \int_\Omega v_{\eta i}\tau_{ij,j}dx = 0 \\ (v'_\eta, w) + \eta[v_\eta, w] + \int_\Omega \sigma_{\eta ij}\varepsilon_{ij}(w)dx - \int_{\Gamma_1} F_N w_N d\Gamma - \int_{\Gamma_1} \sigma_{\eta T} w_T d\Gamma = 0 \end{cases}$$

pour obtenir

$$(65)\quad \begin{cases} \mathcal{A}(\sigma', \tau-\tau_o) + \langle\chi, \tau-\tau_o\rangle + \int_\Omega v_i\tau_{ij,j}dx = 0 & \forall \tau \in \mathcal{U}ad \\ (v', w) + \int_\Omega \sigma_{ij}\ _{ij}(w)dx - \int_{\Gamma_1} F_N w_N d\Gamma - \langle\xi, w\rangle_{\Gamma_1} = 0 & \forall w \in V. \end{cases}$$

Choisissons alors

$$(66)\qquad w_i = \phi_i \in \mathcal{D}(\Omega)$$

dans la 2e équation (65). Il vient

$$(67)\qquad v'_i = \sigma_{ij,j} + f_i$$

ce qui prouve que $\sigma_{ij,j} \in L^2(\Omega)$. Multipliant alors (67) par $w_i \in H^1(\Omega)$, et intégrant par parties sur Ω (ce qui est licite) il vient

$$(68)\quad (v'_i, w_i) + \int_\Omega \sigma_{ij}\varepsilon_{ij}(w)dx - \int_{\Gamma_1} \sigma_N w_N d\Gamma - \int_{\Gamma_1} \sigma_T w_T d\Gamma - \\ - \int_{\Gamma_2} \sigma_{ij} n_j w_i d\Gamma = (f, w)$$

d'où par comparaison avec (65)

$$\int_{\Gamma_1} (\sigma_N - F_N) w_N d\Gamma + \int_{\Gamma_1} \sigma_T w_T d\Gamma - \langle\xi, w\rangle_{\Gamma_1} + \int_{\Gamma_2} \sigma_{ij} n_j w_i d\Gamma = 0 \qquad \forall w \in V$$

ce qui implique facilement que

$$(69) \qquad \begin{cases} \sigma_N = F_N \ \text{ sur } \Gamma_1 \,, \quad \sigma_{ij} n_j = 0 \ \text{ sur } \Gamma_2 \\ \langle \xi, w \rangle_{\Gamma_1} = \int_{\Gamma_1} \sigma_T w_T d\Gamma \qquad \forall w \in V. \end{cases}$$

Il reste à montrer que $\chi = \Lambda(\sigma)$, ce qui s'obient par un raisonnement de monotonie. Quel que soit $\tau \in L^2(\tilde{V})$ nous avons

$$(70) \qquad \int_0^T (\Lambda(\sigma_\eta) - \Lambda(\tau), \sigma_\eta - \tau) dt \geq 0,$$

d'où, en utilisant (64) aménagée,

$$(71) \qquad \int_0^T (\Lambda(\tau), \tau - \sigma_\eta) dt \geq \int_0^T (\Lambda(\sigma_\eta), \tau - \sigma_\eta) dt$$

$$\geq \frac{1}{2} (\sigma_\eta) + \frac{1}{2} (v_\eta)^2 - \int_0^T (f, w) dt - \int_0^T (\sigma'_\eta, \tau) d\tau - \eta[\sigma_\eta, \tau].$$

Prenant la limite inférieure des deux membres, utilisant la semi-continuité inférieure de $\mathcal{A}(\sigma_\eta) + |v_\eta|^2$ et (65), il vient

$$(72) \qquad \int_0^T (\chi - \Lambda(\tau), \sigma - \tau) dt \geq 0.$$

Choisissant alors $\tau = \sigma + \lambda\mu$ où $\lambda > 0$, $\mu \in L^2(\tilde{V})$, il vient après division par λ et passage à la limite $\lambda \to 0$,

$$\int_0^T (\chi - \Lambda(\sigma), \mu) dt \geq 0$$

ce qui implique

$$(73) \qquad \Lambda(\sigma) = \chi \,,$$

ce qui achève la démonstration du théorème 1.

6) Démonstration du théorème 2.

6.1. Unicité. Elle se démontre comme pour le théorème 1.

6.2. Existence. Nous désignons par $\{v_\varepsilon, \sigma_\varepsilon\}$ la solution obtenue au théorème 1 et définie par

$$(74) \qquad \{v_\varepsilon, \sigma_\varepsilon\} \in L^2(0,T \,;\, V \times \tilde{V}) \qquad \sigma_\varepsilon(t) \in \mathcal{U}ad,$$

(75) $\mathcal{A}(\sigma'_\varepsilon,\tau-\sigma_\varepsilon) + \frac{1}{\mu}\,(\beta(\sigma_\varepsilon),\tau-\sigma_\varepsilon) + \frac{1}{\varepsilon}\,(\lambda(\sigma_\varepsilon),\tau-\sigma_\varepsilon)_{\Gamma_1} +$

$$+ \int_\Omega v_{\varepsilon i}(\tau_{ij,j}-\sigma_{\varepsilon ij,j})dx = 0 \qquad \forall \tau \in \mathcal{U}ad$$

(76) $$v'_\varepsilon - \sigma_{\varepsilon ij,j} = f_i$$

Introduisons la forme linéaire $\Lambda_1(\sigma_\varepsilon)$ sur V_o par

$$< \Lambda_1(\sigma_\varepsilon),\ \tau > = \int_{\Gamma_1} \lambda(\sigma_\varepsilon).\tau d\Gamma \ , \qquad \forall \tau \in \tilde{V}_o \ .$$

Il résulte alors des estimations du n° 5 que

(77) $\frac{1}{\varepsilon}\ \Lambda_1(\sigma_\varepsilon) \in$ borné de $L^2(0,T\ ;\ \tilde{V}_o)$.

De plus

(78)
$$\begin{cases} \{v_\varepsilon,\ \sigma_\varepsilon\} & \in & \text{borné de} & L^\infty(0,T\ ;\ H\times\tilde{H}) & \\ \{v'_\varepsilon,\sigma'_\varepsilon\} & \in & \text{borné de} & L^\infty(0,T\ ;\ H\times\tilde{H}) & \\ \beta(\sigma_\varepsilon) & \in & \text{borné de} & L^\infty(0,T\ ;\ \tilde{H}) & \\ \sigma_{\varepsilon ij,j} & \in & \text{borné de} & L^\infty(0,T\ ;\ H) & \text{(d'après (76))} \\ \varepsilon_{ij}(v_\varepsilon) & \in & \text{borné de} & L^\infty(0,T\ ;\ \tilde{H}), & \text{(d'après (75)).} \end{cases}$$

Il résulte des estimations (77) et (78) que l'on peut extraire de v_ε, σ_ε une sous suite, encore notée v_ε, σ_ε, telle que

(79)
$$\begin{cases} \{v_\varepsilon,\ \sigma_\varepsilon\} & \to \{v,\sigma\} & \text{dans} & L^\infty(0,T\ ;\ V\times\tilde{V}) & \text{faible étoile} \\ \{v'_\varepsilon,\sigma'_\varepsilon\} & \to \{v',\sigma'\} & \text{dans} & L^\infty(0,T\ ,\ H\times\tilde{H}) & \text{faible étoile} \\ \beta(\sigma_\varepsilon) & \to \xi_1 & \text{dans} & L^\infty(0,T\ ;\ \tilde{H}) & \text{faible étoile} \\ \Lambda_1(\sigma_\varepsilon) & \to 0 & \text{dans} & L^2(0,T\ ;\ V_o) & \text{fort.} \end{cases}$$

Formons alors, $\forall \tau \in L^2(0,T\ ;\ \tilde{V}_o)$,

$$\int_o^T < \Lambda_1(\sigma_\varepsilon) - \Lambda_1(\tau),\ \sigma_\varepsilon - \tau> dt \geq 0$$

et faisons tendre ε vers zéro en utilisant (79) ; il vient

$$\int_o^T < \Lambda_1(\tau), \sigma - \tau > \ dt \geq 0$$

d'où l'on déduit que $\Lambda_1(\sigma) = 0$, d'où

(80) $\qquad |\sigma_T| \leq g \quad \text{sur } \Gamma_1.$

Choisissons alors $\tau \in L^2(0,T ; \tilde{V})$, $\tau(t) \in \mathcal{U}ad$, $|\tau_T| \leq g$ sur Γ_1 ;

en tenant compte de

$$< \Lambda_1(\sigma_\varepsilon), \tau - \sigma_\varepsilon > \ \leq \ 0$$

l'égalité (75) donne

(81) $\qquad \mathcal{A}(\sigma'_\varepsilon, \tau-\sigma_\varepsilon) + \frac{1}{\mu}(\beta(\sigma_\varepsilon), \tau-\sigma_\varepsilon) + \int_\Omega v_{\varepsilon i}(\tau_{ij,j} - \sigma_{\varepsilon ij,j})dx \geq 0$

et en intégrant sur (0,T) et utilisant (76)

(82) $$\int_o^T \frac{1}{\mu}(\beta(\sigma_\varepsilon), -\sigma_\varepsilon)dt = \frac{1}{2}\mathcal{A}(\sigma_\varepsilon(T)) + \frac{1}{2}|v_\varepsilon(T)|^2 - \int_o^T (\sigma'_\varepsilon, \tau)dt -$$
$$- \frac{1}{\mu}\int_o^T (\beta(\sigma_\varepsilon), \tau)dt - \int_o^T v_{\varepsilon i}\tau_{ij,j}dt - \int_o^T (f, v_\varepsilon)dt.$$

Utilisant la semi-continuité inférieure de $\mathcal{A}(\sigma_\varepsilon(T))$ et $|v_\varepsilon(T)|^2$

on peut passer à la limite inférieure des deux membres pour obtenir,

(83) $$\liminf \int_o^T \frac{1}{\mu}(\beta(\sigma_\varepsilon), \sigma_\varepsilon)dt \geq \frac{1}{2}\mathcal{A}(\sigma(T)) + \frac{1}{2}|v(T)|^2 -$$
$$- \int_o^T \mathcal{A}(\sigma', \tau)dt - \frac{1}{\mu}\int_o^T (\xi_1, \tau)dt - \int_o^T v_i \tau_{ij,j}dt - \int_o^T (f, v)dt.$$

On peut passer à la limite dans (76) pour obtenir

(84) $\qquad v' - \sigma_{ij,j} = f_i.$

On en déduit alors pour (83) que

(85) $$\liminf \int_o^T \frac{1}{\mu}(\beta(\sigma_\varepsilon), \sigma - \sigma_\varepsilon)dt \geq - \int_o^T \mathcal{A}(\sigma', \tau - \sigma)dt -$$
$$- \int_o^T \frac{1}{\mu}(\xi_1, \tau - \sigma)dt - \int_o^T v_i(\tau_{ij,j} - \sigma_{ij,j})dt.$$

Utilisant alors la monotonie de l'opérateur β, il s'ensuit que

(86) $$\int_0^T [\mathcal{A}(\sigma',\tau-\sigma) + \frac{1}{\mu}(\xi_1,\tau-\sigma) + (v_i,\tau_{ij,j}-\sigma_{ij,j})]dt \geq 0$$

ce qui, par application d'un théorème de Lebesgue, fournit

(87) $$\mathcal{A}(\sigma',\tau-\sigma) + \frac{1}{\mu}(\xi_1,\tau-\sigma) + (v_i,\tau_{ij,j}-\sigma_{ij,j}) \geq 0$$

$$\forall \tau \in \mathcal{U}ad, \quad |\tau_T| \leq g \quad \text{sur} \quad \Gamma_1.$$

Par ailleurs il est clair, compte tenu des estimations (78), que

$$\sigma_{ij}n_j = 0 \quad \text{sur} \quad \Gamma_2 \quad , \qquad \sigma_N = F_N \quad \text{sur} \quad \Gamma_1 .$$

Il reste donc seulement à montrer que $\xi_1 = \beta(\sigma)$. On utilise pour cela la monotonie de l'opérateur β en procédant comme il a été fait dans le n° 5 pour montrer que $\chi = \Lambda(\sigma)$. Ceci achève la démonstration du théorème 2.

7) Démonstration du théorème 3.

Elle utilise exactement les mêmes techniques que celle du théorème 2 et les estimations obtenues aux n° 5 et 6 qui étaient indépendantes de μ.

REFERENCES

G. DUVAUT [1]. Existence et unicité de la solution d'un problème dynamique en élasto-visco-plasticité et plasticité parfaite avec conditions de frottement à la frontière. Séminaire de plasticité de l'Ecole Polytechnique. Sept. 1972

G. DUVAUT et J.L. LIONS [1]. Les inéquations en mécanique et en physique. Dunod 1972.

[2]. Un problème d'élasticité avec frottement. Journal de Mécanique. Vol. 10, n° 3, Sept. 1971 pp. 409-420.

P. GERMAIN [1]. Cours de mécanique des milieux continus. Tome 1. Masson Paris 1972.

J.L. LIONS [1]. Quelques méthodes de résolution des problèmes aux limites non linéaires. Dunod-Gauthier-Villars 1971

G. MANDEL [1]. Séminaire de Plasticité. Ecole Polytechnique 1971. Publication Sc. et Tech. du Ministère de l'Air n° 116.

J. MOREAU [1]. Fonctionnelles convexes. Collège de France. 1966-1967.

T. ROCCAFELLAR [1]. Convex analysis. Princeton University Press 1970.

G. Duvaut

QUATRIEME CONFERENCE

PLAQUE EN FORTE FLEXION SOUMISE A DES CONDITIONS UNILATERALES

1) Introduction.

Nous nous proposons l'étude de la déformation d'une plaque en forte flexion soumise à des conditions à la frontière unilatérales. La théorie non linéaire des plaques en forte flexion qui conduit aux équations dites de Von Karman est établie dans le livre de Landau et Lifschitz [1] auquel le lecteur pourra se reporter. Une étude générale de ces types de problèmes a été faite par G. Duvaut et J.L. Lions [1] avec certains types de conditions aux limites. D'autres types de conditions aux limites conduisent à des problèmes unilatéraux non linéaires du type semi-coercifs (cf J.L. Lions et G. Stampacchia [1]) et ont été résolus par M. Potier [1]. Lorsqu'on applique à la frontière de la plaque des forces situées dans le plan de cette dernière on peut observer des phénomènes de flambement de la plaque et sur le plan mathématique on est conduit à des problèmes de bifurcation (cf Do [1]).

Ici nous nous proposons de donner une introduction à ces types de problèmes en donnant les équations générales, la formule de Green adaptée et la formulation variationnelle et les résultats d'existence et d'unicité (limitée à des sollicitations assez faibles) dans le cas d'un problème particulier simple. Pour les démonstrations nous renverrons à l'article plus complet de G. Duvaut et J.L. Lions [1] cité précédemment.

2) Problème physique et équations.

Nous considérons une plaque mince qui dans son état non déformé est assimilé à sa trace Ω sur un plan Ox_1x_2. La région Ω est supposée ouverte et bornée dans Ox_1x_2. Sa frontière $\partial\Omega$ est supposée réguliére et constituée de deux parties Γ_1 et Γ_2 avec

(1) $$\text{mes } \Gamma_1 > 0 .$$

La plaque est encastrée le long de Γ_1. Le long de Γ_2 une butée rigide interdit les déplacements vers les x_3 négatifs (le repère $Ox_1x_2x_3$ est orthonormé direct). Nous désignons par $\{u_1,u_2\xi\}$ le vecteur déplacement des points de la plaque et par $\{\sigma_{\alpha\beta}\}$ le tenseur symétrique des contraintes planes, α et β prenant les valeurs 1 et 2. De plus la plaque est soumise à une densité de forces surfaciques f(x) portées par Ox_3.

Les équations d'équlibre et de comportement des plaques en théorie non linéaire s'écrivent,

(2) $$D\Delta^2\xi - h \frac{\partial}{\partial x_\beta} (\sigma_{\alpha\beta} \frac{\partial\xi}{\partial x_\alpha}) = f$$

(3) $$\frac{\partial}{\partial x_\beta} \sigma_{\alpha\beta} = 0$$

(4) $$\sigma_{\alpha\beta} = a_{\alpha\beta\gamma\delta} \left[\varepsilon_{\gamma\delta}(u) + \frac{1}{2} \frac{\partial\xi}{\partial x_\gamma} \frac{\partial\xi}{\partial x_\delta}\right]$$

(5) $$\varepsilon_{\alpha\beta}(u) = \frac{1}{2} (\frac{\partial u_\alpha}{\partial x_\beta} + \frac{\partial u_\beta}{\partial x_\alpha}) , \qquad u = \{u_1,u_2\} ,$$

où h est l'épaisseur de la plaque, D le module de rigidité à la flexion, $a_{\alpha\beta\gamma\delta}$ des coefficients d'élasticité satisfaisants à

$$(6) \quad \begin{cases} a_{\alpha\beta\gamma\delta} = a_{\beta\alpha\gamma\delta} = a_{\gamma\delta\alpha\beta} \\ a_{\alpha\beta\gamma\delta}\, \varepsilon_{\alpha\beta}\varepsilon_{\gamma\delta} \geq \alpha_o \varepsilon_{\alpha\beta}\varepsilon_{\alpha\beta}\,, \quad \alpha_o = \text{Cste} > 0, \quad \forall \varepsilon_{\alpha\beta} = \varepsilon_{\beta\alpha} \end{cases}$$

Les conditions aux limites du problème sont

$$(7) \qquad \xi = \frac{\partial \xi}{\partial n} = 0 \qquad \text{sur } \Gamma_1$$

où $n = \{n_1, n_2\}$ est la normale extérieure unitaire à $\partial\Omega$,

$$(8) \qquad u_1 = u_2 = 0 \qquad \text{sur } \Gamma_1$$

$$(9) \qquad \sigma_{\alpha\beta} n_\beta = 0 \qquad \text{sur } \Gamma_2$$

$$(10 \qquad \xi \geq 0, \quad F \geq 0, \quad \xi . F = 0 \qquad \text{sur } \Gamma_2$$

où F représente la densité linéique de forces extérieures portées par Ox_3 et appliquées le long de Γ_2. Nous serons amenés dans un premier temps à remplacer la condition (10) par une condition régularisée

$$(11) \quad \begin{cases} \xi > 0 \implies F = 0 \\ \xi \leq 0 \implies -F = \frac{1}{\varepsilon}\xi \end{cases}$$

le coefficient positif ε étant donné. La condition (11) exprime que la butée qui limite les déplacements $\xi_{|\Gamma_2} < 0$ est élastique au lieu d'être rigide. Le cas rigide sera obtenu par passage à la limite $\varepsilon \to 0$.

3) <u>Formulation variationnelle.</u>

A partir des équations (2) (3) (4) on établit (cf G. Duvaut et J.L. Lions [1] [2])la formule de Green suivante,

(12) $$a(\xi,z) = \int_\Omega h\sigma_{\alpha\beta}\frac{\partial\xi}{\partial x_\alpha}\frac{\partial z}{\partial x_\beta}dx = \int_\Gamma h\sigma_{\alpha\beta}n_\beta\frac{\partial\xi}{\partial x_\alpha}zd\Gamma +$$
$$+ \int_\Gamma F(\xi)zd\Gamma - \int_\Gamma M(\xi)\frac{\partial z}{\partial n}d\Gamma + \int_\Omega fzdx$$

où $z = z(x_1,x_2)$ est une fonction régulière sur $\bar{\Omega}$, à (ξ,z) la forme bilinéaire

(13) $$a(\xi,z) = D\int_\Omega \left[\frac{\partial^2\xi}{\partial x_1^2}\frac{\partial^2 z}{\partial x_1^2} + \frac{\partial^2\xi}{\partial x_2^2}\frac{\partial^2 z}{\partial x_2^2} + \nu\left(\frac{\partial^2\xi}{\partial x_1^2}\frac{\partial^2 z}{\partial x_2^2} + \frac{\partial^2\xi}{\partial x_2^2}\frac{\partial^2 z}{\partial x_1^2}\right) +\right.$$
$$\left. + 2(1-\nu)\frac{\partial^2\xi}{\partial x_1\partial x_2}\frac{\partial^2 z}{\partial x_1\partial x_2}\right]dx_1dx_2$$

où ν est le coefficient de Poisson du matériau ; $F(\xi)$ et $M(\xi)$ représent nt les densités linéiques de forces et de moments sur $\partial\Omega$. Ce sont des expressions linéaires par rapport à ξ dont l'expression n'est pas nécessaire ici.

Introduisons les espaces

(14) $$\begin{cases} Z_o = L^2(\Omega), \quad Z = \{z | z \in H^2(\Omega), \quad z = \frac{\partial z}{\partial n} = 0 \text{ sur } \Gamma_1\} \\ V_o = (L^2(\Omega))^2, \quad V = \{v | v \in (H^1(\Omega))^2, \quad v = 0 \text{ sur } \Gamma_1\}. \end{cases}$$

Les espaces Z et V sont de HIlbert quand on les munit des produits scalaires respectifs de $H^2(\Omega)$ et de $(H^1(\Omega))^2$. On montre (cf G. Duvaut J.L. Lions [1]) que $a(\xi,z)$ est coercive sur Z et que la forme bilinéaire $\mathcal{A}(\varepsilon(u),\varepsilon(v))$ donnée par

(15) $$\mathcal{A}(h,k) = \int_\Omega a_{ijkh}h_{ij}k_{kh}, \quad \varepsilon(v) = \{\varepsilon_{ij}(v)\},$$

est coecive sur V, grâce à l'hypothèse (1).

Si $\{\xi,u\}$ est une solution régulière du problème (2)-(10) (resp (2)-(9)(11) nous aurons

(16) $\{\xi,u\} \in Z \times V\ ,\quad \xi_{|\Gamma_2} \geq 0$

(17) $a(\xi,z-\xi) + h\mathcal{A}(\varepsilon(u) + \frac{1}{2}M(\xi),M(\xi,z-\xi)) \geq \int_\Omega f(z-\xi)dx$

$\forall z \in Z\ ,\quad z_{|\Gamma_2} \geq 0$

(18) $\mathcal{A}(\varepsilon(u) + \frac{1}{2}M(\xi),\ \varepsilon(v)) = 0 \qquad \forall v \in V$

où on a posé

(19) $M(z,\xi) = \{\frac{\partial z}{\partial x_\alpha}\ \frac{\partial \xi}{\partial x_\beta}\}\ ,\quad M(z) = M(z,z),$

(respectivement

(19) $\{\xi,u\} \in Z \times V$

(20) $a(\xi,z) + h\,\mathcal{A}(\varepsilon(w) + \frac{1}{2}M(\xi),\ M(\xi,z)) - \int_{\Gamma_2} \frac{1}{\varepsilon}\,\xi^-\ zd\Gamma = \int_\Omega f,zdx$

$\forall z \in Z$

(21) $\mathcal{A}(\varepsilon(u) + \frac{1}{2}M(\xi),\ \varepsilon(v)) = 0 \qquad \forall v \in V).$

Les propriétés (16)(17)(18) (resp (19)(20)(21)) constituent les formulations variationnelles des problèmes envisagés. On montre qu'inversement les solutions de (16)(17)(18) (resp (19)(20)(21)) sont solutions, au moins en un sens affaibli des problèmes envisagés.

4) Résultats.

On démontre (cf G. Duvaut et J.L. Lions [1]) les théorèmes suivants :

Théorème 1. Sous les hypothèses (1)(6) et

(22) $f \in L^2(\Omega)$

il existe au moins une solution u_ε, ξ_ε satisfaisant (19)-(21).

Théorème 2. Sous les hypothèses (1)(6)(22) il existe au moins une solution u, ξ satisfaisant (16)-(18). De plus il existe une sous suite de u_ε, ξ_ε qui converge vers u, ζ dans V × Z faible.

Théorème 3. Si f est de la forme

(23) $f = \eta f_o , \quad f_o \in L^2(\Omega), \quad \eta > 0 ,$

il existe une constante positive η_o telle que si $\eta \leq \eta_o$ la solution u, ξ est unique. Si alors on pose

(24) $u = \eta u_\eta , \quad \xi = \eta \xi_\eta$

on a

(25) $\lim_{\eta \to o} u_\eta = 0 , \quad \lim_{\eta \to o} \xi_\eta = \xi_o \quad$ dans V × Z

où ξ_o est l'unique solution du problème de plaque linéaire

(26) $\begin{cases} \xi \in Z \\ a(\xi, z-\xi) \geq \int_\Omega f(z-\xi)dx \end{cases} \quad \forall z \in Z.$

Les résultats analogues s'obtiennent avec la solution régularisée $u_\varepsilon, \xi_\varepsilon$.

REFERENCES

G. DUVAUT et J.L. LIONS [1]. Problèmes unilatéraux dans la théorie des plaques en forte flexion. A paraître au Journal de Mécanique.

L. LANDAU et LIFSCHITZ.[1] Théorie de l'élasticité. Ed. Mir, Moscou 1967.

Cl. DO [1]. Résultat non publié.

M. POTIER [1]. Thèse de 3e cycle. A paraître.

G. Duvaut

CINQUIEME ET SIXIEME CONFERENCES

RESOLUTION D'UN PROBLEME DE STEFAN

(Fusion d'un bloc de glace à 0°)

1) Problème Physique : On considère un bloc de glace à 0° occupant la région ouverte Ω de $\mathbb{R}^3$ de frontière Γ régulière. On suppose Γ composée de trois parties Γ_1, Γ_2, Γ_3 sans points communs et dont la réunion compose Γ. On suppose de plus que Γ_1 et Γ_3 n'ont pas de frontière commune et que Γ_1 est de mesure strictement positive.

On cherche l'évolution de ce bloc de glace lorsque sa frontière Γ_1 est le siège d'un flux de chaleur, les parties Γ_2 et Γ_3 étant respectivement de flux de chaleur nul et de température 0°. On néglige la variation de volume due à la fusion et on suppose que l'eau fournie reste en place. On désigne par $\mathcal{L}(t)$ la surface de fusion d'équation $t = \ell(x)$, inconnue à priori $(x = (x_1,x_2,x_3))$. On désigne par k la chaleur latente de la glace.

2) Mise en équations : On cherche un champ de température $\theta(x,t)$, $x \in \overline{\Omega}$, $t \in [0,T]$, $T > o$, satisfaisant,

(1) $\frac{\partial\theta}{\partial t} - \Delta\theta = o$ dans $t > \ell(x)$

(2) $\theta(x,t) = o$ dans $t \leqslant \ell(x)$

(3) $\text{grad } \theta \cdot \text{gard } \ell = -k$ ([1]) pour $t = \ell(x)$

(4) $-\dfrac{\partial\theta}{\partial n} = b(\theta-\theta_1)$, $(b > o$ donné$)$ sur Γ_1

(5) $\dfrac{\partial\theta}{\partial n} = o$ sur Γ_2

(6) $\theta = o$ sur Γ_3

(7) $\theta(x,o) = o$ pour $x \in \Omega$.

3) <u>Formulation variationnelle</u>

Suivant C. Baiocchi [1] on introduit :

$$(8)\quad \begin{cases} u(x,t) = \displaystyle\int_{\ell(x)}^{t} \theta(x,\tau)\, d\tau & \text{si } t > \ell(x) \\ u(x,t) = o & \text{si } t \leqslant \ell(x)\ . \end{cases}$$

On pose :

$$(9)\quad \begin{cases} H = L^2(\Omega), \quad V = \{v \mid v \in H^1(\Omega), \quad v = o \text{ sur } \Gamma_3\} \\ K = \{v \mid v \in V \ , \quad v \geqslant o \text{ dans } \Omega\}\ . \end{cases}$$

$$(10)\quad \begin{cases} a(u,v) = \displaystyle\int_\Omega \text{grad } u \text{ grad } v\, dx\ , \quad L(v) = -k \int_\Omega v\, dx \\ (u,v) = \displaystyle\int_\Omega u\, v\, dx\ , \quad \forall\, u,\ v \in V. \end{cases}$$

([1]) Soit $\vec{n}$ la normale à $\mathcal{L}(t)$ dirigée vers la région où $\theta = 0$. On écrit que la flux de chaleur $-\text{grad}\theta\cdot\vec{n}$ à travers la surface $\mathcal{L}(t)$ est la chaleur nécessaire à la fusion du volume V_n de glace où V_n est la vitesse normale, mesurée sur $\vec{n}$, de déplacement de $\mathcal{L}(t)$. D'où :

$$-\,\vec{\text{grad}}\ \theta \cdot \vec{n} = k\, V_n\ , \qquad \text{d'où (3)}\ .$$

La fonction $u(x,t)$ satisfait à :

(11) $u(t) \in K \quad , \quad \forall t \in [0,T]$

(12) $u = o$, $\text{grad } u = o$ pour $t = \ell(x)$,

(13) $u = o$ pour $t < \ell(x)$,

(14) $\frac{\partial u}{\partial t} - \Delta u = -k$ pour $t > \ell(x)$,

(15) $-\frac{\partial u}{\partial x} = b(u - \theta_1 t)$ sur Γ_1 ,

(16) $\frac{\partial u}{\partial n} = 0$ sur Γ_2 .

On déduit de (11) - (16) que u satisfait :

$$(17) \begin{cases} u(t) \in K \, , \quad \forall t \in [0,T] \quad \text{p.p.} \, , \\ (u',v-u)+a(u,v-u)+b \displaystyle\int_{\Gamma_1} (u-\theta_1 t)(v-u)\,d\Gamma \geq L(v-u), \quad \forall v \in K \, , \\ u(x,o) = 0 \, . \text{ [1]} \end{cases}$$

<u>Variante</u> :

Si on remplace la condition (4) par :

(4bis) $\theta = \theta_1$ sur Γ_1 ,

les autres conditions étant inchangées, on est conduit, pour u défini par (8) , à la formulation variationnelle suivante :

on introduit $K_1(t)$ par :

(18) $K_1(t) = \{u | v \in V \, , \; v|_{\Gamma_1} = \theta_1 t \, , \quad v \geq o \quad \text{dans } \Omega\}$

On a alors :

---

[1] On a posé :

$$\frac{\partial}{\partial t} X = X' \, .$$

$$
(19)\begin{cases} u(t) \in K_1(t), & \forall t \in [0,T], \\ (u'(t), v-u(t)) + a(u(t), v-u(t)) \geq L(v-u(t)), & \forall v \in K_1(t) \\ u(o) = 0. \end{cases}
$$

4) Enoncé des résultats

Théorème 1

Pour tout $b > o$, donné, il existe un unique u_b, solution de (17) dans la classe,

$$
(20) \qquad u_b\,,\ u'_b \in L^2(0,T\,;\,V) \cap L^\infty(0,T\,;\,H)\,.
$$

Propriété 1

La solution u_b est telle que, $\forall\ o \leq b_2 \leq b_1$

$$
(21) \qquad o \leq u_{b_2}(x,t) \leq u_{b_1}(x,t) \leq \textcircled{H}\,t
$$

où $\textcircled{H}(x)$ est la fonction définie par :

$$
(22) \qquad \Delta\textcircled{H} = o \text{ dans } \Omega\,,\ \textcircled{H} = \theta_1 \text{ sur } \Gamma_1,\ \textcircled{H} = o \text{ sur } \Gamma_3\,,\quad \frac{\partial\textcircled{H}}{\partial n} = o \text{ sur } \Gamma_2\,.
$$

Propriété 2

La solution u_b est telle que, $\forall t \in [0,T[$,

$$
o \leq \frac{\partial u_b}{\partial t} \leq \textcircled{H}\,,
$$

où $\textcircled{H}$ est définie par (22).

Théorème 2

Il existe u unique solution de (19) dans la classe

$$
u \in L^2(0,T\,;\,V) \cap L^\infty(0,T\,;\,H)
$$

$u' \in L^2(0,T\ ;\ H)$.

De plus u_b tend vers u dans $L^2(0,T\ ;\ H)$ fort et dans $L^2(0,T\ ;\ V)$ faible lorsque b tend vers $+\infty$. De plus, u'_b tend vers u' dans $L^\infty[\Omega \times (0,T)]$ faible étoile.

5) Démonstration du théorème 1

L'unicité est évidente. Pour l'existence on procède par pénalisation, estimation à priori et passage à la limite.

i) Pénalisation : Soit $\varepsilon > o$ et soit $\beta(v)$ défini par :

$$\left.\begin{array}{lll} \beta(v) = o & \text{si} & v(x) \geq o \\ \beta(v) \doteq o & \text{si} & v(x) < o \end{array}\right\} \quad \forall v \in H .$$

On introduit alors $u_\varepsilon(x,t)$ solution (cf. [2]) de

$$(23) \quad \begin{cases} u_\varepsilon \in L^2(0,T\ ;\ V) , \quad u'_\varepsilon \in L^2(0,T\ ;\ H) \\ (u'_\varepsilon,v) + a(u_\varepsilon,v) + \frac{1}{\varepsilon}(\beta(u_\varepsilon),v) + b \int_{\Gamma_1} (u_\varepsilon - \theta_1 t) v \, d\Gamma = L(v), \quad \forall v \in V \\ \dot{u}_\varepsilon(o) = 0 . \end{cases}$$

ii) Estimations à priori I : Choisissant $v = u_\varepsilon(t)$ dans :

(23) on obtient (*) (page suivante)

$$\frac{d}{dt} \frac{|u_\varepsilon|^2}{2} + a(u_\varepsilon) + \frac{1}{\varepsilon}(\beta(u_\varepsilon),u_\varepsilon) + b \int_{\Gamma_1} u_\varepsilon^2 \, d\Gamma = L(u_\varepsilon) + \theta_1 t\, b \int_{\Gamma_1} u_\varepsilon \, d\Gamma$$

d'où il résulte que :

$$(24) \qquad u_\varepsilon \in \text{borné de } L^\infty(0,T\ ;\ H) \cap L^2(0,T\ ;\ V) .$$

iii) Estimations à priori II : Soit $t \in [0,T[$ et $h > o$ tel que $t + h \in [0,T]$.

Dans l'égalité (23) écrite à l'instant t (resp. $(t+h)$) choisissons $v = u(t+h) - u(t)$ (respectivement $v = u(t) - u(t+h)$) et ajoutons membre à membre les égalités obtenues ; il vient, en posant :

(25) $$w(t) = u(t+h) - u(t) ,$$

(26) $$-\left(\frac{dw}{dt}, w\right) - a(w) - b\int_{\Gamma_1} w^2\, d\Gamma \geq -\theta_1\, h\, b \int_{\Gamma_1} w\, d\Gamma .$$

Divisant (26) par $-h^2$ et intégrant sur $(0,t)$, on obtient :

$$\frac{1}{2}\left|\frac{w(t)}{h}\right|^2 + \int_0^t a\left(\frac{w}{h}\right) dt + b\int_0^t\int_{\Gamma_1}\left(\frac{w}{h}\right)^2 d\Gamma\, dt \leq \theta_1 b\int_0^t\int_{\Gamma_1}\left(\frac{w}{h}\, d\Gamma\right)dt + \frac{1}{2}\left|\frac{w(o)}{h}\right|^2$$

Mais $u_\varepsilon(t)$ étant dérivable en t, on a, lorsque $h \to o$,

$$\frac{1}{2}|u'_\varepsilon(t)|^2 + \int_o^t a(u'_\varepsilon(\tau))\, d\tau + b\int_o^t\int_{\Gamma_1} u'^2_\varepsilon\, d\Gamma\, d\tau \leq$$

$$\leq \theta_1\, b\int_o^t\int_{\Gamma_1} u'_\varepsilon\, d\Gamma\, d\tau + \frac{1}{2}|u'_\varepsilon(o)|^2 .$$

Appliquant (23) pour $t = o$ on a :

$$|u'_\varepsilon(o)| \leq \text{Cste}$$

d'où il résulte immédiatement que :

(27) $$u'_\varepsilon \in \text{borné de } L^\infty(0,T\,;\,H) \cap L^2(0,T\,;\,V) .$$

Revenant à (23) on obtient :

(28) $$\frac{1}{\varepsilon}\,\beta(u_\varepsilon) \in \text{borné de } L^2(0,T\,;\,V')\ (^{**}) .$$

(*) On a posé $(u,v) = \int_\Omega u\, v\, dx$, $|u| = (u,u)^{1/2}$, $a(v) = a(v,v)$.

(**) L'espace V' est le dual de V quand on identifie H à son dual.

iv) Passage à la limite : Il résulte des estimations (24) (27) (28) qu'il existe une sous suite, encore notée u_ε, telle que :

$$u_\varepsilon \to u \quad \text{dans} \quad L^2(0,T\ ;\ V) \quad \text{faible}\ ,$$

$$u'_\varepsilon \to u' \quad \text{dans} \quad L^\infty(0,T\ ;\ H) \quad \text{faible étoile}$$

$$u'_\varepsilon \to u' \quad \text{dans} \quad L^2(0,T\ ;\ V) \quad \text{faible}$$

$$\beta(u_\varepsilon) \to o \quad \text{dans} \quad L^2(0,T\ ;\ V') \quad \text{fort}\ .$$

Soit alors $v \in L^2(0,T\ ;\ K)$. Remplaçons dans (25) la fonction test par $v(t) - u_\varepsilon(t)$; il vient, compte tenu de ce que $\beta(v)=o$ et de la monotonie de β,

$$(29) \quad \left(\frac{du_\varepsilon}{dt}, v - u_\varepsilon\right) + a_1(u_\varepsilon, v-u_\varepsilon) \geqslant L(v-u_\varepsilon) + \theta_1 t\ b \int_{\Gamma_1} (v-u_\varepsilon)\ d\Gamma$$

où on a posé :

$$a_1(u,v) = a(u,v) + b \int_{\Gamma_1} uv\ d\Gamma\ .$$

Intégrant sur $(0,T)$ et prenant la limite inférieure des deux membres, on trouve à la fin :

$$(30) \quad \left(\frac{du}{dt}, v-u\right) + a_1(u,v-u) > L(v-u) + \theta_1 t\ b \int_{\Gamma_1} (v-u)\ d\Gamma$$

$$\forall\, v \in K\ , \quad \forall\, t \in (0,T).$$

On vérifie immédiatement que $u(o)$ est limite dans H faible de $u_\varepsilon(o)$. Il en résulte, compte tenu de $\beta(u_\varepsilon) \to o$ dans $L^2(0,T\ ;\ V')$ fort qui entraîne $u(t) \in K$, et de (30) que u est solution de (17) .

Remarque 1

Si on analyse la démonstration du théorème 1, on constate que les conclusions (20)ont pu être obtenues grâce au fait que la forme linéaire du second membre de (17), soit :

$$L_1(t)\ (v) = L(v) + b\ \theta_1\ t \int_{\Gamma_1} v\ d\Gamma$$

telle que :

$$(31)\quad \begin{cases} L_1(t) \in L^2(0,T\ ;\ V') \\ L_1(o) \in H\ , \quad L_1'(t) \in L^2(0,T\ ;\ V'). \end{cases}$$

Les conclusions du théorème 1 sont donc valables sous les hypothèses générales (31) sur le second membre.

6) Démonstration de la propriété 1

i) Montrons que $u(x,t) \leq Ⓗ\, t$.

Nous posons :

$$w(x,t) = (u(x,t) - Ⓗ\, t)^+$$

et nous choisissons :

$$v = u(t) \pm w(x,t)$$

dans (17), ce qui est loisible car $o \leq w(x,t) \leq u(x,t)$.

Il vient alors :

$$(u'(t),w) + a(u(t),w) + b \int_{\Gamma_1} (u-\theta_1 t)\cdot w\ dx = -k \int_\Omega w\ dx \leq 0$$

d'où :

$$(32)\qquad (w'(t)\ ,\ w(t)) \leq 0$$

car :

$$\int_{\Gamma_1} (u - \theta_1 t)\ w\ d\Gamma = \int_{\Gamma_1} w^2\ d\Gamma \geq 0\ ,$$

et

$$a(\textcircled{H}, w) = \int_{\Gamma_1} \frac{\partial \textcircled{H}}{\partial n} w \, d\Gamma \geq 0$$

du fait que $o \leq \textcircled{H} \leq \theta_1$ d'après le principe du maximum ce qui entraîne $\dfrac{\partial \textcircled{H}}{\partial n} \geq 0$.

Il en résulte alors de (32) que $w(t) = 0$ et par conséquent :

$$u(x,t) \leq \textcircled{H}\, t \ , \quad \text{p.p. dans } \Omega \times (0,T).$$

ii) <u>Montrons que</u> $b_1 \leq b_2$ <u>entraîne</u> $u_{b_1} \leq u_{b_2}$.

Pour simplifier l'écriture nous posons :

$$u_i = u_{b_i} \quad , \qquad i = 1 \text{ et } 2 .$$

Dans l'inégalité (17) relative à u_1 (respect. u_2) nous choisissons $v = u_1 - (u_2 - u_1)^-$ (respect. $v = u_2 + (u_2 - u_1)^-$) .

Ajoutant membre à membre les deux inégalités obtenues, il vient :

(33) $$(w',w) + a(w,w) + \int_{\Gamma_1} w\left[b_1(u_1 - \theta_1 t) - b_2(u_2 - \theta_1 t)\right] d\Gamma \leq 0$$

où on a posé :

$$w = (u_1 - u_2)^+$$

Dans (33) l'intégrale sur Γ_1 ne porte en fait que sur l'ensemble des points de Γ_1 où $u_1 > u_2$. Mais alors, tenant compte de i) on a :

(34) $$u_2 - \theta_1 t < u_1 - \theta_1 t \leq 0$$

ce qui entraîne, compte tenu de $o < b_1 < b_2$,

(35) $$w\left[b_1(u_1 - \theta_1 t) - b_2(u_2 - \theta_1 t)\right] \geq 0 .$$

Intégrant alors (33) sur (0,T), on en déduit que :

$$w(t) = 0$$

donc :

$$u_1 \leqslant u_2 \qquad \text{sur} \qquad \Omega \times (0,T) \quad \text{p.p.}$$

7) <u>Démonstration de la propriété 2</u>

i) <u>Démonstration de</u> $u'(t) \geqslant 0$.

Soient $t \in [0,T[$, et $h > o$ tel que $t + h \in [0,T]$.

Nous posons :

(36) $$w(x,t) = [u(x,t+h) - u(x,t)]^- .$$

Dans l'inégalité (17) relative à l'instant t (resp. $t+h$) nous choisissons $v = u(t) - w(t)$ (resp. $v = u(t+h) + w(t)$) et ajoutons membre à membre les deux inégalités obtenues. Il vient :

(37) $$(w(t), (w(t))') + a_1(w(t)) \leqslant - b\, \theta \int_{\Gamma_1} w(t)\, d\Gamma \leqslant 0 .$$

En intégrant (37) sur (0,T) on a :

$$\frac{1}{2} |w(t)|^2 + \int_o^t a_1(w(t)) \leqslant \frac{1}{2} |w(o)|^2 .$$

Mais :

(38) $$w(o) = [u(x,h) - u(x,o)]^- = [u(x,h)]^- = 0 ,$$

d'où il résulte que :

$$w(t) = 0$$

c'est-à-dire :

(39) $$u(x,t+h) - u(x,t) \geqslant 0$$

et par conséquent :

$$u'(x,t) \geqslant 0 \qquad \forall (x,t) \in \Omega \times (0,T) \quad \text{p.p.}$$

ii) Démonstration de $u'(t) \leqslant Ⓗ$

Nous introduisons $\dot{u}_1(x,t)$ par :

(40) $$u_1(x,t) = Ⓗ\, t - u(x,t) \quad ,$$

et nous savons que :

(41) $$o \leqslant u_1(x,t) \leqslant Ⓗ\, t \quad , \qquad u_1'(x,t) \leqslant Ⓗ \; .$$

Il résulte de la propriété 1 et de (17) que u(x,t) satisfait :

(42) $$\begin{cases} u(t) \in \hat{K}(t) = \{v | v \in K, \quad v \leqslant Ⓗ\, t\} \; , \\ (u'(t), v-u(t)+a(u(t),v-u(t)+b\int_{\Gamma_1}(u(t)-\theta_1 t)(v-u(t))d\Gamma \geqslant -k\int_{\Omega}(v-u(t))dx \\ \qquad\qquad \forall v \in \hat{K}(t) \; , \\ u(o) = 0 \; . \end{cases}$$

Par le changement de fonction inconnue (40) et les propriétés (41) et (42) on voit que $u_1(x,t)$ satisfait :

(43) $$\begin{cases} u_1(t) \in \hat{K}(t) \\ (u_1'(t), v_1-u_1(t))+a(u_1(t)-Ⓗ t, v_1-u_1(t))+b\int_{\Gamma_1} u_1(t)(v_1-u_1(t))d\Gamma \geqslant \\ \qquad \geqslant \int_{\Omega}(k+Ⓗ)(v_1-u_1(t))dx \; , \qquad \forall\, v_1 \in \hat{K}(t) \; , \\ u_1(o) = 0. \end{cases}$$

Notre problème est alors de montrer que $u_1'(t) \geqslant 0$. Soit un instant $t \in [0,T[$ et un nombre positif h tel que $t+h \in [0,T]$. Posons :

(44) $$w(x,t) = [u_1(x,t+h) - u_1(x,t)]^-$$

et choisissons dans l'inégalité (46) relative à l'instant t+h (resp. t) l'élément v_1 par :

$$v_1 = u(t+h) + w(t)$$

(resp. par :

$$v_1 = u(t) - w(t)$$

ces choix étant licites. Si nous ajoutons membre à membre les deux inégalités obtenues, nous avons :

(44) $$(w',w) + a(w) + b \int_{\Gamma_1} w^2 \, d\Gamma + h \, a(\text{Ⓗ}, w) \leq 0.$$

Mais, compte tenu de la définition de Ⓗ, il vient $a(\text{Ⓗ},w) \geq 0$, d'où il résulte que :

$$(w', w) \leq 0$$

et en intégrant sur $(0,t)$,

$$|w(t)|^2 \leq |w(o)|^2 = 0 ,$$

ce qui établit la propriété souhaitée.

8) Démonstration du théorème 2

8.1 - Unicité : Elle est immédiate. En effet si $u(x,t)$ et $u^*(x,t)$ sont deux solutions de (19), on choisit $v = u^*$ (resp. $v = u$) dans l'inégalité (19) relative à u (resp. relative à u^*) et on ajoute membre à membre les inégalités obtenues, ce qui donne classiquement, en posant $w = u - u^*$,

(46) $$(w', w) + a(w, w) \leq 0$$

d'où par intégration sur $(0,t)$, $w(t) = 0$, c'est-à-dire

$$u(t) = u^*(t) .$$

8.2 - Existence : Nous allons montrer que la solution u_b de (17) tend vers un élément u solution de (19) lorsque b tend vers $+\infty$. Pour cela nous utilisons les propriétés 1 et 2 établies précédemment et des estimations à priori sur u_b.

i) Estimations à priori

Nous choisissons, dans l'inégalité (17), $v = Ⓗt$, où Ⓗ a été défini par (40). Il vient :

$$(47) \qquad (u_b', Ⓗt - u_b) + a(u_b, Ⓗ - u_b) + b\int_{\Gamma_1}(u_b-\theta_1 t)^2\, d\Gamma \geqslant$$
$$\geqslant -k\int_\Omega (Ⓗt - u_b)\, dx .$$

puis :

$$(48) \qquad (u_b' - Ⓗ, u_b - Ⓗt) + a(u_b - Ⓗt) + b\int_{\Gamma_1}(u_b-\theta_1 t)^2\, d\Gamma \leqslant$$
$$\leqslant \int_\Omega (Ⓗ + k)\,(Ⓗt - u_b)\, dx - a(Ⓗt, u_b - Ⓗt).$$

Intégrant (48) sur $(0,t)$, on obtient :

$$(49) \qquad \frac{1}{2}|u_b - Ⓗt|^2 + \int_0^t a(u_b - Ⓗ\tau)\, d\tau + \int_0^t\int_{\Gamma_1}(u_b-\theta_1\tau)^2 d\Gamma\, d\tau \leqslant$$
$$\leqslant C\int_0^t ||u_b - Ⓗ\tau||\, d\tau ,$$

où C est une constante indépendante de b. Comme b est destiné à tendre vers $+\infty$, on peut supposer $b > 1$. Une norme équivalente à la norme de v dans $H^1(\Omega)$ est donnée par :

$$||v|| = (a(v) + \int_{\Gamma_1} v^2\, d\Gamma)^{1/2}$$

puisque Γ_1 est de mesure strictement positive.

Il s'en suit que (49) donne :

(50) $$\frac{1}{2}|u_b - \textcircled{H}t|^2 + \frac{1}{2}\int_0^t ||u_b - \textcircled{H}\tau||^2 d\tau + (b-1)\int_0^t\int_\Gamma (u_b - \theta_1\tau)^2 d\Gamma d\tau \leq C_1$$

où C_1 est une constante indépendante de b.

On déduit de (50) que :

(51) $u_b \in$ borné de $L^\infty(0,T ; H) \cap L^2(0,T ; V)$

(52) $$(b-1)\int_0^t\int_{\Gamma_1} (u_b - \theta_1\tau)^2 \, d\Gamma \, d\tau \leq C_1 .$$

Nous savons de plus d'après la propriété 2 que :

(53) $u_b' \in$ borné de $L^\infty(\Omega \times (0,T))$

et d'après la propriété 1 que :

(54) $u_b(x,t)$ est une fonction croissante en b, majorée par $\theta_1 t$(1)

ii) <u>Passage à la limite</u>

Il résulte des estimations précédentes que, lorsque b tend vers $+\infty$, au moins pour une sous-suite :

(55) u_b tend vers u dans $L^\infty(0,T ; H)$ faible étoile

(56) u_b tend vers u dans $L^2(0,T ; V)$ faible

(57) u_b' tend vers u' dans $L^\infty(\Omega \times 0,T)$ faible étoile

(58) $u_b(x,t)$ tend vers $u(x,t)$ dans $\mathbb{R}$ ponctuellement en $(x,t) \in \Omega \times (0,T)$.

---

(1) D'après le lemme 2, elle est même majorée par $\textcircled{H}t$.

De plus (54) implique que :

(59) $$\int_0^T \int_\Omega (u_b(x,t))^2\,dx\,dt \to \int_0^T \int_\Omega (u(x,t))^2\,dx\,dt$$

et donc :

(60) u_b tend vers u dans $L^\infty(0,T\,;\,H)$ fort.

Il résulte également des estimations que :

(61) u_b tend vers u dans $L^2(0,T\,;\,L^2(\Gamma_1))$ fort

et d'après (52) nous avons donc :

(62) $$u\big|_{\Gamma_1} = \theta_1 t\ .$$

On voit alors immédiatement que :

(63) $$u(t) \in K_1(t)$$

Choisissons alors $v \in K_1(t)$ dans (17) ,

$$(u_b',\, v - u_b) + a(u_b,\, v-u_b) + b\int_{\Gamma_1} (u_b-\theta_1 t)(\theta_1 t-u_b)\,d\Gamma \geq -k\int_\Omega (v-u_b)\,dx.$$

L'intégrale sur Γ_1 étant négative, on a aussi ,

(64) $$(u_b',v-u_b)+ a(u_b,v-u_b) \geq -k\int_\Omega (v-u_b)\,dx, \quad \forall v \in K_1(t)\ .$$

On peut alors déduire (19) de (64) comme il est indiqué dans [2] .

<u>Remarque 1</u>

On peut obtenir l'existence dans le théorème 2 sans utiliser la propriété 2 par un changement de fonction inconnue et utilisation d'un résultat de H. Brézis [3].

Introduisons :

$$K_2(t) = \{v_2 \mid v_2 \in V\ , \quad v_2\big|_{\Gamma_1} = 0\ , \quad 0 \geq v_2 \geq -\Theta t\}$$

$$u_2(t) = u(t) - (H) t.$$

Compte tenu de la propriété 1 la fonction u_2 doit satisfaire

$$(65) \begin{cases} (u_2', v_2-u_2) + a(u_2, v_2-u_2) \geqslant -\int_\Omega ((H)+k)(v_2-u_2)dx, & \forall v_2 \in K_2(t) \\ u_2(t) \in K_2(t) \\ u_2(o) = 0 . \end{cases}$$

On voit alors facilement que (65) possède une solution unique [3] dans la classe :

$$u_2 \in L^2(0,T ; V) , \qquad u_2' \in L^2(0,T ; H) .$$

Remarque 2 ([1])

On peut également obtenir l'existence dans le théorème 2 par un procédé de pénalisation partielle. Introduisons à cet effet l'opérateur de pénalisation β de H dans H défini par :

$$\begin{cases} \beta(v) = 0 & \text{si} \quad v \geqslant 0 \\ \beta(v) = v & \text{si} \quad v < 0 \end{cases}$$

et l'espace affine :

$$V_1(t) = \{v \mid v \in V , \quad v = \theta_1 t \text{ sur } \Gamma_1 \} .$$

Le scalaire ε étant positif, destiné à tendre vers zéro, soit:

$u_\varepsilon(x,t)$ la solution de l'équation

([1]) Cette remarque a été suggérée à l'A. par J.L. LIONS.

$$
(66)\begin{cases} u_\varepsilon(t) \in V_1(t) , \quad \forall t \in [0,T] \quad \text{p.p.} \\ (u'_\varepsilon, v-u_\varepsilon) + a(u_\varepsilon, v-u_\varepsilon) + \frac{1}{\varepsilon}(\beta(u_\varepsilon), v-u_\varepsilon) = -k\int_\Omega (v-u_\varepsilon)dx, \ \forall v \in V_1(t). \\ u_\varepsilon(o) = 0 . \end{cases}
$$

Choisissant $v = \textcircled{H}t$ dans l'égalité (66) on obtient aisément les majorations ([1])

(67) u_ε ε borné de $L^\infty(0,T ; H) \cap L^2(0,T ; V)$

Choisissant ensuite $v = u_\varepsilon(t)$, $t \in [0,T[$ dans l'égalité (66) relative à l'instant $t+h$, $h > o$, $t+h \in [0,T]$, et $v = u_\varepsilon(t+h)$ dans l'égalité (66) relative à l'instant t. Ajoutant membre à membre les deux égalités et utilisant la monotonie de β on obtient que :

(68) u'_ε ε borné de $L^\infty(0,T ; H) \cap L^2(0,T ; V)$

car $u'_\varepsilon(o)$ est borné dans H .

Les estimations (67) et (68) permettent alors de passer à la limite $\varepsilon \to o$ et d'obtenir l'existence dans le théorème 2.

Remarque 3

Nous n'avons pas utilisé les raisonnements indiqués aux remarques 1 et 2 pour démontrer l'existence dans le théorème 2 car nous avons préféré nous servir des propriétés 1 et 2 qui mettent en évidence certains caractères physiques de la solution. Le théorème 2 montre de plus que la solution de (19) est limite de

---

([1]) On utilise le fait que $\beta(\textcircled{H}t) = o$ et que β est monotone.

solution de (17) lorsque le coefficient b de transmission de la paroi tend vers $+\infty$, ce qui traduit également un fait concret.

Remarque 4

Bien que nous ne l'ayons pas énoncé comme théorème, il est évident d'après la démonstration du théorème 2 que la solution u de (19) satisfait aux propriétés 1 et 2 énoncées pour u_b .

9) Conclusion

Les raisonnements que nous avons mis en oeuvre dans le cas de la glace peuvent naturellement s'adapter à d'autres phénomènes de fusion et complémentairement à des phénomènes de cristallisation. La caractéristique de chacun des cas est qu'initialement la température du milieu continu à transformer doit être uniformément soit à sa température de fusion, soit à sa température de cristallisation. Les conditions aux limites annexes peuvent être diverses suivant les conditions physiques extérieures.

BIBLIOGRAPHIE

[1] C. BAIOCCHI. Note aux C.R.A.S. Déc. 1971.

[2] J.L. LIONS - Quelques méthodes de résolution des problèmes aux limites non linéaires - Dunod Gauthier Villars (1969)

[3] H. BREZIS. Comptes rendus, 274, série A, 1972, p. 310.

CENTRO INTERNAZIONALE MATEMATICO ESTIVO

(C.I.M.E.)

REMARKS ABOUT THE FREE BOUNDARIES OCCURRING IN VARIATIONAL INEQUALITIES

DAVID KINDERLEHRER

Corso tenuto a Bressanone dal 17 al 26 giugno 1973

REMARKS ABOUT THE FREE BOUNDARIES OCCURRING IN VARIATIONAL INEQUALITIES

David Kinderlehrer

1. This discussion concerns two variational inequalities with interior constraints or obstacles. Each of these problems has given rise to questions about the smoothness of solutions and the nature of the free boundary determined by coincidence with the obstacle. The present theme is to show how smoothness of the solution implies smoothness of the free boundary provided that certain geometric conditions, which vary from problem to problem, are satisfied. The emphasis, therefore, is on formulation and illustration rather than demonstration. Our applications are to a linear problem which arises in the study of stationary fluid flow through a porous medium and to a nonlinear problem related to minimizing a functional in a set of functions constrained to lie above a concave obstacle.

Notations

Ω open connected set in R^n with smooth boundary $\partial\Omega$

g smooth in $\bar{\Omega}$

ψ smooth in $\bar{\Omega}$ and $\psi \leq g$ on $\partial\Omega$ (the obstacle)

We may assume for the time being that $g = 0$ since our interest is restricted to interior properties.

$K = K_\psi = \{v \in H^{1,\infty}(\Omega) : v \geq \psi$ in Ω and $v = g$ on $\partial\Omega\}$

$f \in C^1(\bar{\Omega})$

<u>Problem 1</u> Let $a_{ij} \in H^{2,s}(\Omega)$, $1 \leq i,j \leq n$, $a_{ij} = a_{ji}$, $s>n$ and satisfy

$$a_{ij}\,\xi_i\,\xi_j \geq \nu\,|\xi|^2 \quad \text{for} \quad \xi \in R^n \text{ and } x \in \bar{\Omega}\,.$$

$$(1.1)\qquad u \in K : \int_\Omega a_{ij}\, u_{x_i}\,(v-u)_{x_j}\,dx \geq \int_\Omega f(v-u)dx \qquad v \in K$$

<u>Problem 2</u> Let $a(p) = (a_1(p),\ldots,a_n(p)) \in C^2(R^n)$ satisfy: for each C compact in R^n, there exists a $\nu = \nu(C) > 0$ such that

$$(a(p)-a(q))(p-q) \geq \nu|p-q|^2 \qquad \text{for} \quad p,q \in C$$

$$(1.2)\qquad u \in K : \int_\Omega a_j(Du)D_j(v-u)dx \geq \int_\Omega f(v-u)dx \qquad v \in K$$

The solutions to problems 1 and 2 determine a coincidence set

$$(1.3)\qquad I = I(u) = \{x \in \Omega : u(x) = \psi(x)\}$$

whose boundary ∂I, called the curve of separation, is the free boundary we wish to study. To recapitulate, we shall achieve this by exploiting the smoothness of u.

We begin with some very brief informal remarks about the regularity of the solutions to Problems 1 and 2. The first results in this direction for Problem 1 are due to H. Lewy and G.Stampacchia [14] who showed that

$$u \in H^{2,p}(\Omega) \quad \text{when} \quad L\psi \in L^p(\Omega),\ f = g = 0,\ n < p < \infty,$$

$$L\psi = -D_j(a_{ij} D_i \psi)$$

About the same time, Brezis and Stampacchia [5] considered this problem for more general f and g and operators L. It is possible to consider the solution u as the minimum of supersolutions to the equation $Lw = -D_j(a_{ij} D_i w)$ which lie above ψ in Ω. This approach has been studied by H. Lewy and G.Stampacchia [15] and recently by U.Mosco and G.M. Troianiello [20].

The smoothness of the solution to Problem 2 has been associated to its existence. Again, H.Lewy and GStampacchia [16] were able to prove that, for u satisfying (1.2),

$$u \in H^{2,p}(\Omega) \quad \text{when} \quad \psi \in C^2(\bar{\Omega}),\ f = g = 0,\ \text{all } p \in [1,\infty).$$

For the special case of minimum area, we refer to M.Giaquinta and L.Pepe [9]. In general, existence of solutions to Problem 2 depends on the relationship among $\partial\Omega$, $a(p)$, and f.

S.Mazzone has considered this question both from the point of view of the coerciveness of $a(p)$ and the geometrical relationship among $\partial\Omega$, $a(p)$, f, [18];[19]. C.Gerhardt has also treated this problem [7].

It is clear, by considering one dimensional examples, that the solution u to (1.1) or (1.2) need not be in $C^2(\Omega)$. It was thought by some, including the author, that $H^{2,p}$ smoothness was the best generally possible even though for important two dimensional examples ([10],[11]), the second derivatives were bounded. However, J.Frehse [6] recently showed that $u \in H^{2,\infty}(\Omega)$ for $a_{ij} = \delta_{ij}$. By a different method, this conclusion was obtained for general Problems 1 and 2 by H.Brezis and the author [4]. Independently, C. Gerhardt [8] has obtained this result.

The smoothness of the curve of separation ∂I has been related to geometric conditions. We suppose henceforth that $n = 2$. Lewy and Stampacchia [14] observed that if Ω convex, ψ (real) analytic and concave, and $a_{ij} = \delta_{ij}$, then ∂I is an analytic Jordan curve. Recently it has been shown that this conclusion is valid under the assumptions above about Ω and ψ if

$$a_i(p) = \frac{p_i}{\sqrt{1+p^2}}$$

The proof of this result [12] motivates the present work. Questions related to linear analytic equations which share some of the characteristics of the present work are discussed by H.Lewy [17].

2. To show that a given curve is smooth, we shall show it has a smooth representation as the boundary values of a conformal mapping. We formulate a problem in complex variables. Let ω be a simply connected domain in the $z=x_1+ix_2$ plane, which for our purposes we may as well assume to be a closed Jordan domain, and $\Gamma \subset \partial\omega$ be an (open) Jordan arc. Denote by G the upper semidisc $\{|t| < 1, \operatorname{Im} t > 0\}$ in the $t = t_1+it_2$ plane. We consider the conditions below:

(2.1)
ω admits a conformal mapping
$\phi : G \to \omega$ with the properties
$\phi \in H^{1,q}(G)$ for some $q > 2$ and
ϕ is a $1:1$ mapping of $-1 < t < 1$ onto Γ

Let $F \in C^2(U)$, for a neighborhood U of ω, $\omega \subset\subset U$, satisfy

(2.2) $\frac{\partial F}{\partial \bar{z}}(z_o) \neq 0$ for a $z_o \in \Gamma$

Finally, let $f \in H^{1,\infty}(\omega)$ satisfy

(2.3)
$$\frac{\partial f}{\partial \bar{z}}(z) = \mu(z) \quad \text{a. e. in } \omega \text{ where}$$
$$|\mu(z)| \leq \mu_o |z-z_o| < 1 \text{ for } z \in \omega \text{ and some } \mu_o > 0 .$$

Theorem 1 Suppose that Ω, F, f satisfy (2.1), (2.2), (2.3) respectively and that $\phi(t_o) = z_o$. Suppose that

$$f = F \quad \text{on} \quad \Gamma$$

Then for $n = 1$ or 2, there exists a $c \neq 0$ such that

(2.4)
$$\left| \frac{\phi(t)-\phi(t_o)}{(t-t_o)^n} - c \right| \leq c_1 |t-t_o|^{\lambda} \text{ in } \bar{G}, \lambda = 1-\frac{2}{q}$$

Moreover, c and c_1 depend continuously on

$||\phi||_{H^{1,q}(G)}$, $||F||_{C^2(U)}$, and $||f||_{H^{1,\infty}(\omega)}$.

Corollary 2.1 With the hypotheses and notations of Theorem 1,

(2.5)
$$\lim_{z \to z_o} \frac{f(z)-f(z_o)}{z-z_o} = F_z(z_o)+F_{\bar{z}}(z_o)\frac{\bar{c}}{c}, \quad z \in \omega ,$$

where c is defined in (2.4).

It is of interest to know conditions for which $n \doteq 1$.

<u>Corollary 2.2</u> <u>With the hypotheses and notations of Theorem 1, suppose that</u>

(2.6) $U-\omega$ <u>contains a sector of positive angle with vertex</u> z_o <u>or</u>

(2.7) $f(\omega \cup \Gamma) \subset F(U-\omega)$ <u>and</u> $\left|\frac{\partial F}{\partial \bar{z}}(z_o)\right| \neq \left|\frac{\partial F}{\partial z}(z_o)\right|$.

<u>Then</u> (2.4) <u>holds for</u> $n = 1$.

We remark that the validity of (2.4) for each $t_o \in (-1,1)$ with the modulus of c and c_1 independent of t_o implies that $\phi \in C^{1,\lambda}(\bar{G} \cap B_R)$ for each $R < 1$, $B_R = \{|t| < R\}$. This is a known fact in the theory of functions and follows by estimating Cauchy's Integral of the analytic function

$$\frac{\phi(t)-\phi(t_o)}{t-t_o} - c \quad .$$

3. At this point we shall study the curve of separation for some particular examples. The proof of Theorem 1 will be given in §5. Rather than considering the most general conditions, let us discuss the free boundary which arises as the

water level in the stationary flow between two resevoirs separated by a dam of non-homogenous porous medium. For details about the history and formulation of this problem we refer to the lectures of C.Baiocchi in this volume. The particular formulation here follows V.Benci [2]. To agree with the literature, we introduce different notations.

Let R be the interior of a rectangle in the $z = x_1+ix_2$ plane and suppose that α, h, g are assigned smooth functions in $\bar{R}$ satisfying

$$\alpha(z) \geq \alpha_0 > 0 \quad \text{in } \bar{R} \text{ for some } \alpha_0 > 0$$

(3.1) $$h(z) \neq 0 \quad \text{in } \bar{R}$$

$$g(z) \geq 0 \quad \text{on } \partial R$$

Although α, h, and g are particular functions, their form does not concern us here since we shall assume the conclusions of [1],[2]. Denote by

$$K = \{v \in H^1(\Omega) : v \geq 0 \text{ in } R \text{ and } v = g \text{ on } \partial R\}$$

suppose that

(3.2) $$w \in K : \int_R \alpha\, D_i w\, D_i(v-w)\,dx \geq \int_R h(v-w)\,dx \ , \ v \in K .$$

We know that $w \in H^{2,\infty}_{loc}(R)$ by [4]. Let

$$I = \{z \in R : w(z) = 0\} \quad \text{and}$$

$$\Omega = \{z \in R : w(z) > 0\} = R-I$$

According to [1],[2] the free boundary

$$\Gamma = \partial I \cap R$$

is a simple curve connecting two points on opposite parallel sides of the rectangle ∂R. Indeed, assuming that the sides of ∂R form angles of $\pi/4$ with the x_1, x_2 axes, we may assert

(3.3) $\Gamma : x_2 = g(x_1) \quad , \quad |g'(x_1)| \leq M$ a.e. ,

for an appropriate range of x_1.

<u>Theorem 2</u> <u>Let</u> $\Gamma = \partial I \cap R$ <u>be the curve of separation of the solution</u> w <u>to</u> (3.2). <u>Then</u> Γ <u>is a</u> $C^{1,\lambda}$ <u>curve</u>.

<u>Proof</u> We shall verify the conditions (2.1), (2.2), (2.3) in a neighborhood of any $z_o \in \Gamma$. Let $z_o = 0 \in \Gamma$ and $\rho > 0$ such that $B_\rho(0) \subset R$. First we verify (2.2) and (2.3). Observe that

$$(3.4) \qquad \begin{aligned} -D_i(\alpha\, D_i\, w) &= h \quad \text{in} \quad \Omega \\ D_i\, w &= 0 \quad \text{in} \quad I \end{aligned}$$

(since the C^1 function w attains there its minimum).Rewriting (3.4) we see that

$$\Delta w + \frac{1}{\alpha} \sum D_i \alpha \, D_i \, w = - \frac{h}{\alpha} \quad \text{in} \quad \Omega$$

Let v be a solution to the equation

$$\Delta v = - \frac{h}{\alpha} \quad \text{in} \quad B_\rho$$

Since $-h/\alpha$ is smooth, v is in $C^3(B_\rho)$. Define

$$f(z) = D_1(w-v) - i \, D_2(w-v)$$

(3.5)

$$F(z) = -(D_1 v - i \, D_2 v)$$

Then

$$\frac{\partial f}{\partial \bar{z}} (z) = \frac{1}{2} \Delta(w-v) = - \frac{1}{2\alpha} \sum D_i \; \alpha \; D_i \; w \text{ in } \Omega \quad B_\rho$$

Let $z_o \in \Gamma \quad B_\rho$. Since $D_j \, w(z_o) = 0$ and $D_j w$ is Lipschitz in $\bar{B}_\rho$, we see that

$$|D_j w(z) - D_j w(z_o)| \leq \text{const.} |z - z_o| \quad z \in B_\rho, \; z_o \in \Gamma \cap B_\rho$$

Hence

$$\left| \frac{\partial f}{\partial \bar{z}} (z) \right| \leq \text{const.} |z - z_o| \quad \text{for} \quad z \in B_\rho, \; z_o \in \Gamma \cap B_\rho$$

Furthermore, $\frac{\partial F}{\partial \bar{z}} (z_o) = - \frac{1}{2} \Delta v(z_o) = -h(z_o)/2\alpha(z_o) \neq 0$ by (3.1).

Hence, (2.2) and (2.3) are valid for each point $z_o \in \Gamma \cap B_{\bar\rho}$, with constants independent of z_o, and f, F defined by (3.5).

To verify the condition (2.1), we employ a quasi-conformal reflection. Suppose that

$$\Omega \cap B_\rho = \{z : x_2 < g(x_1),\ |z| < \rho\}$$

Then

$$J(z) = x_1 + i(2g(x_1) - x_2)$$

maps $\Omega \cap B_\rho$ onto a portion of I and $J(z) = z$ for $z \in \Gamma$. Furthermore

$$\left| \frac{\partial J}{\partial z} \Big/ \frac{\partial J}{\partial \bar z} \right| = \left| \frac{g'(x_1)}{\sqrt{1+g'(x_1)^2}} \right| \leq \frac{M}{\sqrt{1+M^2}} < 1 .$$

Now let $z = \phi(t)$ be a conformal mapping of $G = \{|t|<1,\ \mathrm{Im}\, t > 0\}$ onto $\Omega \cap B_\rho = \omega$ and set

$$\phi^*(t) = \begin{cases} \phi(t) & t \in \bar G \\ J(\phi(\bar t)) & \mathrm{Im}\, t < 0,\ |t| < 1 \end{cases}$$

Then $\frac{\partial}{\partial \bar t} \phi^* = \mu \frac{\partial}{\partial t} \phi^*$ where

$$|\mu| = \begin{cases} 0 & t \in G \\ \left| \dfrac{g'(x_1)}{\sqrt{1+g'(x_1)^2}} \right| & \operatorname{Im} t < 0, \ |t| < 1 \end{cases} < 1$$

It follows by the "observation of Boyarskii", cf.[3] for example, that

$$\phi^* \in H^{1,q}(G \cap B_R) \quad \text{for a} \quad q > 2 \quad \text{and each} \quad R < 1.$$

Replacing t by $R^{-1} \cdot t$ we may assume that the above holds for $R = 1$. In particular, the restriction of ϕ^* to G, namely ϕ, is in $H^{1,q}(G)$.

Theorem 1 may be applied at <u>each</u> point $z_0 \in B_\rho \cap \Gamma$ from which it follows that

$$\phi \in C^{1,\lambda}(G \cap B_R) \quad \text{for } \lambda = 1 - \frac{2}{q} \quad \text{and each} \quad R < 1.$$

It also follows that the integer $n = 1$ since Γ satisfies an exterior segment condition for each $z_0 \in \Gamma$. (cf.(2.6)).

4. In this paragraph we state an application of Theorem 1 to Problem 2. For this we assume that

$$
(4.1)\qquad
\begin{aligned}
&\Omega \ \text{ is a strictly convex domain in } \ R^2 \\
&\psi \ \text{ strictly concave, in } \ C^3(\bar{\Omega}) \\
&\qquad \psi < 0 \ \text{ on } \ \partial\Omega, \text{ and } \ \max_{\Omega} \psi > 0 \\
&f = 0
\end{aligned}
$$

In this case it is known that ∂I is a Jordan Curve.

Theorem 3 *Under the hypotheses of* (4.1), *the curve of separation* ∂I *associated to the solution* u *of* (1.2) *has a Hölder continuous tangent.*

We refer to [13].

5. We remark on the proof of Theorem 1. It follows the argument of Theorem 1 [13]. One notes that it is not necessary to know that

$$|f_{\bar{z}}/f_z| \le \text{const.}|z-z_0|$$

used in [13] but only that

$$|f_{\bar{z}}| \le \text{const.}|z-z_0|$$

that is, our (2.3).

D. Kinderlehrer

References

[1] C.Baiocchi, Su un problema di frontiera libera connesso a questioni di idraulica, Ann di Mat. P. ed Appl. XCII (1922) 107-127.

[2] V.Benci , on the filtration problem though a porous medium (to appear)

[3] L. Bers , F.John, M.Schechter, Partial Differential Equations Interscience, N.Y. 1962.

[4] H.Brezis and D.Kinderlehrer, The Smoothness to solutions non linear variational inequalities, to appear in Indiana Journal of March.

[5] H.Brezis and G.Stampacchia, Sur la régularité de la solution d'inequations elliptiques, Bull. Soc. Math. France 96 (1968) 153-180.

[6] J.Frehse, On regularity of the solution of a second order variational inequality, Boll. UML (IV)(1972), 312-315.

[7] C.Gerhardt, Hypersurfaces of prescribed mean curvature over obstacles Math. Zeit.

[8] ———— , Regularity of solutions of non linear variational inequalities, Arch. Rat. Mech. and Anal.

[9] M.Giaquinta and L.Pepe, Esistenza e regolarità per il problema dell'area minima con ostacoli in n variabili Ann. Sc. N. S. di Pisa 25 (1971),481-506.

[10] D.Kinderlehrer, The coincidence set of solution to certain variational inequalities, Arch. Rat. mech. and Anal., 40, (1971) 231-250.

[11] ———— , The regularity of the solution to a certain variational inequality, Proc. Symp. P. and Appl. Math. 23 AMS Providence, R.I.

[12] ———— , How a minimal surface leaves an obstacle, Acta Math. (130) 1973.

[13] ———— , Some questions related to the coincidence set in variational inequalities, Symp. Math. (to appear).

[14] H.Lewy and G.Stampacchia, On the regularity of the solution to a variational inequality, C.P.A.M. 22(1969) 153-188.

[15] ——————————,On the smoothness of superharmonics which solve a minimum problem,). d'Analyse Math. 23 (1970) 227-236.

[16] ——————————, On the existence and smoothness solution of some noncoercive variational inequalities, Arch. Rat. Mech. and Anal. 41 (1971) 241-253.

[17] H. Lewy, On the nature of the boundary separating two domains with different regimes (to appear)

[18] S. Mazzone, Existence and regularity of the solution of certain non linear variational inequalities with an obstacle Arch. Rat. Mech. and Anal.

[19] ——————, Un problema di disequazioni variazionali per superficie di curvatura media assegnata, Boll. UML VII (1973) 318-329.

[20] U. Mosco and G.M.Troianello, On the smoothness of solutions of the unilateral Dirichlet problem, Boll. UML.

CENTRO INTERNAZIONALE MATEMATICO ESTIVO
(C.I.M.E.)

TORSION ELASTOPLASTIQUE D'ARBRES CYLINDRIQUES: PROBLEMES OUVERTS

HÉLÈNE LANCHON

Corso tenuto a Bressanone dal 17 al 26 giugno 1973

TORSION ELASTOPLASTIQUE D'ARBRES CYLINDRIQUES

PROBLEMES OUVERTS

H. Lanchon. Références [10] [11]

1. INTRODUCTION: (Les notations adoptées sont les mêmes que celles de G. Duvaut dans ses conférences 1.2.3).

1.1 Problème concret

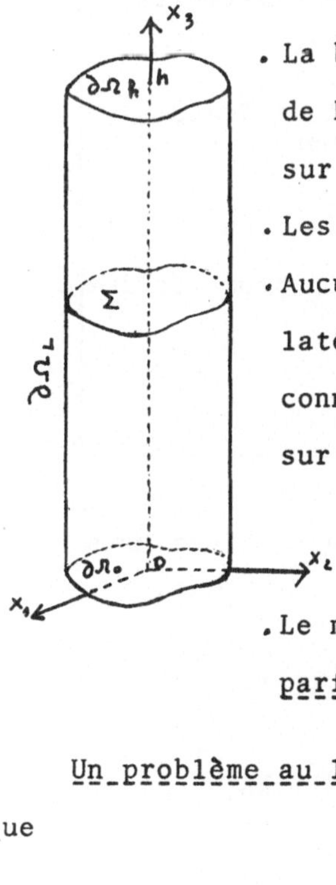

- La barre cylindrique occupe le domaine Ω de R^3; des couples opposés sont exercés sur ses deux bases $\partial\Omega_o$ et $\partial\Omega_h$.
- Les forces volumiques sont supposées nulles.
- Aucun effort n'est exercé sur la surface latérale; (dans le cas d'une section multi-connexe, cette condition est encore vraie sur la frontière latérale des cavités):

$$F = 0 \quad \text{sur} \quad \partial\Omega_L$$

- Le matériau est supposé "élastoplastique, parfaitement plastique".

Un problème au limite associé est: trouver (σ,u) tels que

$$\sigma_{ij,j} = 0 \text{ dans } \Omega$$

$$\sigma_{ij}n_j = 0 \text{ sur } \partial\Omega_2$$

$$\sigma_{33} = 0,\ u_1 = u_2 = 0 \text{ sur } \partial\Omega_o$$

$$\sigma_{33} = 0; \; u_1 = - \alpha(t) h x_2; \; u_2 = \alpha(t) h x_1 \text{ sur } \partial\Omega_h$$

(Une rotation globale est imposée à la base supérieure. $\alpha(t)$ est l'angle de torsion. Par la suite, on supposera $h = 1$.)

De plus, σ et u doivent être reliés par la loi de comportement. Nous envisagerons ici les deux lois suivantes:

$\mathcal{F}(\sigma) \leq 0$ dans Ω (critère de plasticité)

(A) Loi de Henchy	ou	(A)' Loi de Prandtl-Reuss
$\varepsilon_{ij}(u) = A_{ijkh}\sigma_{kh} + \lambda_{ij}$		$\dot{\varepsilon}_{ij}(u) = A_{ijkh}\dot{\sigma}_{kh} + \lambda_{ij}$
(bon modèle mathématique)		(beaucoup plus réaliste)

$\lambda_{ij}(x)\left(\tau_{ij} - \sigma_{ij}(x)\right) \leq 0$ dans Ω $\forall \tau \in C$

avec : . $\mathcal{F}(\sigma) = \frac{1}{2}\sigma_{ij}\sigma_{ij} - \frac{1}{6}(\sigma_{ii})^2 - g^2$: critère de Von Mises (g constante > 0)

. $\varepsilon_{ij}(u) = \frac{1}{2}\left(u_{i,j} + u_{j,i}\right)$: déformation linéarisée

. $A_{ijkh} = \frac{1+\nu}{2E}\left[\delta_{ik}\delta_{jh} + \delta_{ih}\delta_{jk}\right] - \frac{\nu}{E}\delta_{ij}\delta_{kh}$: tenseur d'élasticité de la loi de Hooke.

. λ_{ij} est la déformation plastique.

- $\dot{\phi} = \frac{\partial \phi}{\partial t}$, et la convention de sommation sur les indices répétés est adoptée

- $C = \{\tau \in R^9,\ \tau_{ij} = \tau_{ji};\ \mathfrak{F}(\tau) \leq 0\}$

Nous sommes intéressés par:

- l'existence et l'unicité de la solution (σ, u);
- la détermination pour chaque valeur de $\alpha(t)$ des régions élastiques et plastiques;

$\mathcal{E}_\alpha = \{x/x \in \Omega,\ \mathfrak{F}[\sigma(x)] < 0\}$,

(en effet dans $\mathcal{E}_\alpha$, $\lambda_{ij} = 0$)

$\mathcal{P}_\alpha = \{x/x \quad \Omega,\ \mathfrak{F}[\sigma(x)] = 0\}$

Le problème (A) avec Hencky conduit à un problème statique.
Le problème (A)' avec Prandtl-Reuss conduit à un problème quasistatique d'évolution.

1.2 Méthode et plan de l'exposé

a) Formulation et résultats pour un problème général avec Hencky: Pb(B) .

b) Résolution de (A) comme cas particulier de (B).

c) Formulation et résultats pour un problème général avec Prandtl-Reuss: Pb(C) .

d) Etude de (A)' comme cas particulier de (C). Identité des solutions de (A) et (A)' lorsque $\dot{\alpha}(t) \geq 0$.

e) Généralisation aux cas des sections multiconnexes. Problèmes ouverts.

L'exposé est fait en vue de:

. Souligner les résultats mathématiques importants.

. Montrer comment l'on peut tirer partie d'une solution abstraite pour obtenir des résultats concrets.

. Présenter les problèmes ouverts.

(Par manque de temps, les points c), d) et e) ne seront évoqués que très brièvement.)

2. PROBLEME GENERAL AVEC LA LOI DE HENCKY

2.1 Description du problème (B)

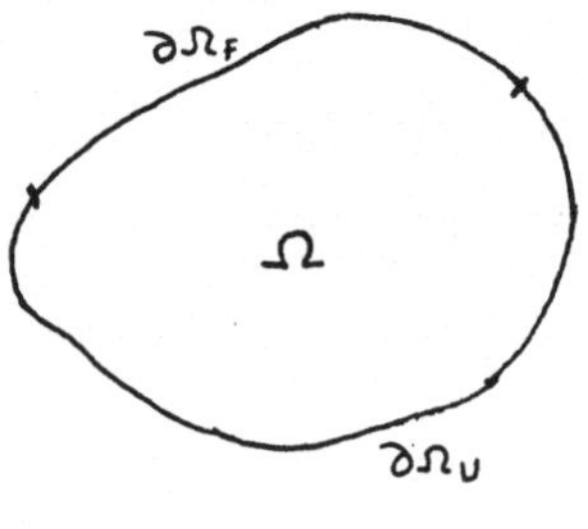

. Ω: domaine arbitraire de R^3.

. données: $\left\{\begin{array}{l} f \text{ dans } \Omega \\ F \text{ sur } \partial\Omega_F \\ U \text{ sur } \partial\Omega_U \end{array}\right.$

. inconnues: (σ,u) tels que,

$$\left.\begin{array}{lll} (1) & \sigma_{ij} = \sigma_{ji} & \text{dans } \Omega \\ (2) & \sigma_{ij,j} + f_i = 0 & \text{dans } \Omega \end{array}\right\} \text{équilibre}$$

$$\left.\begin{array}{lll} (3) & \sigma_{ij}n_j = F_i & \text{sur } \partial\Omega_F \\ (4) & u_i = U_i & \text{sur } \partial\Omega_U \end{array}\right\} \text{conditions aux limites}$$

$$\left.\begin{array}{ll} (5) & \mathcal{F}(\sigma) \leq 0 \\ (6) & \varepsilon_{ij}(u) = A_{ijkh}\sigma_{kh} + \lambda_{ij} \\ (7) & \lambda_{ij}(\tau_{ij} - \sigma_{ij}) \leq 0 \quad \forall\, \tau \in C \end{array}\right\} \begin{array}{l} \text{dans } \Omega. \\ \text{Loi de comportement} \\ \text{de Hencky} \end{array}$$

avec:

(8) $\mathcal{F} : R^9 \to R$ convexe et continue (critère de plasticité)

(9) A_{ijkh} : borné , symétrique et coercif dans Ω (tenseur d'élasticité. Cf. G. Duvaut, 2ème conférence).

2.2 Principe de Haar Karman et formulation variationnelle

Enoncé du principe: si le problème (B) a une solution $(\sigma^o \mu^o)$, alors σ^o minimise la fonctionnelle

$$(10) \quad J(\sigma) = \frac{1}{2}\int_\Omega A_{ijkh}\,\sigma_{ij}\,\sigma_{kh}\,dx - \int_{\partial\Omega_U} \sigma_{\ell j}\, n_j\, U_\ell\, d\Gamma$$

sur l'ensemble des champs de contraintes permis, c'est à dire, vérifiant (1), (2), (3), (5).

(Il suffit d'écrire la différence $J(\sigma) - J(\sigma_o)$ et de faire une intégration par partie en tenant compte de (1) ... (9) pour montrer ce principe.)

• Le choix des espaces fonctionnels et les hypothèses sur les données étant faits pour donner une signification à (1) ... (10), on suppose:

$$\begin{cases} f_i \in L^2(\Omega) \\ F_i \in H^{-1/2}[\partial\Omega] \\ U_i \in H^{1/2}[\partial\Omega] \end{cases}$$

et l'on pose :

$$(11) \quad \mathcal{H} = \left\{ \sigma \,/\, \sigma \in [L^2(\Omega)]^9 \text{ et vérifiant } (1), (2), (3), (5) \text{ p.p} \right\}$$

σ^o, solution éventuelle de (B) doit donc être solution du

Problème (B_1): minimisation de $J(\sigma)$ sur $\mathcal{K}$.

2.3 Résultats et conclusions pour le problème (B)

- $\mathcal{K}$ est un convexe fermé de $[L^2(\Omega)]^9$
- $\mathcal{A}(\sigma,\tau) = \int_\Omega A_{ijkh}\sigma_{ij}\tau_{kh}\, dx$ est une forme bilinéaire, symétrique et coercive.
- $\int_{\partial\Omega_U} \sigma_{\ell j} n_j U_\ell \, d\Gamma$ est une forme linéaire continue sur $\mathcal{K}$

D'où le résultat classique (cf. réf. 2):

Théorème 21: si $\mathcal{K} \neq \emptyset$ (ensemble vide) alors $\exists\, \sigma^o \in \mathcal{K}$ unique tel que :

(12) $J(\sigma^o) \leq J(\sigma) \quad \forall\, \sigma \in \mathcal{K}$ et, σ^o est caracterisé par

(13) $\mathcal{A}(\sigma^o, \sigma-\sigma^o) \geq \int_{\partial\Omega_U} (\sigma_{\ell j} - \sigma^o_{\ell j}) n_j U_\ell \, d\Gamma \quad \forall \sigma \in \mathcal{K}$

Conclusions

Nous avons un résultat d'unicité pour le champ de contraintes solution de (B).

L'existence d'une solution (σ^o, u^o) de (B) ne sera acquise que si $\mathcal{K} \neq \emptyset$ et, dans ce cas, seulement si l'on sait associer à σ^o un champ de déplacements $u^o \in H^1(\Omega)$ vérifiant (4), par la loi de comportement (6), (7).

3. RESOLUTION DU PROBLEME (A) COMME CAS PARTICULIER DE (B)

3.1 Application de 2 . Formulation et résolution d'un nouveau problème (A_1)

Ici $\mathcal{K} = \{\sigma/\sigma \in [L^2(\Omega)]^9,\ \sigma_{ij} = \sigma_{ji},\ \sigma_{ij,j} = 0,$

$(\sigma) \leqslant 0$ p.p. sur Ω

$\sigma_{ij}n_j = 0$ p.p. sur $\partial\Omega_L$, $\sigma_{33} = 0$ p.p. sur $\partial\Omega_o$

et $\partial\Omega_1\}$

avec

$$\mathcal{F}(\sigma) = \frac{1}{2}\sigma_{ij}\sigma_{ij} - \frac{1}{6}(\sigma_{ii})^2 - g^2 = \frac{1}{2}\sigma^D_{ij}\sigma^D_{ij} - g^2;$$

$$\sigma^D = \sigma - \frac{\text{trace } \sigma}{3}\mathbb{1}$$

donc $\quad 0 \in \mathcal{K} \to \mathcal{K} \neq \emptyset \to \exists$ une solution unique σ^o pour (B_1).

Proposition 3.1: σ^o est tel que

(1) $\sigma^o_{ij} = 0$ p.p. sur Ω excepté $\sigma^o_{13} = \sigma^o_{31}$

et $\sigma^o_{23} = \sigma^o_{32}$;

(2) $\sigma^o_{ij,3} = 0$ p.p. sur Ω

Idées de la démonstration: On introduit un $\overline{\sigma}$ qui vérifie la proposition de la manière suivante:

$\overline{\sigma}_{ij} = 0$ sauf $\overline{\sigma}_{p3} = \overline{\sigma}_{3p}$, $(p = 1,2)$, défini comme distribution sur Σ par

$$\langle\phi,\overline{\sigma}_{p3}\rangle = \int_\Omega \phi(x_1,x_2)\overline{\sigma}_{p3}(x_1,x_2,x_3)dx \qquad \forall\ \phi \in \mathcal{D}(\Sigma)$$

(Σ étant la section droite du cylindre).

Alors $\overline{\sigma} \in \mathcal{K}$ et $J(\overline{\sigma}) \leq J(\sigma^o) \rightarrow \sigma^o = \overline{\sigma}$ ce qui démontre la proposition.

Remarque: σ^o est solution du problème:

$$(B_1) \begin{cases} \sigma^o \in \mathcal{K}_T, \quad J(\sigma^o) \leq J(\sigma) \quad \forall\, \sigma \in \mathcal{K}_T \quad \text{où} \\ \mathcal{K}_T = \{\sigma / \sigma \in \mathcal{K}, \quad \sigma_{ij} \equiv 0 \text{ excepté } \sigma_{p3} = \sigma_{3p}, \\ (p = 1,2)\} \end{cases}$$

Lemme 3.1: $\forall\, \sigma \in \mathcal{K}_T$, $\exists\, \theta \in H^1_0(\Sigma)$ unique tel que

$$\theta,_2 = \sigma_{13}; \quad \theta,_1 = -\sigma_{23} \quad \text{p.p. sur } \Sigma.$$

De plus $|\text{grad}\theta| \leq 0$.

Démonstration: $\sigma_{ij} \in L^2(\Omega)$, $\sigma_{ij,j} = 0$ p.p. sur Ω et $\sigma_{33} = 0$ p.p. sur $\partial\Omega_o$ et $\partial\Omega_1$ impliquent

$$\exists\, \phi \in H^1(\Sigma) \text{ tel que } \phi,_2 = \sigma_{13} \text{ et } \phi,_1 = -\sigma_{23}$$

$\sigma_{ij}n_j = 0$ sur $\partial\Omega_L \rightarrow$ grad $\phi \wedge n = 0$ sur $\partial\Sigma$, soit $\phi = k$ (constante) sur $\partial\Sigma$ (dans le cas d'une section multiconnexe on aura pareillement $\phi = k_i$ sur chaque contour $\partial\Sigma_i$ de cavité).

Posons $\theta = \phi - k$, alors $\theta \in H^1_o(\Sigma)$. ($\theta = k_i - k = C_i$ sur $\partial\Sigma_i$ dans le cas multiconnexe).

(14) $|||\theta|||^2 = \int_\Sigma |\text{grad}\theta|^2 dx$ étant une norme sur $H_0^1(\Sigma)$, et le fait que $|\text{grad}\theta|^2 = \sigma_{13}^2 + \sigma_{23}^2$, entrainent l'unicité du θ associé de cette manière à chaque $\sigma \in \mathcal{K}_T$.

Enfin: $\mathcal{F}(\sigma) \leq 0$ implique $|\text{grad}\theta| \leq g$.

Conséquences du lemme 3.1: $\exists$ une bijection entre $\mathcal{K}_T$ et

(15) $K = \{\phi/\phi \in H_0^1(\Sigma),\ |\text{grad}\phi| \leq g\}$

et, pour deux éléments correspondants:

(16) $J(\sigma) = J_\alpha(\theta) = \int_\Sigma |\text{grad}\theta|^2 dx - 4\mu\alpha \int_\Sigma \theta dx;$

$$\mu = \frac{E}{2(1 + \nu)}$$

Les problèmes (B_1) et $(B_1)'$ sont donc équivalents au problème (A_1): minimiser $J_\alpha(\theta)$ sur K.

Lemme 3.2: $\exists\, f_\alpha$ unique dans $H_0^1(\Sigma)$ tel que:

(17) $2\mu\alpha \int \phi\, dx = \langle\phi, f_\alpha\rangle \ \forall\, \phi \in H_0^1(\Sigma)$

où $\langle\ ,\ \rangle$ est le produit scalaire associé a $|||.|||$

(ceci n'est rien d'autre que le théorème de représentation de Riesz).

Alors:

(18) $J_\alpha(\theta) = |||\theta - f_\alpha|||^2 - |||f_\alpha|||^2$

• Une expression équivalente du problème (A_1) peut être donnée:

Trouver $\theta_\alpha \in K$ tel que:

$$|||\theta_\alpha - f_\alpha||| \leq |||\phi - f_\alpha||| \quad \forall \phi \in K$$

Théorème 3.1: Existence et unicité pour (A_1):

$\exists\, \theta_\alpha \in K$ unique qui minimise $J_\alpha(\phi)$ sur K et, θ_α est caractérisé par:

$$(19) \quad \langle\theta_\alpha - \phi,\ \theta_\alpha - f_\alpha\rangle \leq 0 \quad \forall \phi \in K$$

(K est un convexe fermé de $H_0^1(\Sigma)$ et θ_α n'est autre que la projection de f_α sur K).

3.2 Propriétés de θ_α

3.2.1 Propriétés de K:

• K est l'ensemble des fonctions Lipschitziennes, de module g, de $H_0^1(\Sigma)$. Ces fonctions sont donc uniformément continues sur $\overline{\Sigma}$ et, nulles au sens fort sur $\partial\Sigma$.

• K est compact pour la norme de la convergence uniforme; ceci est une conséquence du théorème d'Ascoli:

$X = \overline{\Sigma}$ est un compact de R^2.

$Y = \bigcup_{\phi \in K} \phi(\overline{\Sigma})$ est un compact de R.

K est un ensemble d'applications équicontinues de X dans Y.

En conséquence $\overline{K}$ est uniformément compact dans $\mathcal{C}(X,Y)$, (ensemble des applications continues de X dans Y) or il est facile de montrer que $K = \overline{K}$.

- Enfin, si ϕ_1 et $\phi_2 \in K$ alors $\sup(\phi_1,\phi_2)$ et $\inf(\phi_1,\phi_2) \in K$

3.2.2 Régularité de θ_α

Les résultats de H. Brézis et G. Stampacchia, cf. réf. [4], nous permettent de conclure:

$$\theta_\alpha \in C^{1,1}(\overline{\Sigma}),$$

(ensemble des fonctions continues à dérivées Lipschitziennes) dans les deux cas suivants : si Σ est convexe et $\partial\Sigma$ Lipschitzienne ou, si Σ est non-convexe mais avec une frontière plus régulière.

Nous pouvons alors définir proprement les parties élastique et plastique de la section Σ.

(20) $E_\alpha = \{x/x \in \Sigma,\ |\mathrm{grad}\theta_\alpha(x)| < g\}$, ouvert de Σ

(21) $P_\alpha = \{x/x \in \Sigma,\ |\mathrm{grad}\theta_\alpha(x)| = g\}$.

3.2.3 Propriétés de f_α dont θ_α est la projection sur K

Proposition 3.2: Soit $\gamma = \dfrac{f_\alpha}{\mu\alpha}$, alors:

1) γ est la solution du problème de Dirichlet:

(22) $\Delta\gamma + 2 = 0$ sur Σ et $\gamma = 0$ sur $\partial\Sigma$

2) $\gamma \in W^{2,p}(\Sigma) \cap C^{\infty}(\Sigma) \cap C^{1,1}(\overline{\Sigma})$, $\forall\ 2 \leq p \leq \infty$.

3) $\gamma > 0$ sur Σ

4) $|\mathrm{grad}\gamma|$ est borné sur Σ et ne peut atteindre son maximum que sur $\partial\Sigma$.

Démonstration: par définition $\langle\phi,\gamma\rangle = 2\int_{\Sigma}\phi dx \quad \forall\ \phi \in H_0^1(\Sigma)$; et ceci est la formulation variationnelle du problème de Dirichlet mentionné en 1).

2) est obtenu par les théorèmes classiques de régularité.

3) et 4) sont des conséquences directes du principe du maximum, en effet: $\Delta\gamma = -2 < 0$ montre que γ est superharmonique et donc ne peut atteindre son minimum que sur la frontière; ce minimum est 0 donc $\gamma > 0$ dans Σ. De même $\Delta|\mathrm{grad}\gamma|^2 = 2\ \mathrm{grad}\Delta\gamma\ \mathrm{grad}\gamma + 2\gamma_{,i\ell}\gamma_{,i\ell} = 2\gamma_{,i\ell}\gamma_{,i\ell} \geq 0$ sur Σ; $|\mathrm{grad}\gamma|^2$ est donc subharmonique et ne peut atteindre son maximum, qui est fini puisque $\gamma \in C^{1,1}(\overline{\Sigma})$, que sur $\partial\Sigma$.

Posons $M = \sup_{\overline{\Sigma}}|\mathrm{grad}\gamma|$ et $\alpha_o = \frac{g}{\mu M}$

3.2.4 Propriétés de θ_α pour $0 \leq \alpha \leq \alpha_o$: solution élastique

Proposition 3.3: Pour $0 \leq \alpha \leq \alpha_o$, $\theta_\alpha = f_\alpha$ ce qui implique que toute la section reste élastique.

En effet, $\alpha \leq \alpha_0 \rightarrow |\text{grad } f_\alpha| \leq g \rightarrow f_\alpha \in K \rightarrow \theta_\alpha = f_\alpha$; par ailleurs: $|\text{grad } f_\alpha| < g$ dans Σ (cf. proposition 3.2, partie 4), donc $E_\alpha = \Sigma$. Les propriétés de f_α dérivent d'ailleurs directement de celles de γ et montrent que f_α est bien la solution classique du problème élastique.

3.2.5 Comportement de θ_α pour $\alpha_0 < \alpha < +\infty$: solution élastoplastique

Proposition 3.4: Pour $\alpha_0 < \alpha < +\infty$, $P_\alpha \neq \Phi$; (Cela signifie qu'il y a une région plastique) et, par ailleurs:

$$\Delta\theta_\alpha + 2\mu\alpha = 0 \text{ sur } E_\alpha \text{ et } \theta_\alpha \in C^\infty(E_\alpha)$$

Démonstration: Posons

$$\Sigma_{g_n} = \{x/x \in \Sigma \text{ t.q. } |\text{grad } \theta_\alpha(x)| \leq g - \frac{1}{n}\}$$

et, à tout $\psi \in \mathcal{D}(\Sigma_{g_n})$ associons:

$$\phi_1 = \begin{cases} \theta_\alpha - \dfrac{\psi}{nk} \text{ sur } \Sigma_{g_n} & \text{avec } k = \sup\limits_{x \in \Sigma_{g_n}} |\text{grad } \psi(x)| \\ \theta_\alpha \text{ sur } \Sigma - \Sigma_{g_n} & \end{cases}$$

Alors $\phi_1 \in K$ ainsi que $\phi_2 = 2\theta_\alpha - \phi_1$; ces deux éléments peuvent donc être choisis comme fonctions tests dans (19) ce qui entraine:

$$\int_{\Sigma_{g_n}} \text{grad}(\theta_\alpha - f) \text{ grad}\psi \, dx = 0 \qquad \forall \psi \in \mathcal{D}(\Sigma_{g_n})$$

Par passage à la limite lorsque $n \to \infty$, nous obtenons:

$$\int_{E_\alpha} \text{grad}(\theta_\alpha - f_\alpha)\text{grad}\psi \, dx = 0 \qquad \forall \psi \in \mathcal{D}(E_\alpha)$$

D'où la conclusion de la proposition.

3.2.6 Comportement de θ_α lorsque $\alpha \to \infty$: solution complètement plastique.

Proposition 3.5: quand $\alpha \to \infty$ alors:

1) θ_α converge uniformement vers $\theta_\infty \in K$
2) θ_∞ est l'enveloppe supérieure des fonctions de K.
3) $\theta_\infty(x) = g\delta(x,\partial\Sigma)$ où $\delta(x,\partial\Sigma)$ désigne la distance de x à $\partial\Sigma$.
4) $|\text{grad}\theta_\infty| = g$. p.p. sur Σ; toute la section est donc plastique.

Démonstration: (19) peut être écrit:

$$\frac{\langle\theta_\alpha,\theta_\alpha\rangle}{\mu\alpha} - \frac{\langle\theta_\alpha,\phi\rangle}{\mu\alpha} - \langle\gamma,\theta_\alpha\rangle + \langle\gamma,\phi\rangle \leq 0 \qquad \forall\phi \in K;$$

Lorsque $\alpha \to \infty$, θ_α décrit un sous ensemble infinie de K. K étant uniformément compact, $\exists$ une sous suite θ_{α_i} qui converge uniformément vers $\theta_\infty \in K$. On peut donc passer à la limite dans l'inéquation précédente compte tenu du fait que

(23) $\quad <\gamma,\phi - \theta_\infty> \leq 0 \qquad \forall\, \phi \in K$

L' unicité vient du fait que s'il existe 2 solutions:
$\theta_\infty^{(1)} \neq \theta_\infty^{(2)}$ alors $\theta_\infty = \sup(\theta_\infty^{(1)},\theta_\infty^{(2)}) \in K$ et satisfait

$$<\gamma,\dot{\theta}_\infty - \theta_\infty^{(p)}> > 0;$$

d'où la contradiction.

(Les points 2, 3, 4 ne présentent pas de difficultés.)

3.2.7 Comportement général de θ_α en fonction de α

Proposition 3.6: si $\alpha < \alpha'$, alors $\theta_\alpha < \theta_{\alpha'}$ sur un sous ensemble de mesure non nulle de Σ. En particulier

$$\theta_\alpha > 0 \text{ partout sur } \Sigma, \forall \alpha > 0.$$

Demonstration: (i) $\theta_\alpha(x) \leq \theta_{\alpha'}(x)$; en effet

soit $\Sigma_o = \{x/x \in \Sigma,\ \theta_\alpha(x) > \theta_{\alpha'}(x)\}$

et $\Sigma_1 = \{x/x \in \Sigma,\ \theta_\alpha(x) \leq \theta_{\alpha'}(x)\}$

En appliquant (19) aux deux fonctions test suivantes :

$\phi_1 = \inf\{\theta_\alpha,\theta_{\alpha'}\}$ et $\phi_2 = \sup\{\theta_\alpha,\theta_{\alpha'}\}$, nous obtenons

$$(\alpha' - \alpha) \int_{\Sigma_o} \left[\theta_{\alpha'}(x) - \theta_\alpha(x)\right] dx \geq 0$$

ce qui implique $\Sigma_o = \phi$.

(ii) si $\theta_\alpha(x) = \theta_\alpha'(x)$ partout sur Σ, alors $\theta_\alpha = \theta_\infty$; supposons en effet $\theta_\alpha = \theta_\alpha' \neq \theta_\infty$, alors $E_\alpha = E_\alpha' \neq \phi$ et $\Delta\theta_\alpha = \Delta\theta_\alpha' = - 2\mu\alpha = - 2\mu\alpha'$ sur E_α, ce qui est impossible puisque $\alpha = \alpha'$.

(iii) Pour α fini, $\theta_\alpha \neq \theta_\infty$; en effet si $\theta_\alpha = \theta_\infty$ alors

$$<\theta_\infty - \phi, \theta_\infty - f_\alpha> \leq 0 \qquad \forall \phi \in K.$$

Soit alors

$$\phi = \begin{cases} k \text{ sur } \Sigma_k = \{x/x \in \Sigma,\ \theta_\infty(x) \geq k\} \\ \theta_\infty \text{ ailleurs} \end{cases}$$

avec $k \in]0, M_\infty[$, $M_\infty = \sup\limits_{x \in \Sigma} \theta_\infty(x)$; $\phi \in K$ et l'inéquation implique $M_\infty - k \geq \frac{g^2}{2\mu\alpha}$ $\forall k \in]0, M_\infty[$ ce qui est faux.

3.3 Apparition et propagation des zones plastiques

3.3.1 Nouvelles définitions des régions élastique et plastique

H. Brézis et M. Sibony (5) ont obtenu le résultat de comparaison suivant: si

(24) $K' = \{\phi/\phi \in H_0^1(\Sigma),\ |\phi| \leq \theta_\infty$ p.p. sur $\Sigma\}$ alors

$K \subset K'$ et les deux problèmes:

(P) $\theta_\alpha \in K; \langle\theta_\alpha - \phi, \theta_\alpha - f_\alpha\rangle \leqslant 0 \quad \forall \phi \in K$

(P') $\theta_\alpha' \in K'; \langle\theta_\alpha' - \phi, \theta_\alpha' - f_\alpha\rangle \leqslant 0 \quad \forall \phi \in K'$

ont même solution: $\theta_\alpha = \theta_\alpha'$.

De là on peut montrer que

(25) $E_\alpha = \{x/x \in \Sigma, \theta_\alpha(x) < \theta_\infty(x)\}$

(26) $P_\alpha = \{x/x \in \Sigma, \theta_\alpha(x) = \theta_\infty(x)\}$

Cette nouvelle définition de E_α et P_α est essentielle pour établir les propriétés concrètes suivantes qui sont intuitives et facile à démontrer.

3.3.2 Contiguité des régions plastiques et de $\partial\Sigma$

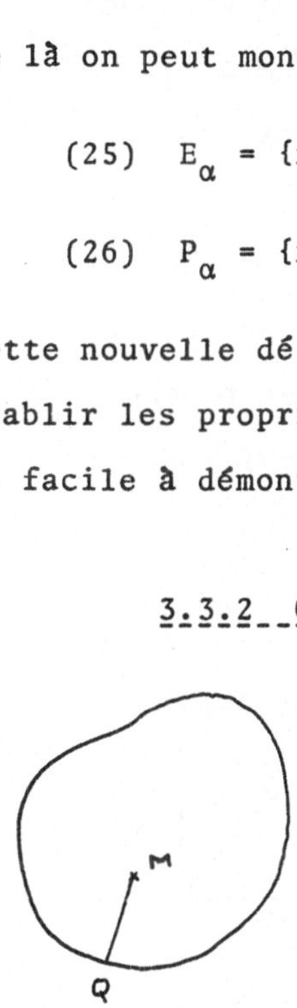

Si $M \in P_\alpha$ et, si $Q \in \partial\Sigma$ est tel que

$$|MQ| = \delta(M, \partial\Sigma)$$

alors $MQ \subset P_\alpha$.

En particulier toute possibilité d'îles plastiques est exclue.

3.3.3 Partition de Σ en zones d'influence

Soit Γ, l'ensemble des points de Σ qui sont équidistants de au moins deux points de $\partial\Sigma$. Γ peut être facilement déterminé pour chaque cas de section.

Nous pouvons prouver que P_α ne peut traverser Γ; en effet la dérivée de θ_∞ subit une discontinuité à la traversée de Γ et ne peut donc être égal à θ_α qui appartient à $C^{1,1}(\overline{\Sigma})$ Ainsi les différents arcs de Γ partagent la section Σ en différentes parties, chacune d'elles étant exclusivement influencée par une partie determinée de $\partial\Sigma$. Nous dirons donc que $\tau_i \subset \Sigma$ est la zone d'influence de $\partial\Sigma_i \subset \partial\Sigma$ si $\forall\, M \in \tau_i$, il existe un unique $Q \in \partial\Sigma$ tel que $\delta(M,\partial\Sigma) = |MQ|$ et si $Q \in \partial\Sigma_i$ (cf. figures ci dessous).

En conclusion:

La plasticité commence à apparaître lorsque $\alpha = \alpha_0$ en un ou plusieurs points de $\partial\Sigma$, elle se propage de proche en proche à partir de là et chaque composante connexe reste enfermée dans une zone d'influence. Pour α fini, il reste donc toujours un voisinage élastique autour de Γ.

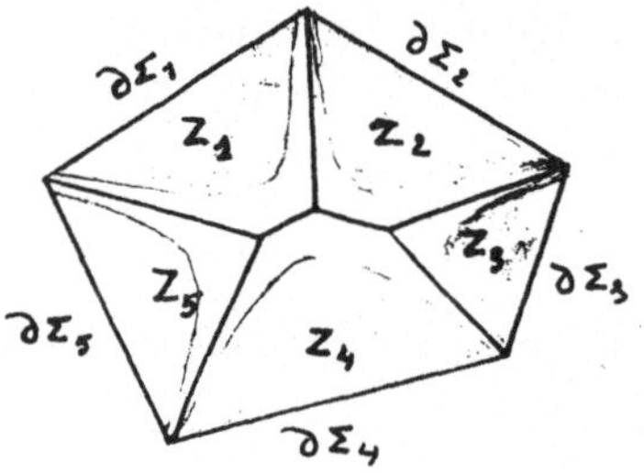

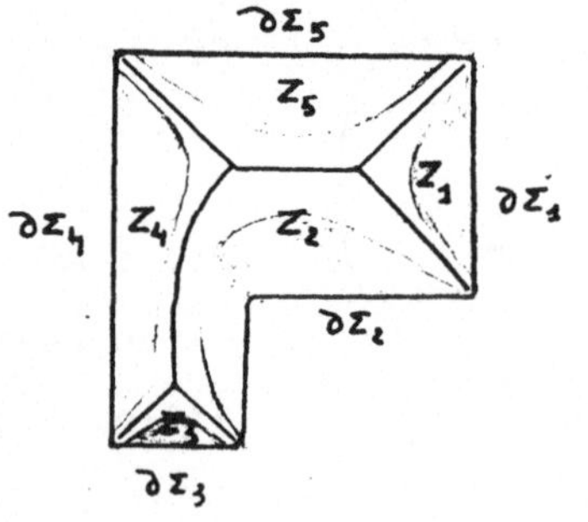

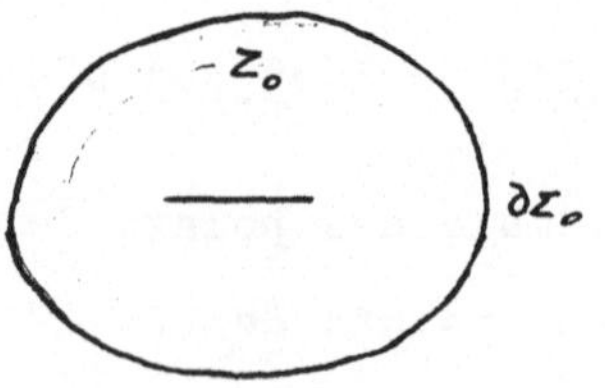

3.4 Détermination du champ de déplacements solution de (A)

Signalons simplement que pour chaque valeur de α on est capable d'associer σ^o soit:

$$\sigma^o_{ij} = 0 \text{ excepté } \sigma^o_{13} = \sigma^o_{31} = \frac{\partial\theta_\alpha}{\partial x_2} \text{ et } \sigma^o_{23} = \sigma^o_{32} = -\frac{\partial\theta_\alpha}{\partial x_1},$$

un champ de déplacements u^o par la loi de comportement (6), (7), cf. [10], et ceci grâce à un résultat de H. Brézis [9]. Nous obtenons l'unicité à une translation globale près, parallèle a Ox_3, ce qui est physiquement tout à fait normal.

Pour cette détermination, la régularité de θ_α ainsi que toutes les propriétés énoncées en 3.2 et 3.3 sont indispensables or la plupart de ces propriétés sont obtenues grâce au théorème de comparaison de H. Brézis et M. Sibony [5].

Finalement le problème (A) est complètement résolu.

4. PROBLEME GENERAL AVEC "PRANDTL-REUSS" ET APPLICATION AU PROBLEME (A')

• On trouvera dans G. Duvaut, J.L. Lions [7] la formulation et les résultats pour un problème au limite général du type (B), mais cette fois avec la loi de comportement de Prandtl-Reuss. Les résultats sont du même style que ceux obtenus en 2 mais, σ^o est ici solution d'une inéquation qui contient à la fois σ et $\dot{\sigma} = \frac{d\sigma}{dt}$, et qui ne correspond pas à la minimisation d'une fonctionnelle.

Pour l'application au problème (A') nous procédons comme en 3.1 pour montrer que σ^o à la forme simple: $\sigma^o_{ij} = 0$ sauf σ^o_{13} et σ^o_{23} et ne dépend que de x_1 et x_2. Les deux différences mentionnées ci dessus apportent tout de même une difficulté supplémentaire car nous devons remplacer l'argument $J(\overline{\sigma}) \leq J(\sigma^o)$ par:

$$\mathcal{A}\ (\overline{\sigma} - \sigma^o, \overline{\sigma} - \sigma^o) \leq 0$$

ce qui est plus délicat mais conduit à la même conclusion.

• Le problème (A_1') se formule alors comme suit: Trouver

(27) $\theta_\alpha(t) \in K$ tel que $<\dot{\theta}_\alpha(t) - f_\alpha(t), \theta_\alpha(t) - \phi> \leq 0$

$\forall \quad \phi \in K.$

Ce problème est résolu par la théorie des opérateurs maximaux monotones et des semi groupes non linéaires de contraction [9].

• Si $\dot{\alpha}(t) \geq 0$, les solutions de (19) et (27) sont les mêmes, ce qui signifie que la loi de Hencky et celle de Prandtl-Reuss donnent même résultat. Ceci est obtenu grâce à un nouveau théorème de comparaison de H. Brézis [6] qui s'appuie encore sur H. Brézis, M. Sibony [5].

5. GENERALISATION AU CAS DES SECTIONS MULTICONNEXES: PROBLEMES OUVERTS.

5.1 Notations et nouvelle formulation

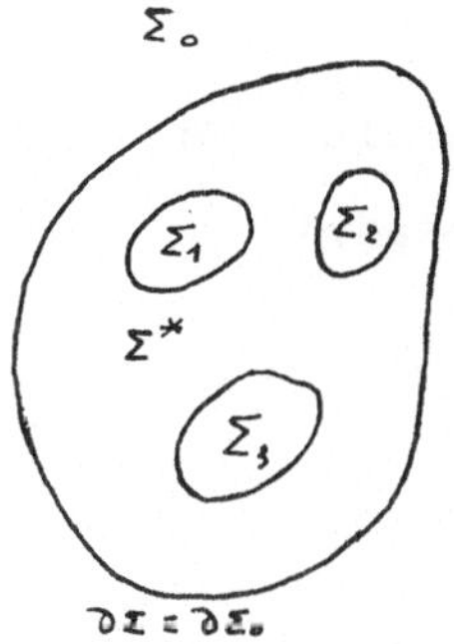

On désigne toujours par Σ la section du cylindre mais par Σ^* la partie multiconnexe effectivement occupée par le matériau.

Soient Σ_1, Σ_2, ...Σ_q les sections des q cavités.

Désignons encore par Σ_o l'extérieur de $\overline{\Sigma}$.

Les résultats de 3.1 sont valables ici mais, comme nous l'avons déjà signalé, l'absence d'efforts sur la frontière latérale se traduit par $\theta_\alpha^* = 0$ sur $\partial\Sigma_o = \partial\Sigma$ et $\theta_\alpha^* = C_i$ (constante) sur $\partial\Sigma_i$ (nous affectons une étoile à tout ce qui correspond à la section multiconnexe Σ^*).

En prolongeant, par C_i, θ_α^* sur chaque Σ_i (et désignant encore par θ_α^* la fonction ainsi prolongée) nous continuons à travailler sur la section Σ toute entière et obtenons θ_α^* comme solution unique du problème (A_1^*): Trouver

(28) $\theta_\alpha^* \in K^*; <\theta_\alpha^* - \phi, \theta_\alpha^* - f_\alpha> \leq 0 \quad \forall \phi \in K^*$ où

(29) $K^* = K \cap E$

(30) $E = \{\phi / \phi \in H_o^1(\Sigma), \ \phi = \text{constante sur chaque } \Sigma_i, \ i = 1, \ldots q\}$

5.2 Résultats et problèmes ouverts (R_1^*) et (R_2^*)

- K^* a les mêmes propriétés que K (cf. paragraphe 3.2.1).

- Nous n'avions pas jusqu'à ces derniers temps de résultat de régularité pour θ_α^*; ce résultat étant simplement espéré comme suit:

$$(R_1^*) \qquad \theta_\alpha^* \in C^{1,\alpha}(\overline{\Sigma}^*)$$

(Il semble cependant que pendant cette session de Bressanone, C. Gerhardt ait obtenu ce résultat avec peut être quelques restrictions sur la forme des Σ_i)

- f_α^*, projection de f_α sur E est la solution élastique du problème.

- Nous introduisons comme en 2.2.3,

$$\alpha_o^* = \frac{\alpha}{\sup\limits_{\overline{\Sigma}} |\mathrm{grad}\ f_\alpha^*|},$$

et montrons que pour $\alpha \leq \alpha_o^*$ toute la section Σ^* reste élastique.

- Lorsque $\alpha \to \infty$, nous montrons que $\theta_\alpha^*(x)$ tend vers:

$$(31) \qquad \theta_\infty^*(x) = g\ \delta^*(x, \partial\Sigma),$$

où $\delta^*(x,\partial\Sigma)$ est en quelque sorte "une distance au bord" généralisée définie comme suit:

$$(32) \qquad \delta^*(x,\partial\Sigma) = \inf_{i \in \{0,1 \ldots, q\}} \{\delta(x,\Sigma_i) + d_i\}$$

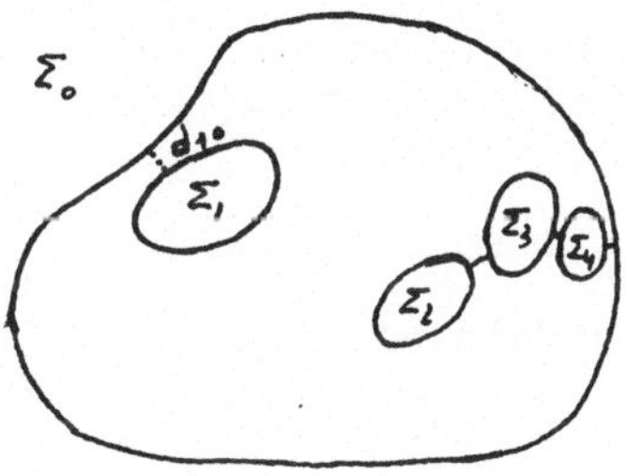

d_i étant le plus court chemin entre $\partial\Sigma_i$ et $\partial\Sigma_o$ si l'on convient que la traversée de chaque Σ_i donne une contribution nulle. Dans l'exemple de la figure ci contre, nous avons

$$d_o = 0, \; d_1 = d_{10}, \; d_2 = d_{23} + d_{34} + d_{40} < d_{20},$$

$$d_3 = d_{34} + d_{40} \leqslant d_{30} . \qquad (d_{ij} = \delta[\Sigma_i, \Sigma_j])$$

• Nous nous heurtons alors à un second problème ouvert:

(R_2^*) $\left\{\begin{array}{l} \text{y a t'il identité entre } \theta_\alpha^* \text{ et, } \theta_\alpha^{*\prime} \text{ solution de:} \\ (33) \quad \theta_\alpha^{*\prime} \in K^{*\prime}, \; <\theta_\alpha^{*\prime} - \phi, \; \theta_\alpha^{*\prime} - f_\alpha> \leqslant 0 \\ \qquad\qquad\qquad\qquad \forall \; \phi \in K^{*\prime} \\ (34) \quad K^{*\prime} = K' \cap E \qquad \text{cf. (24)?} \end{array}\right.$

(Ce résultat serait l'équivalent de celui de H. Brézis, M. Sibony.)

En l'absence de ces résultats (R_1^*) et (R_2^*) nous avons procédé par analogie avec le cas simplement connexe, pour formuler un certain nombre de conjectures relatives à l'apparition et au développement de la plasticité dans Σ^*. L'analyse numérique de certains cas de figures non nécessairement trivial a montré que ces conjectures étaient tout à fait vérifiées [11]; ceci est sûrement un encouragement pour la démonstration rigoureuse de (R_1^*) et (R_2^*).

L'obtention de (R_1^*) et (R_2^*) permettrait en outre de montrer l'existence d'un champs de déplacements associé a θ_α^* et de comparer les résultats donnés par Hencky avec ceux obtenus par Prandtl-Reuss.

Signalons enfin que l'analyse numérique avec le convexe $K^{*\prime}$ est plus aisée: les temps de calcul sont beaucoup plus courts et les résultats plus précis.

H. Lanchon

REFERENCES

"Espaces de Sobolev"

[1] J.L. LIONS - E. MAGENES

"Problèmes aux limites non homogènes et applications". Dunod, Paris (1968).

"Minimisation d'une fonctionnelle"

[2] J.L. LIONS

"Sur le contrôle optimal de systèmes gouvernées par des équations aux derivées partielles". Dunod, Gauthier Villars (1969).

"Inéquations variationnelles"

[3] J.L. LIONS et G. STAMPACCHIA

"Variational inequalities". Comm. Pure Appl. Math. 20 (1967) p.493-519.

[4] H. BREZIS et G. STAMPACCHIA

"Sur la régularité de la solution d'inéquations elliptiques". Bull. Soc. Math. France Vol. 96 (1968) p.153-180.

[5] H. BREZIS et M. SIBONY

"Equivalence de deux inéquations variationnelles et applications". Arch. Rat. Mech. Anal. Vol. 41 (1971) p.254-265.

[6] H. BREZIS

"Problèmes unilatéraux". Thèse Université Paris VI (1971).

[7] G. DUVAUT - J.L. LIONS

"Les inéquations en mécanique et en physique". Dunod, (1972).

"Détermination d'un champ de déplacements"

[8] H. BREZIS

"Multiplicateur de Lagrange en torsion élastoplastique" Arch. Rat. Mech. Anal. Vol. 49 No. 1 (1972) p.32-41.

"Théorie des opérateurs maximaux monotones et des semi groupes non linéaires de contraction"

[9] H. BREZIS

"Opérateurs maximaux monotones et semi-groupes de contractions dans des espaces de Hilbert". Cours Université Paris VI et CNRS ERA 215 (1971).

"Torsion élastoplastique"

(a) Théorie

[10] H. LANCHON

"Torsion élastoplastique d'un arbre cylindrique de section simplement ou multiplement connexe". Thèse Université Paris VI. Département de Mécanique (1972).

(b) Etude numérique

[11] R. GLOWINSKI et H. LANCHON

"Torsion élastoplastique d'une barre cylindrique de section multiconnexe". Journal de Mécanique Vol. 12 No. 1 (Mars 1973).

CENTRO INTERNAZIONALE MATEMATICO ESTIVO

(C.I.M.E.)

DUALITE EN CALCUL DES VARIATIONS

JEAN MICHEL LASRY

Corso tenuto a Bressanone dal 17 al 26 giugno 1973

J. M. Lasry

Résumé :

Soit P un problème de calcul des variations :

$$\text{(P) Minimiser } \ell(\lambda u) + \int_\Omega L(\omega,u(\omega),\Lambda u(\omega))d\omega$$

où Λ est un opérateur différentiel linéaire (gradient, laplacien,....) et λ un opérateur de trace (ex : λu = restriction de u à $\partial\Omega$) où L est une intégrande normale convexe et où ℓ est une fonctionnelle convexe. Nous associons à P le problème dual

$$(P^*) \text{ Minimiser } \ell^*(\mu p) + \int_\Omega L^*(\omega,Mp(\omega),p(\omega))d\omega$$

où ℓ^* et $L^*(\omega,.,.)$ sont les fonctions duales de ℓ et $L(\omega,.,.)$ où M est l'opérateur différentiel adjoint de $-\Lambda$ (divergence, laplacien,...) et où μ est l'opérateur de trace pour lequel on a la formule de Green $< u,Mp > + < \Lambda u,p > + < \lambda u, \rho p > = 0$. Nous étudions la dualité entre P et P^* ($\inf P = -\inf P^*$, conditions d'extrémalité liant des solutions $\bar{u}$ et $\bar{p}$ de P et P^* respectivement existence de $\bar{p}$). En particularisant $\Omega =]0,T[$ et $\Lambda = \frac{d}{dt}$ on retrouve les problèmes de Bolza convexes étudiés par Rockafellar. D'autrepart en particularisant ℓ et L on retrouve la dualité pour les opérateurs aux dérivées partielles développée par R. Temam.

J. M. Lasry

INTRODUCTION

1. Dans une série d'articles R.T. Rockafellar a étudié la dualité pour les problèmes de Bolza convexes, c-à-d pour les problèmes du type :

$$(P) \text{ Minimiser } \left[\ell(x(0), x(T)) + \int_0^T L(t,x(t),\dot{x}(t))dt \right] \text{ pour } x \in A$$

où A est l'ensemble des fonctions absolument continues, ℓ une fonction convexe s.c.i. et L une intégrande normale convexe (ℓ et L sont à valeurs dans $]-\infty, +\infty]$).

D'autre part R. Temam a développé la dualité pour des problèmes de cacul des variations, notamment le problème de Plateau

$$\underset{u \in H_0^1(\Omega)}{\text{Minimiser}} \int_\Omega \sqrt{1+(\mathrm{grad}(u-u_0))^2}\, d\omega$$

et de nombreux autres exemples faisant intervenir des opérateurs aux dérivées partielles (gradient, laplacien,...).

Nous allons donner une formulation commune de ces problèmes, et ainsi étudier la dualité pour des problèmes de calcul des variations faisant intervenir à la fois les intégrandes normales convexes et les opérateurs linéaires aux dérivées partielles.

On va retrouver dans le cas général des développements usuels en dualité : construction d'un problème dual P^* tel que $\inf P = -\inf P^*$ et tel que les conditions d'extrémalité fournissent un système de conditions nécessaire et suffisant d'optimalité. Le résultat le plus marquant est, suivant l'idée de T. Rockafellar (1,2) déjà reprise par R. Temam (1,2), que l'existence d'une solution optimal du problème dual est liée à la stabilité du problème primal vis à vis de certaines perturbations. Ce qui permet de ramener un problème d'existence à une étude de stabilité.

. Pour obtenir un résultat d'existence sur P on échange les rôles de P et P^*(ce qui est générallement possible).

2. La dualité entre P et P^* repose dans sa formulation même sur la formule de Green. Dans la démonstration celle-ci intervient sous la forme d'une propriété analogue au lemme d'Euler-Lagranga, lemme que l'on peut dans l'esprit de la dualité écrire ainsi :

si $x \in L^1(0,T)$ et $y \in L^1(0,T)$ et si

$$\sup_{y \in C^\infty(0,T)} \left\{ [x(0) - \alpha]\, y(0) + [x(T) - \beta]\, y(T) + \int_0^T [z(t)y(t) + x(t)\dot{y}(t)]\, dt \right\}$$

est fini alors $x(0) = \alpha$, $x(T) = \beta$ et $z(t) = x(0) + \int_0^t x(\tau)dt$.

Cette analogie n'est pas fortuite puisque les relations d'extrémalité obtenues par dualité coïncident pour des Lagrangiens réguliers avec les équations d'Euler.

3. Nous renvoyons pour une bibliographie complète et un historique de la dualité en calcul des variations aux articles de Rockafellar cités et au livre de I. Ekeland et R. Temam ou on trouvera aussi de nombreux exemples d'applications à des problèmes classiques.

* *

NOTATIONS

On désigne par Ω un ouvert borné régulier de $\mathbb{R}^n$, $\Gamma(= \partial\Omega)$ son bord. Pour $i \in \mathbb{N}$ et $1 \leq \delta \leq \infty$ on note $L_i^\delta(\Omega,\mathbb{R}^i)$ l'espace des fonctions δ-sommables de Ω dans $\mathbb{R}^i$; $D_i(\Omega) = D_i(\Omega,\mathbb{R}^i)$ l'espace des fonctions C^∞ de Ω dans $\mathbb{R}^i$ à support compact; $D_i(\overline{\Omega})$ et $D_i(\Gamma)$ les espaces $C^\infty(\overline{\Omega},\mathbb{R}^i)$ et $C^\infty(\Gamma,\mathbb{R}^i)$ respectivement; $D_i'(\Omega)$ les distributions dans Ω à valeurs dans $\mathbb{R}^i$.

1. Espaces fonctionnels, opérateurs, formule de Green

Nous donnons d'abord (A) le formalisme qui recouvre les trois exemples principaux (B).

A On désigne par $\mathbb{U}$ un espace fonctionnel tel que $D_m(\overline{\Omega}) \subset \mathbb{U} \subset L^\alpha(\Omega,\mathbb{R}^m)$ où $\alpha \in [1, +\infty]$ (en général $\mathbb{U}$ sera un espace de Sobolev; $\mathbb{U} = H^\rho(\Omega)$, $\mathbb{U} = W^{s,\rho}(\Omega),\ldots$).

On désigne par Λ un opérateur différentiel linéaire défini pour $\phi = (\phi_1,\ldots,\phi_m) \in D_m(\overline{\Omega})$ par

$$\Lambda\phi = \sum_{j=1}^{k} \sum_{|\rho| \leq N} a_{\rho,j} \frac{\partial}{\partial x_1^{\rho_1}} \cdots \frac{\partial}{\partial x_n^{\rho_n}} \phi_j \quad \text{avec } a_\rho \in D_k(\overline{\Omega}).$$

On suppose que pour $u \in \mathbb{U}$, Λu appartient à L_k^β où β est donné dans $[1, +\infty[$.

On désigne par λ un opérateur linéaire de $\mathbb{U}$ dans un espace de Banach E tel que l'on ait $D_g(\Gamma) \subset E \subset D_g'(\Gamma)$ (injections continues). On suppose que λ vérifie

$\lambda(D_m(\Omega)) = 0$ et $\lambda(D_m(\overline{\Omega})) = D_g(\Gamma)$.

Soit M l'ajoint de $-\Lambda$ au sens des distributions.

On désigne par $\mathbb{P}_o$ l'espace

$$\mathbb{P}_o = \{p \in L_k^{\beta'} \ ; \ Mp \in L_m^{\alpha'}\}$$

où α' et β' sont les exposants conjugués de α et β ($1/\alpha' + 1/\alpha = 1$; $1/\beta' + 1/\beta = 1$). On suppose que sur $\mathbb{P}_o$ est défini un opérateur linéaire $\mu : \mathbb{P}_o \longrightarrow D'_g(\Gamma)$ tel que la "formule de Green" suivante :

$$\int_\Omega p\,\Lambda\,u + \int_\Omega uMp + <\lambda u, \mu p> = 0$$

est vérifiée pour $(u,p) \in D_m(\bar{\Omega}) \times \mathbb{P}_o$ et pour $(u,p) \in \mathbb{U} \times \mathbb{P}$ avec

$$\mathbb{P} = \{p \in \mathbb{P}_o \ ; \ \mu p \in E'(\text{dual de } E)\}.$$

B Exemples

exemple 1. Prenons pour $\mathbb{U}$ l'espace de Sobolev $H^1(\Omega)$, pour Λ l'opérateur gradient et pour λ l'opérateur de restriction à Γ. Si $u \in H^1(\Omega)$, on a $u \in L^2(\Omega)$, $\Lambda u \in L^2(\Omega,\mathbb{R}^n)$ et $\lambda u \in H^{1/2}(\Gamma)$. (On a donc $\alpha = \beta = 2$, $m = 1$, $k = n$, $E = H^{1/2}(\Gamma)$).

L'adjoint de $-\Lambda$ est l'opérateur M = divergence. L'espace $\mathbb{P}_o$ est donc l'espace

$$V = \{p \in L^2(\Omega,\mathbb{R}^n) \ , \ \text{div}\ p \in L^2(\Omega)\}.$$

Pour $p \in \mathbb{P}_o$ on peut définir le "flux sortant" $\vec{p}.\vec{\nu}$. (où ν est la normale extérieure à Γ. Il vient : $\mathbb{P} = \{p \in \mathbb{P}_o$, $\mu p \in E' = H^{-1/2}(\Gamma)\} = \mathbb{P}_o = V$.

On a donc $\mathbb{P} = \mathbb{P}_o$ dans cet exemple.

Enfin, la formule de Green s'écrit ici

$$\int_\Omega p.\mathrm{grad}u + \int_\Omega u\,\mathrm{div}p + < U_\Gamma, - p.\nu > = 0$$

(où $U_\Gamma = \lambda u$ est la trace de u sur Γ) et elle est vérifiée pour $(u,p) \in H^1(\Omega) \times V$ (cf. Temam (3) ou Lions-Magenes (1)).

<u>exemple 1'</u> : On peut échanger les rôles de $\mathbb{U}$ et $\mathbb{P}$ dans l'exemple précédent (ce qui n'est pas toujours possible). Prenons pour $\mathbb{U}$ l'espace V de l'exemple 1, pour Λ l'opérateur divergence et pour λ l'opérateur $\lambda u = u.\nu$ (= flux sortant). Il vient alors M = gradient et :

$$\mathbb{P}_o = \{p \in L^2(\Omega);\ \mathrm{grad}\ p \in L^2(\Omega,\mathbb{R}^n)\} = H^1(\Omega).$$

μ est l'opérateur de restriction, $\mathbb{P} = \mathbb{P}_o = H^1(\Omega)$ et la formule de Green est la même que dans l'exemple 1.

<u>exemple 2</u>. Prenons pour $\mathbb{U}$ l'espace A_m des fonctions absolument continues de $[0,T]$ dans $\mathbb{R}^m$ (donc $\Omega =]0,T[$ et $\Gamma = \{0,T\}$), pour Λ l'opérateur de dérivation $\frac{d}{dt}$, et pour λ l'opérateur $\lambda u = (u(0), u(T))$ à valeurs dans $\mathbb{R}^m \times \mathbb{R}^m$. On vérifie alors que $M = -\frac{d}{dt}$ que $\mathbb{P}_o = \mathbb{U}$, que μ est l'opérateur défini par : $\mu p = p(0), -p(T))$, que $\mathbb{P} = \mathbb{P}_o$ et que la formule de Green se réduit à la formule d'intégration par parties pour les fonctions absolument continues (avec $\dot{x} = dx/dt$) :

$$\int_0^T u\,\dot{p} + \int_0^T \dot{u}\,p + (u(0)p(0) - u(T)p(T)) = 0.$$

<u>exemple 3</u> Prenons pour $\mathbb{U}$ l'espace

$\mathbb{U} = \{u \in H^1(\Omega),\ \Delta u \in L^2(\Omega)\}$

pour Λ l'opérateur Δ (laplacien). Sur $\mathbb{U}$ on peut définir l'opérateur de trace λ par

$$\lambda u(u_{|\Gamma}, \frac{\partial u}{\partial \nu}) \in H^{1/2}(\Gamma) \times H^{-1/2}(\Gamma)$$

(donc $\alpha = \beta = 2$, $m = k = 1$, $E = H^{1/2}(\Gamma) \times H^{1/2}(\Gamma)$, $g = 2$)

L'opérateur M adjoint de $-\Lambda$ est $-\Delta$ et :

$$\mathbb{P}_o = \{p \in L\ (\Omega);\ \Delta p \in L^2(\Omega)\}$$

Sur $\mathbb{P}_o$ on peut définir l'opérateur de trace μ par (cf. Lions Magenès (1))

$$\mu p = (\frac{\partial p}{\partial \nu}\ ,\ -p_{|\Gamma}) \in H^{-3/2}(\Gamma) \times H^{-1/2}(\Gamma)$$

D'où, par définition (puisque $E' = H^{-1/2}(\Gamma) \times H^{1/2}(\Gamma)$)

$$\mathbb{P} = \{p \in \mathbb{P}_o\ ;\ \frac{\partial p}{\partial \nu} \in H^{-1/2}(\Gamma) \text{ et } p_{|\Gamma} \in H^{1/2}(\Gamma)\}$$

On montre que $\mathbb{P} = \mathbb{U}$ et que l'on a la formule de Green

$$\int_\Omega p\,\Delta u - \int_\Omega u\,\Delta p + \langle u_{|\Gamma}, \frac{\partial p}{\partial \nu}\rangle - \langle p_{|\Gamma} \frac{\partial u}{\partial \nu}\rangle = 0$$

pour $(u,p) \in \mathbb{U} \times \mathbb{P}$ (cf. Lions Magenès (1))(et pour $(u,p) \in \mathcal{D}(\overline{\Omega}) \times \mathbb{P}_o$)

2. Fonctionnelles, énoncés des problèmes

A) On désigne par $L : \Omega \times \mathbb{R}^m \times \mathbb{R}^k \to]-\infty, +\infty]$ une fonction borélienne telle que, pour presque tout $\omega \in \Omega$, la fonction $L(\omega,.,.)$ est convexe s.c.i. propre (c-à-d $L(\omega,.,.) \not\equiv +\infty$). Autrement dit L est une intégrande normale convexe au sens de Rockafellar(3).

On désigne par $L^* : \Omega \times \mathbb{R}^m \times \mathbb{R}^k \to]-\infty, +\infty]$ la fonction définie par

$$L^*(\omega,c,d) = \text{Sup}\,\{ac + bd - L(\omega,a,b);(a,b) \in \mathbb{R}^m \times \mathbb{R}^k\}$$

C'est encore une intégrande normale convexe et $L^{**} = L$ (cf. idem)

<u>Notation</u> : <u>Si</u> u <u>et</u> v <u>sont des fonctions mesurables de</u> Ω <u>dans</u> $\mathbb{R}^m$ <u>et</u> $\mathbb{R}^k$ <u>respectivement, on notera</u> $L(u,v)$ <u>la fonction mesurable</u> $\omega \to L(\omega,u(\omega), v(\omega))$ <u>et</u> $L^*(u,v)$ <u>la fonction mesurable</u> $\omega \to L^*(\omega,u(\omega),v(\omega))$. <u>Par exemple on écrira</u> $\int L(u,v)$ <u>ou</u> $\int_\Omega L(u,v)$ <u>pour</u> $\int_\Omega L(\omega,u(\omega),v(\omega))d\omega$.

B) <u>On se donne une fonctionnelle</u> $\ell : E \to]-\infty, +\infty]$ <u>convexe s.c.i.</u> propre (c-à-d $\ell \not\equiv +\infty$) et sa duale $\ell^* : E' \to]-\infty, +\infty]$ définie par

$$\ell^*(y) = \text{Sup}\,\{<x,y> - \ell(x);\ x \in E\}$$

Dans la pratique E sera souvent un espace fonctionnel (cf. exemple 1 : $E = H^{1/2}(\Gamma)$) et ℓ sera de la forme

$$\ell(\omega) = \int_\Gamma f(\xi,\omega(\xi))\,d\xi \qquad \text{pour } \omega \in H^{1/2}(\Gamma).$$

Si E est un espace de Sobolev on trouvera le calcul de ℓ^* dans Brezis (1).

C) <u>Hypothèse de finitude</u> (H_o). <u>On suppose qu'il existe</u> $u_o \in \mathbb{U}$ <u>et</u> $p_o \in \mathbb{P}$ <u>tels que les quantités suivantes sont finies</u> :

$$\ell(\lambda u_o), \int_\Omega |L(u_o, \Lambda u_o)|, \ell^*(\mu p_o), \int_\Omega |L(Mp_o,p_o)|$$

Cette hypothèse implique en particulier que pour presque tout $\omega \in \Omega$ et tout $(a,b) \in \mathbb{R}^m \times \mathbb{R}^k$ on a

$$L(\omega,a,b) \geq a.Mp_o(\omega) + b.\,p_o(\omega) - L^*(\omega,Mp_o(\omega),p_o(\omega))$$

$$L^*(\omega,a,b) \geq a.\,u_o(\omega) + b.\,\Lambda u_o(\omega) - L(\omega,u_o(\omega),\Lambda u_o(\omega)).$$

On déduit de la première inégalité que si $(u,v) \in L_m^\alpha \times L_k^\beta$, on a

$$L(\omega,u(\omega),\ v(\omega)) \geq A(\omega) \quad \text{avec} \quad A \in L\,(\Omega),$$

donc que l'intégrale $\int_\Omega L(u,v)$ est bien définie à valeur dans $]-\infty,+\infty]$.

De même on déduit de la deuxième inégalité que pour $(p,q) \in L_k^{\beta'} \times L_m^{\alpha'}$ l'intégrale $\int_\Omega L^*(q,p)$ est bien définie à valeur dans $]-\infty\ ,\ +\infty]$.

D. Les considérations précédentes permettent de définir les problèmes suivants :

$$(P)\ \underset{u \in \mathbb{U}}{\text{Minimiser}}\ [\,\ell(\lambda u) + \int_\Omega L(\omega),u(\omega),\Lambda u(\omega))d\omega]$$

$$(P^*)\ \underset{p \in \mathbb{P}}{\text{Minimiser}}\ [\,\ell^*(\mu p) + \int_\Omega L^*(\omega,Mp(\omega),p(\omega))d\omega]\,.$$

<u>exemple 1</u> (suite) Dans ce cas les problèmes s'écrivent (compte tenu du choix de $\mathbb{U}$,A,.. voir ci-dessus 1.B)

$$(P_1)\ \underset{u \in H^1(\Omega)}{\text{Minimiser}}\ [\,\ell(u_{|\Gamma}) + \int_\Omega L(\omega,u(\omega),\ \text{grad}u(\omega))d\omega]$$

$$(P_2^*)\ \underset{p \in L^2(\Omega,\mathbb{R}^n)\ ,\text{div}p \in L^2}{\text{Minimiser}}\ [\,\ell^*(-p.\nu) + \int_\Omega L^*(\omega,\text{div}p(\omega),p(\omega))d\omega]$$

<u>exemple 2</u> (suite) Il vient

$$(P_2)\ \underset{u \in A_m}{\text{Minimiser}}\ [\,\ell(u(0),u(T)) + \int_0^T L(t,u(t),\dot{u}(t))dt]$$

$$(P_2^*)\ \underset{p \in A_m}{\text{Minimiser}}\ [\,\ell^*(p(0),\ -p(T)) + \int_0^T (t,\dot{p}(t),p(t))dt]$$

avec $A_m = \{x \in L^1(0,T;\mathbb{R}^m)\ ;\ \dot{x} \in L^1(0,T;\mathbb{R}^m)\}$

Ce sont les problèmes de Bolza convexes étudiés par Rockafellar (1,2).

exemple 3 Il vient avec $\mathbb{B} = \{u \in H^1(\Omega),\ \Delta u \in L^2(\Omega)\}$

$$(P_3) \quad \underset{u \in \mathbb{B}}{\text{Minimiser}} \left[\ell(u_{|\Gamma}, \frac{\partial u}{\partial \nu}) + \int_\Omega L(\omega, u(\omega), \Delta u(\omega))d\omega \right]$$

$$(P_3^*) \quad \underset{p \in \mathbb{B}}{\text{Minimiser}} \left[\ell(\frac{\partial p}{\partial \nu}, -p_{|\Gamma}) + \int_\Omega L^*(\omega, -\Delta p(\omega), p(\omega))d\omega \right]$$

3. Premières propriétés

Posons pour $u \in \mathbb{U}$ et $p \in \mathbb{P}$

$$I(u) = \ell(\lambda u) + \int_\Omega L(u, \Lambda u)$$

$$J(p) = \ell^*(\mu p) + \int_\Omega L^*(Mp; p)$$

A) Pour tout $(u,p) \in \mathbb{U} \times \mathbb{P}$ on a

$$I(u) + J(p) = \ell(\lambda u) + \ell^*(\mu p) + \int_\Omega (L(u,\Lambda u) + L^*(Mp,p))$$

$$\geq \langle \lambda u, \mu p \rangle + \int_\Omega (u.Mp + \Lambda\, u.p)$$

Donc $I(u) + J(p) \geq 0$ d'après la formule de Green. D'où :

$$\inf P + \inf P^* = \inf_{u \in \mathbb{U}} I(u) + \inf_{p \in \mathbb{P}} J(p) \geq 0$$

D'après l'hypothèse H_o on a $I(u_o) < +\infty$ et $J(p_o) < +\infty$. Donc :

$$-\infty < -\inf P^* \leq \inf P < +\infty.$$

B) Conditions d'extrémalité

L'égalité $I(u) + J(p) = 0$ a lieu si et seulement si

$$S \begin{cases} \ell(\lambda u) + \ell^*(\mu p) = \langle \mu u, \mu p \rangle \\ \int_\Omega (L(u,\Lambda u) + L^*(Mp,p)) = \int_\Omega (u.Mp + \Lambda\, u.p) \end{cases}$$

La deuxième condition du système S est équivalent à

$$L(\omega,u(\omega),\Lambda u(\omega)) + L^*(\omega,Mp(\omega),p(\omega)) = u(\omega).Mp(\omega) + \Lambda u(\omega).p(\omega)$$

pour presque tout $\omega \in \Omega$. (puisque $f \geq g$ et $\int f = \int g$ impliquent $f = g$ presque partout)

Notons $\partial\ell(\lambda u)$ (resp.$\partial\ell^*(\mu p)$) le sous gradient de ℓ (respℓ^*) au point λu (resp. μp).

Notons $\partial_{u,\Lambda u} L(\omega,u(\omega), \Lambda u(\omega))$ le sous gradient de la fonction $(a,b) \to L(\omega,a,b)$ au point $(u(\omega), \Lambda u(\omega))$.

Notons $\partial_{Mp,p} L^*(\omega,Mp(\omega),p(\omega))$, le sous gradient de la fonction $(a,b) \to L^*(\omega,a,b)$ au point $(Mp(\omega),p(\omega))$.

Le système (S) équivalent à chacun des quatre systèmes (équivalents) $S_{1,3}$, $S_{1,4}$, $S_{2,3}$, $S_{2,4}$ formés à partir de deux des quatre conditions suivantes

(1) $\lambda u \in \partial\ell^* \mu$

(2) $\mu p \in \partial\ell\ (\lambda u)$

(3) $(u(\omega), \Lambda u(\omega)) \in \partial_{Mp,p} L^*(\omega,Mp(\omega),p(\omega))$

pour presque tout $\omega \in \Omega$

(4) $(Mp(\omega), p(\omega)) \in \partial_{u,\Lambda u} L(\omega,u(\omega), \Lambda u(\omega))$

pour presque tout $\omega \in \Omega$

Supposons que $(\omega,a,b) \to L(\omega,a,b)$ est différentiable par rapport à (a,b) ; notons $\frac{\partial L}{\partial a}$ et $\frac{\partial L}{\partial b}$ les dérivées partielles par rapport à la deuxième et la troisième variable. Alors la conditions (4) s'écrit

$$p(\omega) = \frac{\partial L}{\partial b}(\omega,u(\omega), \Lambda u(\omega))$$

$$Mp(\omega) = \frac{\partial L}{\partial a}(\omega,u(\omega), \Lambda u(\omega))$$

En éliminant p on trouve l'équation "d'Euler" du problème (P) :

$$M \frac{\partial L}{\partial b}(\omega,u(\omega), \Lambda u(\omega)) = \frac{\partial L}{\partial b}(\omega,u(\omega), \Lambda u(\omega))$$

Par exemple dans l'exemple 2 cette équation s'écrit

$$\frac{d}{dt}\frac{\partial L}{\partial \dot{x}}(t,x(t), \dot{x}(t)) = \frac{\partial L}{\partial x}(t,x(t), \dot{x}(t))$$

C'est l'équation d'Euler du problème (P_1) (cf. § 1.B et § 2.D).

4. Théorème

D'après les paragraphes 3.A et 3.B ci-dessus on voit que pour un couple $(\overline{u},\overline{p}) \in \mathbb{U} \times \mathbb{P}$ les propriétés sont équivalentes :

i) le couple $(\overline{u},\overline{p})$ vérifie le système $S_{i,j}$ $(i = 1,2;\ J = 3,4)$

ii) le couple $(\overline{u},\overline{p})$ vérifie $I(\overline{u}) + J(\overline{p}) = 0$

iii) on a $\inf P + \inf P^* = 0$ et $\overline{u}$ et $\overline{p}$ sont des solutions optimales de P et P^* respectivement.

Nous allons voir que l'égalité $\inf P + \inf P^* = 0$ et l'existence d'une solution optimale de P^* sont liées à la stabilité de P vis à vis de certaines perturbations.

A. Fonction de stabilité

Pour $(v,w) \in L_k^\beta \times E$ on pose

$$\gamma(v,w) = \inf_{u\in \mathbb{U}} [\ell(w + \lambda u) + \int_\Omega L(\omega,u(\omega), v(\omega) + \Lambda u(\omega))d\omega]$$

En particulier $\gamma(0,0) = \inf P$: on dit que γ est la fonction de stabilité de P(par rapport aux perturbations (v,w)).

Montrons d'abord à l'aide de H_o et de la formule de Green que

$\gamma(v,w) > -\infty$. Il vient

$$\ell(w + \lambda u) + \int_\Omega L(u,v + \Lambda u)$$

$$\geq \langle w + \lambda u, \mu p_o \rangle - \ell^*(\mu p_o) + \int_\Omega [u.Mp_o + (v + \Lambda u).p_o] - \int_\Omega L^*(Mp_o,p_o)$$

$$\geq \langle w, \mu p_o \rangle + \int_\Omega v.p_o - \ell^*(\mu.p_o) - \int_\Omega L^*(Mp_o,p_o)$$

D'où : $\gamma(v,w) \geq \langle w,\mu p_o \rangle + \int_\Omega v.p_o + c$ (avec $c \in \mathbb{R}$)

D'autre part on vérifie (en utilisant la convexité de ℓ et de $L(\omega,.,.)$ et la linéarité de λ et de Λ) que γ <u>est une fonction convexe</u>.

B) Enoncé

Théorème : On fait l'hypothèse (H_1) que pour tout $u \in L^\alpha_m$ il existe $v \in L^\beta_k$ tel que $L(u,v) \in L^1(\Omega)$. Alors :

1) on a $\inf P = -\inf P^*$ si et seulement si γ est s.c.i. en $(0,0)$. Dans ce cas le problème P^* a une solution optimale si et seulement si γ est sous différentiable en $(0,0)$.

2) Si γ est majorée dans un voisinage fort de $(0,0)$ dans $L^\beta_k \times E$, alors $\inf P = -\inf P^*$ et P^* a une solution optimale.

3) Si $\inf P = -\inf P^*$, il est équivalent de dire que $(\bar{u},\bar{p}) \in \mathbb{U} \times \mathbb{P}$ vérifie l'un des systèmes $S_{1,3}$, $S_{1,4}$, $S_{2,3}$, $S_{2,4}$ ou que $\bar{u}$ et $\bar{p}$ sont des solutions optimales de P et P^* respectivement.

Démonstration Le point 3 ne fait que reprendre les considérations du paragraphe 3.

Le point 2 résulte du point 1. En effet, si la fonction convexe γ est majorée dans un voisinage (fort) de (0,0) elle est continue (0,0) (pour la topologie forte) et sous-différentiable.

Reste le point 1? On calcule γ^* et on trouve pour $(p,r) \in L_k^{\beta'} \times E'$:

i) Si $Mp \in L_m^{\alpha'}$ et si $\mu p = r$:

ii) $\gamma^*(p,r) = \ell^*(\mu p) + L^*(Mp,p)$ sinon $\gamma^*(p,r) = +\infty$.

Il en résulte que $\inf P^* = \inf \gamma^*$. Comme $\inf \gamma^* = -\gamma^{**}(0)$ d'une part, et comme $\gamma(0,0) = \inf P$ d'autre part, l'égalité $\inf P + \inf P^* = 0$ a lieu si et seulement si $\gamma(0) = \gamma^{**}(0)$, c'est à dire si et seulement si γ est s.c.i. en (0,0) (pour la topologie forte ou pour la topologie faible).

Enfin comme $\inf P^* \in \mathbb{R}$, il est équivalent de dire que γ^* atteind son minimum ou que P^* a une solution optimale. Si γ est s.c.i. en (0,0), il est équivalent de dire que γ^* atteind son minimum ou que γ est sous-différentiable en (0,0). On remarque qu'on a le résultat plus précis :

$(p,r) \in \partial\gamma(0,0) \Longleftrightarrow [\, p \in \mathbb{P}$, $r = \mu p$, p est une solution optimale de $P^*]$

Le théorème est démontré modulo le calcul de γ^* qui est la partie difficile de la démonstration (on trouvera ce calcul, dans le cas de l'exemple 1, dans Berliocchi et Lasry (2). Le calcul dans le cas général est à peu près le même).

<u>Remarque 1</u>. Pour obtenir des résultats d'existence de solutions optimales pour le problème P, on étudie la stabilité de P^*. Plus préci-

sément on cherche à appliquer le théorème après avoir échangé les rôles de P et P^*: on utilise $L^{**} = L$, $\alpha'' = \alpha$, $\beta'' = \beta$,... Cependant on ne peut pas toujours échanger $\mathbb{U}$ et $\mathbb{P}$; c'est tout de même possible dans les exemples envisagés ici.

Remarque 2. Le point 2 du théorème est le plus utile car c'est le plus facile à appliquer.

§ 5. Exemples (suite et fin)

exemple 1. Dans le cadre de l'exemple 1 (cf. § 1 et 2) on trouve notamment les problèmes variationnels classiques de Neumann et de Dirichlet (on trouvera d'autres exemples dans Ekeland-Temam(1)).

On prend

$$L(\omega,x,y_1,\ldots,y_n) = a_o(\omega)x^2 + xz(\omega) + \Sigma\, a_{ij}(\omega)y_i y_j$$

avec $\Sigma\, a_{i,j}\, \xi_i\xi_j \geq a\,(\xi_1^2 + \ldots + \xi_n^2)$, $\forall\xi \in \mathbb{R}^n$, $\forall\omega \in \Omega$; $a_{i,j} \in L^\infty(\Omega)$;

$a_o(\omega) \geq \alpha$; $a_{i,j} = a_{j,i}$; $z \in L^2(\Omega)$.

Les a_{ij} sont donc les coefficients d'un opérateur fortement elliptique dans Ω.

On calcule l'intégrande duale. Il vient :

$$L^*(\omega,x^*,y_1^*,\ldots,y_n^*) = \frac{1}{4a_o(\omega)}\,[\, z(\omega) - x^*]^2 + \frac{1}{4}\,\Sigma\, b_{ij}(\omega)y_i^* y_j^*$$

où les $(b_{ij}(\omega))$ sont les éléments de la matrice inverse de la matrice $(a_{ij}(\omega))$

Soient ℓ_1 et ℓ_2 les fonctionnels convexes sur $H^{1/2}(\Gamma)$:

$$\ell_1(u) = 0 \qquad \forall u \in H^{1/2}(\Gamma)$$

$$\ell_2(u) = 0 \qquad \text{si} \quad u = 0$$

$$\ell_2(u) = +\infty \qquad \text{si} \quad u \neq 0$$

On vérifie sans difficulté que le point 2 du théorème s'applique aux problèmes $P_{1,1}$ (Neumann) et $P_{1,2}$ (Dirichlet) suivants :

$$(P_{1,1}) \quad \underset{u \in H^1(\Omega)}{\text{Minimiser}} \quad \{\ell_1(u_{|\Gamma}) + \int_\Omega \{a_o\, u^2 + uz + \sum_{i,j} a_{ij} \frac{\partial u}{\partial x_i} \frac{\partial u}{\partial x_j}\}$$

$$(P_{1,2}) \quad \underset{u \in H^1(\Omega)}{\text{Minimiser}} \quad \{\ell_2(u_{|\Gamma}) + \int_\Omega \{a_o\, u^2 + uz + \Sigma\, a_{ij} \frac{\partial u}{\partial x_i} \frac{\partial u}{\partial x_j}\}$$

On obtient les problèmes duaux suivants

$$(P^*_{1,i}) \begin{cases} \text{Minimiser} \quad \frac{1}{4} \int_\Omega \frac{1}{a_o} (z - \operatorname{div} p)^2 + \frac{1}{a} \Sigma\, b_{ij}\, p_i\, p_j \\ \text{pour} \quad p \in L^2(\Omega, \mathbb{R}^n)\, ,\ \operatorname{div} p \in L^2(\Omega) \\ \text{si} \quad i = 1,\ p.\nu = 0;\ \text{si} \quad i = 2, \text{ pas de condition sur } p.\nu. \end{cases}$$

Les relations d'extrémalité s'écrivent par exemple pour p_2

$$p_i(\omega) = 2\, \Sigma\, a_{ij}(\omega) \frac{\partial u}{\partial x_i}(\omega)$$

$$\operatorname{div} p(\omega) = 2\, a_o(\omega) + z(\omega)$$

$$p.\nu = 0$$

ce qui donne en éliminant p le système elliptique usuel.

<u>exemple 2</u> (suite, cf. § 1 et § 2).

On trouvera de nombreux exemples de problèmes de Bolza convexes dans T. Rockafellar (1 et 2).

exemple 3. Nous allons voir dans un cas particulier comment appliquer le formalisme général.

Soit $z \in L^2(\Omega)$ donné. Considérons le problème

$$P \begin{cases} \text{Minimiser } \int_\Omega |y - z|^2 + \int_\Gamma (\frac{\partial y}{\partial \nu})^2 \\ \text{pour } y \in H^1(\Omega),\ y = \Delta y \text{ dans } \Omega,\ y \geqslant 0 \text{ sur } \Gamma. \end{cases}$$

On introduit le Lagrangien $L : \Omega \times \mathbb{R}^2 \to]-\infty, +\infty]$

$$L(\omega,a,b) = \begin{cases} (a - z(\omega))^2 & \text{si } a = b \\ +\infty & \text{si } a \neq b \end{cases}$$

et le coût frontière $\ell : H^1(\Gamma) \times L^2(\Gamma) \to [0, +\infty]$

$$\ell(v,w) \begin{cases} \|w\|_2 & \text{si } v \geqslant 0 \\ +\infty & \text{sinon} \end{cases}$$

On pose $\mathbb{U} = \{u \in H^{3/2}(\Omega);\ \Delta y \in L^2(\Omega)\}$.

Sur $\mathbb{U}$ est défini un opérateur linéaire continu de trace $\lambda : \mathbb{U} \to H^1(\Gamma) \times L^2(\Gamma)$ par $\lambda u = (u_{|\Gamma}, \frac{\partial u}{\partial \nu})$ (cf. Lions-Magenes (1)). On pose $\Lambda = \Delta : \mathbb{U} \to L^2(\Omega)$. Le problème P s'écrit alors (l'équation $y = \Delta y$ est "dans" L).

$$\text{(P) Minimiser } \ell(\lambda u) + \int_\Omega L(\omega, u(\omega), \Lambda u(\omega))d\omega \quad \text{pour } u \in \mathbb{U}.$$

Montrons que le point 2 du théorème s'applique.

Soit $\gamma : L^2(\Omega) \times H^1(\Gamma) \times L^2(\Gamma) \to]-\infty, +\infty]$ la fonction de stabilité. La définition (§ 4.A) donne ici

$$\gamma(u,v,w) = \begin{cases} \inf \int_\Omega |y - z|^2 + \int_\Gamma (\frac{\partial y}{\partial \nu} + w)^2 \\ \text{pour } y \in \mathbb{U},\ y + v \geqslant 0 \text{ sur } \Gamma,\ y - \Delta y = w \text{ dans } \Omega. \end{cases}$$

Soit $y_{v,w}$ la solution du système

$$y - \Delta y = w \text{ dans } \Omega, \; y + v = 0 \text{ sur } \Gamma.$$

On montre (cf. Lions Magenès (1)) que $(v,w) \to y_{v,w}$ est continue de $L^2(\Omega) \times H^1(\Gamma)$ dans Ш. En reportant $y_{v,w}$ dans la définition de γ on trouve :

$$\gamma(u,v,w) \leqslant a + b(\|u\|_2 + \|v\|_{H^1} + \|w\|_2) \text{ (avec } a,b \in \mathbb{R})$$

Le problème dual a donc une solution optimale. En calculant

$$(P^*) \begin{cases} \text{Minimiser } \frac{1}{4} \int_\Gamma p^2 + \int_\Omega [(\frac{p-\Delta p}{2})^2 + (p - \Delta p)z] \\ p \in L^2(\Omega), \; \Delta p \in L^2(\Omega), \; \frac{\partial p}{\partial \nu} \leqslant 0 \end{cases}$$

Enfin une condition nécessaire et suffisante pour que (u,p) soit un couple de solution de P et P^* respectivement est que

i) $y = \Delta y$

ii) $\frac{1}{2}(p - \Delta p) = y + z$

iii) $y \geqslant 0$ sur $\Gamma, \frac{\partial p}{\partial \nu} \leqslant 0$ sur Γ

$$\int_\Gamma y \frac{\partial p}{\partial \nu} = 0$$

iv) $p = 2\frac{\partial y}{\partial \nu}$

On montre sans difficultés l'existence d'une solution optimale de P par les méthodes usuelles (on peut aussi étudier la stabilité de P^*, ce qui n'est pas non plus difficile). On peut aussi montrer l'existence d'un état adjoint p vérifiant les conditions i, ii, iii, iv, par les méthodes plus classiques (que la dualité) de Lions (1).

J. M. Lasry

Bibliographie

Berliocchi H. et Lasry J.-M.,

1) C.R. Acad. Sc., Paris, 1973, t. 276, p. 1611-1614

2) Annalles de l'université de Bordeaux II, 231, Cours de la libération, Talence, à paraître début 1974.

Brézis H., Intégrales convexes dans les espaces de Sobolev, (Sym. Eq. der. Part.) Jerusalem, 1972.

Ekeland I. et Temam R., Analyse convexe et problèmes variationnels Paris, Dunod, à paraître.

Lions J.-L., Contrôle optimal, Dunod, 1971.

Lions J.-L. et Magenes E., Problèmes aux limites.., Dunod, 1969.

Moreau J.-J., Séminaire Leray, Collège de France, Paris, 1967.

Rockafellar T.

1) Conjugate convexe functions in optimal control and the calculus of variations, J. of Math. appl. 32,1970, p. 174-222.

2) Existence and duality theorems for convex problems of Bolza. Trans. Amer. Math. Soc. vol. 159, sept. 1971.

3) Intégrals which are convex functionals, I, Pacific J. Math, 24 1968, p. 525-540.

4) Duality and stability in extremum problems involoving convex functions, Pac. J. Math 24, 1968, 525-539.

5) Convex analysis, Princeton Univ. Press, Princeton, N.J.USA, 1969.

J. M. Lasry

Temam R.,

Solutions généralisées de certaines équations du type hyper-surfaces minima, Arch. Rac. Mech. and Anal., 44, 1971, pp.121-126.

Henri Berliocchi, 11, rue Croix Bosset, Sèvres -92-

Jean-Michel Lasry,18, avenue de la Motte Piquet, Paris -7ème-

Equipe de recherche associée au CNRS n° 249

Mathématiques de la décision, Université de Paris IX-Dauphine

75016 - PARIS.

CENTRO INTERNAZIONALE MATEMATICO ESTIVO

(C. I. M. E.)

ON UNILATERAL CONSTRAINTS, FRICTION AND PLASTICITY

JEAN JACQUES MOREAU

Corso tenuto a Bressanone dal 17 al 26 giugno 1973

ON UNILATERAL CONSTRAINTS, FRICTION AND PLASTICITY

J. J. MOREAU

CONTENTS

J. J. Moreau

Chapter 3. SUPERPOTENTIALS AND PERFECT CONSTRAINTS

3. a Configurations and forces
3. b Statical laws
3. c Frictionless bilateral constraints
3. d Perfect unilateral constraints
3. e Superpotentials
3. f Dual minimum properties
3. g Saddle-point property
3. h One-dimensional examples
3. i An example of compound system
3. j Various treaments of the equilibrium problem

Chapter 4. LAWS OF RESISTANCE

4. a Velocities and forces
4. b Pseudo-potentials
4. c Viscous resistance
4. d Velocity constraints
4. e Friction and plasticity
4. f Dissipation function
4. g Superposition of resistance laws

Chapter 5. MOVING SETS

5. a Hausdorff distance and variation
5. b The case of convex sets in a normed space
5. c Intersection of two moving convex sets
5. d Distance and penalty function in a Hilbert space
5. e Moving convex set in a Hilbert space
5. f The sweeping process
5. g Existence theorem
5. h Discretization algorithm

J. J. Moreau

Chapter 6. QUASI-STATIC EVOLUTION OF AN ELASTOPLASTIC SYSTEM

J. J. Moreau

I INTRODUCTION

1. a ORIENTATION

Three intermingled themes run in all the following : variational statements, the duality in paired linear spaces, the convexity of sets or functions. These are precisely three leading themes of Optimization Theory, as it has been developed for several decades ; in fact the study of optimization problems started many progresses of modern convexity theory, in which duality plays an essential part.

In Mechanics these three themes have been present for more than two centuries. There is no need to recall the importance of variational ideas in the development of Analytical Dynamics. Observe, however, that these ideas often served as a mere scaffolding, to be removed before the end of the construction. Lagrange equations arose from the variational properties of a mechanical system subject to frictionless constraints and conservative forces only ; but actually Analytical Dynamics has a much wider scope, so that some modern treatises on the subject may develop it in the framework of Differential Geometry, without reference to any properly variational fact. Variational calculus acted here in suggesting some mathematical structure which eventually supplanted it. In another domain a similar evolution took place quite recently when the variational approach of partial differential equations gave rise to the theory of Variational Inequalities which have not much to do with extremum problems.

The classical Calculus of Variations, developed in the context of differentiability, automatically involves the duality of linear spaces, possibly without formalizing it. In Statics, for instance, it is usual to characterize the equilibrium configurations of a "frictionless" system with finite freedom, by equalling to zero the partial derivatives of the potential energy. This induces to consider these partial derivatives as the "components" of mechanical actions or "forces", in a genral sense ; in fact this constitutes the correct way to formulate calculation rules about forces, which are preserved under the change of variables ; for example if some evolution of the system takes place, one obtains a simple expression for the work or the power of forces. This benefit in calculation (and also the possible connection with Thermodynamics) promoted the use of energy methods in many domains ; however these methods may have been a hindrance when they happened to prevent scientists from considering phenomena which could not be described by means of potential functions. Here again one improves by forgetting the variational stimulus and considering respectively displacements and forces as the elements of two linear spaces placed in duality by the bilinear form "work". Such was already the underlying idea of the traditional method of virtual work.

About convexity, on the other hand, it must be noted that Mechanics was probably the first physical domain to make use of this concept ; this was in formulating the equilibrium condition of a heavy solid body lying on a horizontal plane : the vertical line drawn from the centre of mass must meet the convex hull of the points of support. This is typically a result concerning unilateral constraints. In fact the study of dynamical problems for systems of finite or infinite freedom with unilateral constraints (e.g. the inception of cavitation in a perfect incompressible fluid ; see MOREAU [7], [8], [9]) initially motivated the part taken by the author in the development of convexity theory. It must be stressed that convexity is involved in the theory of unilateral constraints in an essential way ; it is not used as a convenience assumption made to facilitate mathematical treatment, as it often happens, for instance, in Optimization.

These lectures do not deal with dynamics, but only with equilibrium or quasi-static evolution, i.e. evolution problems where inertia is negligible. The motion of a system is studied when resistance phenomena, such as friction or the resistance of a plastic system to yielding, are taken into account. Here again convexity is involved from the stage of formulating the resistance law itself. Many mechanists feel that the occurrence of convexity in this connection is essential, probably with some thermodynamical significance.

Classical Coulomb's law of friction enters into our general scheme of resistance laws admitting a (convex) pseudo-potential. It will be objected that this law gives only a rather rough approximation of the friction phenomena ; experimentally, when the sliding velocity increases from zero the friction coefficient begins with decreasing, while the existence of a superpotential would only allow it to increase. The author's position in this matter is the following.

Traditional physics almost always starts from linear laws as first approximations to which improvements have possibly to be added by taking terms of "higher order" into account. The common habit of assuming differentiabity in formulations is connected with the same tendency, as the meaning of differentials is precisely to describe some "tangent" linear mappings. On the contrary Coulomb's law of friction is radically nonlinear and nondifferentiable ; nevertheless there is no doubt that this law agrees with the fundamental features of the friction phenomenon and as such it is always used in practice as the first approximation, possibly subject to further improvements. For instance the augmented friction when the sliding velocity is small or vanishes is frequently explained as a sort of welding which takes place between the bodies in contact, and has to be broken when sliding occurs.

Let us suggest that, in plasticity as well as in friction, our pseudo-potential formalism describes the primary phenomenon exactly as in other domains of physics the primary phenomena admit linear formulations. This causes no conceptual difficulty ; on the other hand, the considerable amount of work which has been devoted in recent decades to optimization techniques makes now available the computational methods permitting to deal numerically with "subdifferential calculus" and convex analysis.

1. b SUMMARY OF CHAPTER 2

The preparatory Chapter 2 presents the elements of the duality theory of convex functions and the subdifferentials of such functions. The articulation of the concepts is sufficiently detailed but the proofs of the main statements are not given. Except otherwise indicated the reader may find them in MOREAU [10], a multigraph report. Some are also given in the recent book of P.J. LAURENT [1], which devotes a chapter to this subject. Of course, the book of R.T. ROCKAFELLAR [2], yet restricted to finite dimensional spaces, supply much of the fundamental informations.

J. J. Moreau

The setting is that of a pair of real linear spaces, say (X,Y), placed in duality by a bilinear form denoted as $\langle .,. \rangle$. This duality is spposed separating, i.e. the two linear forms defined on X by $x \mapsto \langle x,y \rangle$ and $x \mapsto \langle x,y' \rangle$ are identical only if the elements y and y' of Y are equal, and the symmetric assumption is made whith exchanging the roles of the two spaces. Therefore, if one of the two spaces has a finite dimension, the dimension of the other is the same ; in this case, every linear form defined on one of the two spaces can be represented in the preceding way and is continuous with regard to the natural topology of finite dimensional linear spaces. The situation is more complicated for infinite dimensional spaces. Recall in that case that each of the two spaces, say X for instance, may be endowed with various locally convex topologies which are compatible with the duality (X,Y) in the sense that relatively to any of them, the continuous linear forms are exactly the functions $x \mapsto \langle x,y \rangle$ with arbitrary y in Y. By the separation assumption made above, these topologies are Hausdorff ; it is a classical fact that among them the weak topology $\sigma(X,Y)$ is the coarsest and the Mackey topology $\tau(X,Y)$ is the finest. Observe that, by usual separation arguments, the closed convex sets are the same relatively to all these topologies, thus in the following we shall sometimes refer to closed convex sets without specifying the topology. Same remark for the lower semi-continuous convex functions.

1. c SUMMARY OF CHAPTER 3

Chapter 3 takes up Mechanics by the study of material systems whose set of possible configurations, denoted by $\mathcal{U}$, is endowed with a linear space structure. Such is in particular the case, due to the use of linear approximation, in many practical situations where it is supposed that the considered system presents only "infinitely small deviations" from some reference state which constitutes the zero of the linear space $\mathcal{U}$. By the bilinear form "work" the linear space $\mathcal{U}$ is placed in duality with another linear space $\mathcal{F}$ whose elements represent, in a general sense, forces applied to the system. An example in § 3. a shows why this duality may be supposed separating.

In this framework a statical law is a relation, arising from the study of some of the physical processes in which the system is involved, formulated between the possible configuration, say $u \in \mathcal{U}$, of the system and some, say $f \in \mathcal{F}$, among the forces it experiences if it happens to come through this configuration. Such a relation may depend on time. The concept of a statical law which admits a potential function is recalled.

At this stage it is stressed that the word constraint

possesses in Mechanics a stricter sense than it receives, for instance, in Optimization (observe that the French mechanical term is "liaison", while "côntrainte" has other meanings). Describing a mechanical constraint requires fundamentally more information than defining some set of permitted configurations ; some precisions must be given about the confining process, in the formulation of which the force of constraint or reaction is involved. Paragraphs 3. c and 3. d emphasize, in the linear framework of this Chapter, that frictionless constraints, bilateral or unilateral, are statical laws. Precisely they come into the general class of the statical laws which possess a superpotential, i.e. the relations between u and f which can be written under the form $-f \in \partial \phi(u)$, where ϕ denotes a convex numerical function, possibly taking in some part of the space $\mathcal{U}$ the value $+\infty$. The classical laws possessing a potential function also belong to this class, as far as the potential function is convex.

If all the mechanical actions experienced by the system (possibly excepting forces which vanish in any expected equilibrium) are represented by the conjunction of statical laws admitting time-independent superpotentials, the equilibrium configurations trivially possess some extremum properties in the space $\mathcal{U}$. Paragraph 3. f supposes that all these mechanical actions have been grouped in order to be summarized as the conjunction of two statical laws admitting the respective superpotentials ϕ_1 and ϕ_2 ; then $u \in \mathcal{U}$ is an equilibrium configuration if and only if there exists $f_1 \in \mathcal{F}$ such that $- f_1 \in \partial \phi_1 (u)$ and $f_1 \in \partial \phi_2 (u)$. The determination of f_1 prior to that of u is classically called a statical approach to the equilibrium problem ; the duality theory of convex functions immediately yields some extremum formulation for this problem. This involves the respective dual function ϕ_1^* and ϕ_2^* of ϕ_1 and ϕ_2 , generalizing the so-called complementary energy of linear elastostatics. Similar correspondances between extremum problems formulated in two paired linear spaces are a familiar feature in convex optimization, as well is familiar the connection of such a pair of problems with a saddle-point property concerning some function called a Lagrangian. In fact, Paragraph 3. g gives a simultaneous characterization of u and f_1 as a saddle-point in the product space $\mathcal{U} \times \mathcal{F}$. As all the preceding pattern may usually be applied to each definite mechanical system in several different ways, it is able to generate a great number of extremal or saddle-point characterization of equilibrium. The foregoing concepts were first published as a short Note (MOREAU [11]) in which proofs were not given.

Paragraph 3. h illustrates the formalism by some examples of one-dimensional systems. Paragraphs 3. i and 3. j emphasize the application to a lattice of bars ; this introduces two pairs of finite dimensional linear spaces (X,Y) and (E,S), a linear mapping D from X into E and the adjoint mapping D^* from S into Y : this is a very common algebraic pattern in elastostatics. Various ways of exploiting it are presented ; in particular the last one is meant to prepare for the evolution problem of elastoplastics, to be treated in Chapter 6. More details about continuous media and the function spaces involved in their study are given by B. Nayroles in his lectures.

1. d SUMMARY OF CHAPTER 4

This Chapter, devoted to resistance laws does not require a linear space structure for the set of the possible configurations. In fact it is a constant feature in Mechanics to associate with each configuration of a system a real linear space $\mathcal{V}$; the elements of $\mathcal{V}$ constitute, in some sense, the values that may take the velocity of the system if it comes through the considered configuration. A second linear space $\mathcal{F}$ is also associated with each configuration ; the elements of $\mathcal{F}$ form, in a generalized sense, the possible values of forces which may be applied to the system at an instant it happens to have the considered configuration. The spaces $\mathcal{V}$ and $\mathcal{F}$ corresponding to a given configuration are placed in duality by a bilinear form : $\langle v,f\rangle$ denotes the

<u>power</u> of the force $f \in \mathcal{F}$ if the system possesses the velocity $v \in \mathcal{V}$. In the special case of Chapter 3, it turns out that $\mathcal{V}$ may be identified with $\mathcal{U}$ and the same $\mathcal{F}$ is associated with every configuration.

We call in general <u>resistance law</u> a relation formulated between the possible velocity $v \in \mathcal{V}$ and a force say $f \in \mathcal{F}$, arising from some of the physical processes in which the system is involved. This is properly a resistance phenomenon if the relation is dissipative, i.e. if it implies $\langle v,f\rangle \leqslant 0$.

Here again, the case where it exists a function ϕ defined on $\mathcal{V}$, called the <u>pseudo-potential</u> of the resistance law, such that the relation takes the form $-f \in \partial \phi(v)$ deserves special attention. If, in particular $0 \in \partial \phi(0)$, the relation is sure to be dissipative ; the pseudo-potential is called in this special case a <u>resistance function</u> and one may suppose without loss of generality, that $\phi(0) = 0$. An elementary example is that of <u>viscosity laws</u> : then ϕ is a quadratic form, traditionally called the Rayleigh function.

The main application of these ideas concerns <u>dry friction</u> and <u>plasticity</u> ; this corresponds to a function ϕ which is sublinear, i.e. convex and positively homogeneous. Equivalently, ϕ is the support function of a closed convex subset of $\mathcal{F}$, denoted by $-C$, containing the origin. An essential fact in such a case is that the considered resistance law, namely $-f \in \partial\phi(v)$, neither defines f as a single-valued function of v nor v as a single-valued function of f ; to v = 0, in particular, correspond as possible values for f all the points of C . This is a familiar feature of the Coulomb law for dry friction or of the Prandtl - Reuss law for perfect plasticity. In their conventional formulation they may, at first sight, look like a piecing together of heterogeneous empirical data ; the present formulation on the contrary reveals the strong mathematical consistency of each of these laws. The rest of these lectures is meant to display the efficiency of such an approach. The reader will see, on the other hand, in P. GERMAIN [1] how our pseudo-potential formalism may take place in the more familiar setting of a textbook on Continuum Mechanics.

For what concerns Coulomb's law of dry friction it will be objected that, in most practical problems, the normal component of the contact force, which enters here in the expression of ϕ as a constant, is unknown. Our position is to consider this quantity as one of the state variables of the system.

Paragraph 4. d comes back to perfect constraints as they were introduced by Chapter 3. In the present kinematical context, these constraints are manifested as relations between the velocity of the system and some force acting on it, namely the reaction of the constraint. These relationstoo can be represented by means of pseudo-potentials and the same is true for the nonholonomic perfect constraints of traditional Mechanics (actually an extreme case of friction) : we propose to refer to such relations as velocity constraints.

Friction or plasticity laws, as well as viscosity laws, exhibit a very usual property : the corresponding dissipated power $-\langle v,f\rangle$ can be expressed as a single-valued function of the velocity, classically called the dissipation function. There is a priori no reason for this function to be related to the pseudo-potential if it exists ; paragraph 4. f characterizes the resistance laws for which such a relation holds.

The chapter ends with remarks about viscoplasticity : adding some viscosity to a resistance law of the plasticity or friction type described above, amounts to replace the indicator function ψ_C of the set C (the function taking the value 0 on this set and $+\infty$ outside) by a penalty function of the same set.

1. e SUMMARY OF CHAPTER 5

This is a purely mathematical part. The application of the foregoing mechanical formalism to evolution problems requires, in particular, some investigations about the motion of a set.

By means of Hausdorff distance, the classical concept of the variation of a function defined on a real interval is adapted to moving sets in a metric space ; the absolute continuity of such sets is similarly introduced.

As convex subsets of a normed space may be described in terms of their support functions, a special approach of moving sets is developed for this case. In the same setting of normed spaces and convex moving sets, Paragraph 5. c establishes an intersection theorem which formulates sufficient conditions for the intersection of two absolutely continuous convex moving sets to be itself absolutely continuous.

The rest of the Chapter is restricted to Hilbert spaces. Paragraph 5. b considers among other topics the distance from a moving point $t \mapsto z(t)$ to a moving convex set $t \mapsto C(t)$; if both are absolutely continuous the distance is an absolutely continuous numerical function and some inequality involving derivatives is established, as a preparation for the following.

Paragraph 5. c introduces the sweeping process associated with a moving convex set in the Hilbert space H. This gives a fundamental example of an evolution problem under unilateral constraint ; from the mathematical standpoint this process features also as a constituent of several more complicated situations ; in particular it will be met again in the treatment of the elastoplastic problem of Chapter 6. The author has already devoted several studies to this problem, mainly published as multigraph seminar reports (cf. MOREAU [17], [18], [20], [21]). The method used in § 5. g to establish an existence theorem consists in a regularization technique, equivalent in the present context to representing the given moving convex set by penalty functions.

The Chapter ends with an algorithm of time discretization for the solution of the sweeping problem ; the convergence of this algorithm is proved by using again regularization, but with a time-dependent "penalty coefficient".

1. f SUMMARY OF CHAPTER 6

This final Chapter shows how all the foregoing operates when applied to the quasi-static evolution problem for elastoplastic systems. This involves a linear space $\mathcal{U}$ as configuration space and, according to the conventional conception of elastoplasticity, the system is treated as formed by two components : the "visible" or "exposed" component, denoted by $x \in \mathcal{U}$, and the "hidden" or "plastic" component denoted by $p \in \mathcal{U}$. The elastic restoring force depends only on the difference $x-p$. The component x undergoes perfect constraints and loads, both depending on time in a given way. The component p undergoes a resistance related to its "velocity" $\dot{p}$ by a law of the type studied in § 4. This is only perfect plasticity, but a very simple example suggests that strain hardening too could be taken into account by a similar pattern, provided a sufficiently large space would be affected to the "hidden" or "internal" variable p ; this point of view is adopted by several authors.

Great simplification is brought by a notation trick by which the configuration space $\mathcal{U}$ and the force space $\mathcal{F}$ are identified with a single Hilbert space H ; the norm in H is related to the elastic energy.

An existence theorem is proved by reduction to the sweeping process of Chapter 5 ; thereby a time-discretization algorithm is

J. J. Moreau

2 DUALITY AND SUBDIFFERENTIALS OF CONVEX FUNCTIONS

2. a POLAR FUNCTIONS

Let X , Y be a pair of real linear spaces placed in separating duality by the bilinear form $\langle .,. \rangle$. Let f be a function defined, for instance, on X, with values in $\bar{R} = [-\infty, +\infty]$. Consider the affine function defined on X by

$$(2.1) \qquad x \mapsto \langle x,y \rangle - \rho$$

with y fixed in Y, called the <u>slope</u> of this affine function, and ρ fixed in R ; such is the general form of the affine functions which are continuous for some, then for any, locally convex topology on X compatible with the duality.

An usual question is that of determining wether this affine function is a <u>minorant</u> of f ; a trivial necessary and sufficient condition for that is

$$(2.2) \qquad \rho \geqslant \sup_{x \in X} [\langle x,y \rangle - f(x)] \ .$$

Attention is drawn thereby to the function f^* defined on Y by

$$(2.3) \qquad f^*(y) = \sup_{x \in X} [\langle x,y \rangle - f(x)]$$

called the <u>polar function</u> of f.

In particular the equality $f^*(y) = +\infty$, for some $y \in Y$, means that f possesses no affine minorant having y as slope ; such is the case, for instance, whichever is y, if f takes somewhere in X the value $-\infty$.

EXAMPLE. Let A be a subset of X; take as f the indicator function ψ_A of A, i.e.

$$\psi_A(x) = \begin{cases} 0 & \text{if } x \in A \\ +\infty & \text{if } x \notin A . \end{cases}$$

Its polar function

$$\psi_A^*(y) = \sup_{x \in X} [\langle x,y\rangle - \psi_A(x)] = \sup_{x \in A} \langle x,y\rangle$$

is classically known under the (rather improper) name of the support function of A. Take y different from zero in Y and $\rho \in \mathbb{R}$; the affine function (2.1) is a minorant of ψ_A iff the closed half space $\{x \in X : \langle x,y\rangle - \rho \leqslant 0\}$ contains A. In view of condition (2.2) this is possible only if $\psi_A^*(y) < +\infty$; in such a case taking exactly $\rho = \psi_A^*(y)$ yields a half-space which is minimal, with regard to inclusion, among the half-spaces containing A; but that does not mean this half-space is necessary a "supporting half-space"; its boundary hyperplane need not meet A, even when A is closed and convex.

2. b PAIRS OF DUAL FUNCTIONS

For the construction of the supremum in (2.3) one may equivalently consider only the values of x such that $f(x) < +\infty$. Therefore, whichever is f, the function f^* belongs to the set, denoted by $\Gamma(Y,X)$, of the functions on Y which are the pointwise suprema of collections of affine functions like $y \mapsto \langle x,y\rangle - \sigma$, $x \in X$, $\sigma \in \mathbb{R}$. Using Hahn-Banach's theorem, one proves that, besides the constant $-\infty$ (it is the supremum of an empty collection), the set $\Gamma(Y,X)$ consists exactly of the functions on Y, with values in $]-\infty, +\infty]$, which are convex and l.s.c. for some locally convex topology on Y compatible with the duality (Y,X), then l.s.c. for all such topologies.

The spaces X and Y play here symmetric roles ; there is no inconvenience in denoting in the same way by the star * the function defined on X as the polar of a given function on Y. Then the bipolar of f is defined on X by

$$f^{**}(x) = \sup_{y \in Y} [\langle x,y\rangle - f^*(y)] .$$

The construction of this supremum may be equivalently be restricted to the values of y such that $f^*(y)$ is finite ; that means f^{**} is the supremum of the affine functions like (2.1), with ρ verifying equality in (2.2) ; they are the maximal affine minorant of f, so that f^{**} may also be defined as the pointwise supremum of all the affine function of the form (2.1) which minorize f. This supremum is equivalently characterized as the greatest element of $\Gamma(X,Y)$ minorizing f or Γ- hull

For instance, if A is a subset of X, the Γ - hull of the indicator function ψ_A is the indicator function of the closed convex hull of A.

The preceding implies that if it is a priori supposed that $f \in \Gamma(X,Y)$ and $g \in \Gamma(Y,X)$ one has the equivalence

$$f = g^* \iff f^* = g .$$

Then f and g are said mutually polar or conjugate or dual functions. In this way the star * induces a one-to-one correspondance between $\Gamma(X,Y)$ and $\Gamma(Y,X)$; as the constant $+\infty$ corresponds to the constant $-\infty$, the correspondance is also one-to-one between the elements of $\Gamma(X,Y)$ and $\Gamma(Y,X)$ other than these singular constants : these elements are called the proper closed convex functions on X and Y; the sets of them will be denoted by $\Gamma_0(X,Y)$ and $\Gamma_0(Y,X)$ respectively.

From the definition of polarity it immediately follows

$$\forall x \in X \quad , \quad \forall y \in Y \quad : \quad f(x) + g(y) \geqslant \langle x,y \rangle$$

called Fenchel's inequality.

REMARK ON TERMINOLOGY. Most of the words introduced by the preceding definitions are the English transcriptions of French terms currently used by French speaking people after the author's multigraph report of 1966 (MOREAU [10]). This involves but slight discrepancies from the book of R.T. ROCKAFELLAR [2] : following the 1949 initiating paper of

W. FENCHEL [1], Rockafellar prefers the locutions "conjugate functions" to "dual functions". It may be inconvenient to call also conjugate of f, as he does, the function f^* associated by (2.3) with some f which does not necessarily belong to $\Gamma(X,Y)$. As this so called "conjugacy" is no more a symmetric correspondance, the author chose in the 1966 report, to use in this connotation the term polar function. Unfortunately, in the meantime, Rockafellar applied the word polar to another kind of correspondance (cf. Sec. 15 of his book) concerning nonnegative closed convex functions vanishing at the origin, which generalizes some classical conjugacy of gauge functions (see § 2. h below) ; but there does not seem to be much risk of confusion.

2. c IMAGES OF PROPERTIES OR RELATIONS

Many properties or relations concerning functions defined, for instance, on X, imply some properties or relations concerning the polar of them. Here we restrict ourselves to a few of these "images by polarity" considering exclusively functions $f, f_1, f_2, \ldots$ which belong to $\Gamma(X,Y)$ and denoting by $g, g_1, g_2, \ldots$ their polar (i.e. dual) functions.

Easy calculation yields :

1° Homothety. If $\sigma \in R$ is a non zero constant, the identity

$$\forall x \in X \quad : \quad f_1(x) = f_2(\sigma x)$$

is equivalent to

$$\forall y \in Y \quad : \quad g_1(y) = g_2(\frac{1}{\sigma} y) \ .$$

2° <u>Multiplication by a positive constant</u>. If λ is a strictly positive constant, the identity

$$\forall x \in X \quad : \quad f_1(x) = \lambda \, f_2(x)$$

is equivalent to

$$\forall y \in Y \quad : \quad g_1(y) = \lambda \, g_2(\frac{1}{\lambda} y) \ ;$$

the right member is sometimes written as a "right product by λ" : notation $g_1 = g_2 \lambda$.

In particular a function g belonging to $\Gamma(Y,X)$ is the support function of a subset of X (or equivalently the support function of the closed convex hull of this subset) if and only if its dual f is an indicator, i.e. this dual takes only the values 0 and $+\infty$. That means f remains unchanged under the multiplication by any $\lambda > 0$; in view of the preceding, this is equivalent to g being <u>positively homogeneous</u> (i.e. <u>sublinear</u>, due to the assumed convexity of g). A more special situation is that of a function g belonging to $\Gamma(Y,X)$ which at the same time is an indicator function and is sublinear : this happens if and only if f possesses the same properties ; in such a case f and g are respectively the indicator functions of two <u>mutually polar</u>

(closed, convex) cones, P and Q, i.e.

$$Q = \{y \in Y : \forall x \in P , \langle x,y\rangle \leqslant 0\}$$

and symmetrically

$$P = \{x \in X : \forall y \in Q , \langle x,y\rangle \leqslant 0\} .$$

3° Translation. If $a \in X$ and $\alpha \in \mathbb{R}$, the identity

$$\forall x \in X : f_1(x) = f_2(x - a) + \alpha$$

is equivalent to

$$\forall y \in Y : g_1(y) = g(y) + \langle a,y\rangle - \alpha .$$

4° Product spaces. Let (X_i, Y_i), $i = 1, 2, \ldots, n$, be n pairs of real linear spaces placed in duality by n bilinear forms respectively denoted by $\langle .,.\rangle_i$. If $x = (x_1, x_2, \ldots, x_n)$ denotes the generic element of the linear space

$$X = X_1 \times X_2 \times \ldots \times X_n$$

and $y = (y_1, y_2, \ldots, y_n)$ the generic element of the linear space

$$Y = Y_1 \times Y_2 \times \ldots \times Y_n$$

the bilinear form

$$\langle x,y\rangle = \langle x_1,y_1\rangle_1 + \langle x_2,y_2\rangle_2 + \ldots + \langle x_n,y_n\rangle_n$$

places X and Y in duality. For each i, denote by f_i, g_i a pair of functions defined respectively on X_i and Y_i and mutually polar with regard to the bilinear form $\langle .,.\rangle_i$. It is easy to see that the functions f and g defined on X and Y respectively by

$$f(x) = f_1(x_1) + f_3(x_3) + \dots + f_n(x_n)$$

$$g(y) = g_1(y_1) + g_2(y_2) + \dots + g_n(y_n)$$

are mutually polar with regard to $\langle .,. \rangle$.

The following result is less trivial (see proofs in MOREAU [3] or [10]) :

5^0 Continuity. The setting is again that of single pair of linear spaces (X,Y). The function $f \in \Gamma_0(X,Y)$ is finite and continuous at the origin for some locally convex topology on X compatible with the duality (then for the Mackey topology τ (X,Y) which is the finest of them) if and only if the dual function $g \in \Gamma_0(Y,X)$ is inf-compact, i.e. for any $k \in R$ the "level set" or "slice" $\{y \in Y : g(y) \leqslant k\}$ is compact for some (locally convex) topology on Y compatible with the duality (then for the weak topology σ (Y,X) which is the coarsest of them). Note that, due to the convexity of g, a sufficient condition for that is the existence of some $k > \inf g$ such that this compactness holds.

Using translation (cf. 3^0 above) one concludes that the continuity of f at some point $x_o \in X$ is equivalent to the compactness of the "oblique slices of g with slope x_o", i.e. the sets $\{y \in Y : g(y) - \langle x_o, y \rangle \leqslant k\}$.

2. d INF - CONVOLUTION AND THE IMAGE OF ADDITION

Let us denote by $\dot{+}$ the commutative and associative operation extending classical addition to any pair of elements of $\bar{R} = [-\infty, +\infty]$ by putting $(-\infty) \dot{+} (+\infty) = +\infty$ (symmetrically the operations $\underset{\bullet}{+}$ extends classical addition by the convention $(-\infty) \dot{+} (+\infty) = -\infty$).

Let f_1, f_2 be functions defined on the linear space X with values in $\overline{R}$; the function f defined on X by

$$(2.4) \qquad f(x) = \inf_{u \in X} [f_1(u) \dot{+} f_2(x-u)] = \inf_{v \in X} [f_1(x-v) \dot{+} f_2(v)]$$

is called the infimal convolute, or shortly inf-convolute, of f_1 and f_2 ; it is denoted by $f_1 \nabla f_2$ (or also $f_1 \square f_2$, as in ROCKAFELLAR [2], when there is no risk of confusion with the supremal convolute $f_1 \Delta f_2$, which would be symmetrically defined by using "sup" and $\dot{+}$). This operation is commutative and associative ; if f_1 and f_2 are convex, so is $f_1 \nabla f_2$, etc...

Example 1. If f_2 is the indicator function of a singleton $\{a\}$, then $f_1 \nabla f_2$ is a translate of f_1, namely the function

$$x \mapsto f_1(x-a) \quad .$$

Example 2. If A is a subset of X and $\|.\|$ a norm on this linear space, then $(\psi_A \nabla \|.\|)(x)$ is the distance from the point x to the set A.

Example 3. If A and B are two subsets of X, the inf-convolute $\psi_A \nabla \psi_B$ is the indicator function of the set

$$A + B = \{x \in X \ : \ \exists\, a \in A \ , \ \exists\, b \in B \ , \ x = a + b\} \ .$$

Coming back to the setting of the pair of spaces (X,Y) in duality, the computation of polar functions yields easily

$$(f_1 \nabla f_2)^* = f_1^* \dot{+} f_2^* .$$

Suppose now that f_1 and f_2 belong to $\Gamma(X,Y)$ and that g_1 and g_2 are their polar (i.e. dual) functions ; taking the polars of both members of the preceding equality leads to

$$(f_1 \nabla f_2)^{**} = (g_1 \dot{+} g_2)^* . \tag{2.5}$$

Addition $\dot{+}$ is a composition law in $\Gamma(Y,X)$; (2.5) describes the composition law in $\Gamma(X,Y)$ which is the image of it by the one-to-one mapping * ; this composition law is the Γ - <u>hull of inf-convolution</u> (cf. § 2. b above) ; we denote it by $\underline{\nabla}$; it may be called Γ -<u>convolution</u>.

Of practical importance are the cases where $f_1 \nabla f_2$ happens to belong to $\Gamma(X,Y)$ so that the double star may be omitted in (2.5). Let us just formulate here the two most usual of them.

It is still assumed that f_1 and f_2 belong to $\Gamma(X,Y)$.

1° Suppose that the set, denoted by cont f_1, of the points where f_1 is finite and continuous, for some topology compatible with the duality, and the set

$$\mathrm{dom}\, f_2 = \{x \in X \ : \ f_2(x) < +\infty\}$$

are such that

$$\mathrm{cont}\, f_1 + \mathrm{dom}\, f_2 = X .$$

Then $f_1 \nabla f_2$ is either the constant $-\infty$ or is finite and continuous everywhere in X for the considered topology ; therefore

$f_1 \nabla f_2 \in \Gamma(X,Y)$, hence $f_1 \nabla f_2 = f_1 \underline{\nabla} f_2$.

2° Suppose that there exists a point y_o in Y at which both functions g_1 and g_2 are finite, one of them continuous at this point (for some topology compatible with the duality) ; then $f_1 \nabla f_2 \in \Gamma(X,Y)$; furthermore this inf-convolution is exact, i.e., whichever is x, the infimum in (2.4) is a minimum. Note that the hypothesis is equivalent to the following : both functions $x \mapsto f_1(x) - \langle x, y_o \rangle$ and $x \mapsto f_2(x) - \langle x, y_o \rangle$ are bounded from below and one of them is inf-compact for the weak topology $\sigma(X,Y)$ (cf. § 2 c) .

2. e SUBGRADIENTS AND SUBDIFFERENTIALS

Let f denote a function defined on X, with values in $\overline{R}$; an element y of Y is called a subgradient of f at the point $x \in X$ if y is the slope of an affine minorant of f exact at the point x, i.e. taking at this point the same value as f. This requires that the value f(x) is finite and that the expected minorant has the form

$$u \mapsto \langle u - x, y \rangle + f(x) .$$

Using condition (2.2) for an affine function to minorize f, one obtains the following representation for the set, denoted by $\partial f(x)$, of the subgradients of f at the point x

$$\partial f(x) = \{ y \in Y \ : \ f^*(y) - \langle x, y \rangle \leqslant - f(x) \} .$$

This set is called the *subdifferential* of f at the point x. The convexity and the lower semicontinuity of f^* imply that $\partial f(x)$ is a convex, possibly empty, subset of Y, closed for the topologies compatible with the duality (Y,X). If $\partial f(x)$ is not empty the function f is said to be *subdifferentiable* at the point x.

Trivially *the function* f *possesses a finite minimum attained at the point* x *if and only if* $\partial f(x)$ *contains the zero of* Y.

Recall that the function f is said *weakly differentiable*, or Gâteaux-differentiable, at the point x, relatively to the duality (X,Y), if there exists $y \in Y$ (necessarily unique) such that for any $u \in X$, the function $t \mapsto f(x + t\,u)$ of the real variable t possesses for $t = 0$ a derivative equal to $\langle u,y \rangle$; the element y is called the *weak gradient*, or Gâteaux-gradient, of the function f at the point x, relatively to the duality (X,Y). If in addition the function f is *convex*, one easily finds that *the subgradient* $\partial f(x)$ *consists of the single element* y. When X is a normed space, Y its topological dual, all this a fortiori holds if f is *Fréchet-differentiable* at the point x.

Subdifferentiability finds its clearest setting when a pair of dual, i.e. mutually polar functions $f \in \Gamma_o(X,Y)$ and $g \in \Gamma_o(Y,X)$ is considered. Then, for x in X and y in Y *the three following properties are equivalent* :

(2.6) $\qquad y \in \partial f(x)$

(2.7) $\qquad x \in \partial g(y)$

(2.8) $\qquad f(x) + g(y) - \langle x,y \rangle = 0 \quad ;$

observe that, by Fenchel's inequality, the $=$ sign above may equivalently be replaced by $\leqslant$. If these properties hold, the points x and y are said conjugate relative to the pair of mutually polar functions (f,g).

EXAMPLE. Take as f the indicator function ψ_C of a nonempty closed convex subset of X. Then the relation $y \in \partial \psi_C(x)$ is trivially equivalent to the following : the point x belongs to C and the set $\{u \in X : \langle u - x, y \rangle \leqslant 0\}$ contains C. If y differs from the zero of Y this set is a closed half-space whose boundary is a supporting hyperplane of the set C at the point x ; then one classically says that $y \in Y$ is an outward normal vector at the point x of the convex set $C \subset X$. Let us agree to take this locution in a weak sense, by considering also the zero of Y as a normal vector at the point x if it belongs to $\partial \psi_C(x)$; thus the set $\partial \psi_C(x)$ will be called the outward normal cone at the point x. This cone is empty if $x \notin C$; if $x \in C$ it contains at least the zero of Y and reduces to this single element, in particular, when x is an internal point of C (i.e. every straight line drawn to x intersects C along a segment to which x is interior). In terms of the support function ψ_C^* of C, condition (2.8) yields that if x

belongs to C one has

$$\partial \psi_C(x) = \{y \in Y : \psi_C^*(y) = \langle x,y \rangle \}$$

$$= \{y \in Y : \psi_C^*(y) \leqslant \langle x,y \rangle \} .$$

REMARK. For a pair of spaces (X,Y) with finite dimension and convex functions f, g which are differentiable, relations (2.6), (2.7), (2.8) show that the correspondance between f and g reduces to the classical Legendre transform.

Let us come back to the case of an arbitrary f and possibly infinite dimensional spaces. By associating with every $x \in X$ the subset $\partial f(x)$ of Y one defines a multimapping (also called a multifunction, or a multivalued mapping, or a set-valued mapping) from X into Y. Independently of the formalization of subgradients and the "subdifferential calculus" (MOREAU [2] ; similar ideas were also present in Rockafellar's Thesis, Harvard, 1963) this multimapping was considered in G.J. MINTY [1] as the leading example of monotone, possibly multivalued, operator. In fact whichever are x and x' in X, whichever are y in $\partial f(x)$ and y' in $\partial f(x')$, if any, one finds easily

$$\langle x - x', y - y' \rangle \geqslant 0$$

which is, by definition, the monotony property of the multimapping ∂f.

2. f ADDITION RULE

The main calculation rule for subdifferentials concerns addition. If f_1 and f_2 are two numerical functions, defined for instance on X, the inclusion

$$\partial f_1(x) + \partial f_2(x) \subset \partial (f_1 + f_2)(x) \tag{2.9}$$

is trivial. If this inclusion holds as an equality of sets the functions f_1 and f_2 are said to possess the additivity of the subdifferentials at the point x .

Let us indicate two usual sufficient conditions for that :

1° If both functions f_1 and f_2 are convex, one of them weakly differentiable at the point x, inclusion (2.9) holds as an equality of sets.

2° If both functions f_1 and f_2 are convex and if there exists a point x_o in X at which one of them is continuous, with both values $f_1(x_o)$ and $f_2(x_o)$ finite, inclusion (2.9) holds as an equality of sets for every x in X. Continuity must be understood here in the sense of some (locally convex) topology compatible with the duality (X,Y) : thus the less stringent hypothesis is obtained by taking the finest of them, i.e. the Mackey topology τ (X,Y).

EXAMPLE. Make $f_1 = f$, a function defined on X, with values in $]-\infty, +\infty]$ and $f_2 = \psi_C$, the indicator function of a non empty subset C of X. The problem of minimizing the restriction of f to C is clearly equivalent to that of minimizing, over the whole of X, the function $f + \psi_C$; a minimizing point x is characterized by

(2.10) $$0 \in \partial(f + \psi_C)(x)$$

a condition which is implied by

(2.11) $$0 \in \partial f(x) + \partial \psi_C(x) \quad .$$

When the additivity of the subdifferentials holds, conditions (2.10) and (2.11) are equivalent.

Such is the case for instance, by 1° above, if the set C is convex, and the function f convex, everywhere weakly differentiable : then (2.11), written as

(2.12) $$-\operatorname{grad} f(x) \in \partial \psi_C(x)$$

is a necessary and sufficient condition for x to be a solution of our "constrained minimization problem". Make in particular $X = Y = H$, a separated pre-Hilbert space with the inner product $(.|.)$ playing the role of the bilinear form $\langle .,. \rangle$. Let a be an arbitrary element of H ; define the function f by

$$f(x) = \frac{1}{2}(x-a \mid x-a) = \frac{1}{2}\|x-a\|^2 \quad .$$

Elementary calculation proves that this function is convex and weakly differentiable relatively to the duality (H,H), with

$$\operatorname{grad} f(x) = x - a \quad .$$

Then (2.12) yields a necessary and sufficient condition for x to be the nearest point to a in C

$$a - x \in \partial \psi_C(x) \quad ; \tag{2.13}$$

such an x is denoted by $\mathrm{proj}_C(a)$ or $\mathrm{proj}\,(a, C)$, if it exists. Uniqueness of this points results from f being strictly convex ; recall on the other hand that if H is complete, i.e. if it is a Hilbert space, the existence of $\mathrm{proj}_C\, a$ is secured for any $a \in H$.

2. g IMAGES BY LINEAR MAPPINGS

Let (F,G) be another pair of linear spaces, placed in separating duality by a bilinear form denoted by $\langle\langle .,. \rangle\rangle$. Let A be a linear mapping from F into X, weakly continuous (i.e. continuous from F endowed with any topology compatible with the duality (F,G), to X endowed with the weak topology $\sigma\,(X,Y)$). Weak continuity implies the existence of the adjoint (or transpose) of A, i.e. the linear mapping A^* from Y into G such that

$$\forall\, u \in F \quad , \quad \forall\, y \in Y \quad : \quad \langle A\,u, y \rangle \;=\; \langle\langle u, A^* y \rangle\rangle \quad .$$

Let $f \in \Gamma\,(X,Y)$; clearly the function

$$f \circ A \quad : \quad u \mapsto f(A\,u)$$

belongs to $\Gamma\,(F,G)$; one proves (see ROCKAFELLAR [3]) that its dual function $(f \circ A)^*$ is the Γ - hull of the function defined on G by

$$v \mapsto \inf \{ f^*(y) \;:\; A^* y \;=\; v \} \quad . \tag{2.14}$$

If in addition there exists a point in the range of A at which f is finite and continuous (for some topology compatible with the duality (X,Y)) then $(f \circ A)^*$ equals the function (2.14) itself. Under the same assumption, for every $u \in F$, the subdifferential $\partial(f \circ A)(u)$ is the image of $\partial f(A u) \subset Y$ under the mapping A^* ; this may be expressed by writing

(2.15) $$\partial(f \circ A) = A^* \circ \partial f \circ A \quad .$$

2. h CONJUGATE GAUGE FUNCTIONS AND QUASI - HOMOGENEOUS CONVEX FUNCTIONS

The setting is again that of a single pair of spaces (X,Y). Let A be a closed convex subset of X containing the origin ; denote by B the polar set of A, i.e.

$$B = \{y \in Y \;:\; \forall\, x \in A \;,\; \langle x,y\rangle \leqslant 1\} \,.$$

Then A is, symmetrically, the polar set of B. It is easily seen that the gauge function of A, namely the function a defined on X by

$$a(x) = \inf \{\lambda \in \,]0, +\infty[\;:\; \tfrac{1}{\lambda} x \in A\} \,,$$

is the support function of B ; symmetrically the gauge function b of B is the support function of A. We shall refer to this situation by saying that (a, b) is a pair of conjugate gauge functions.

For sake of simplicity let us restrict ourselves here to the case where both functions take only finite values ; this means that A is absorbent in X (i.e. the origin is an internal point) and that it is bounded relatively to the topologies compatible with the duality ; equivalently B possesses the same properties in Y. Such is the case, for instance if X is a given normed space, Y its dual endowed with the usual norm : the respective norms form a pair of conjugate gauge functions and the corresponding mutually polar sets are the closed unit balls of the two spaces.

One finds

$$b(y) = \sup_{a \in X} \frac{\langle x,y \rangle}{a(x)}$$

and the symmetrical relation (this can be extended to possibly infinite valued conjugate gauge functions, under some notational precautions).

Consider on the other hand a mapping ϕ from $[0, +\infty[$ into $[0, +\infty]$ possessing the following properties : ϕ is convex, non decreasing, lower semi continuous and $\phi(0) = 0$ (actually ϕ is continuous on the interior of $\mathrm{dom}\,\phi = \{\xi \in [0, +\infty[\;:\; \phi(\xi) < +\infty\}$). Classically, with such a function is associated its Young conjugate γ defined on $[0, +\infty[$ by

$$\gamma(\eta) = \sup(\xi\,\eta - \phi(\xi))$$

which possesses the same properties ; ϕ is, in turn, the Young conjugate of γ.

Examples :

1° $\phi(\xi) = \frac{1}{p}\xi^p$, $\gamma(\eta) = \frac{1}{q}\eta^q$

where p and q denote two constants in $]1, +\infty[$, such that $1/p + 1/q = 1$.

2° $\phi(\xi) = \lambda\,\xi$, $\gamma(\eta) = \begin{cases} 0 & \text{if } 0 \leqslant \eta \leqslant \lambda \\ +\infty & \text{if } \lambda < \eta < +\infty \end{cases}$

where $\lambda \in [0, +\infty[$ is a constant.

Exclude the singular case where one of the two functions ϕ and γ is the constant zero. Then one proves that the functions $f = \phi \circ a$, $g = \gamma \circ b$ respectively defined on X and Y, i.e.

$$f(x) = \phi\,(a(x)) \quad , \quad g(y) = \gamma\,(b(y))$$

are a pair of dual functions in the sense of the preceding paragraphs.

Each of these functions is said quasi-homogeneous (or gauge-like in ROCKAFELLAR [2]) ; in fact in the special case where $\phi(\xi) = \frac{1}{p}\,\xi^p$, the function f is positively homogenous with degree p. The functions defined in this way, for instance on X, may be characterized as follows : they are the elements of $\Gamma_o\,(X,Y)$ such that the various sets $\{x \in X : f(x) \leqslant k\}$ (the "slices" of f), for $k \in \mathbf{R}$ are homothetic to A. (they are empty for $k < 0$).

Concerning the determination of the subdifferentials of these functions, let us only indicate : Two points $x \in X$ and $y \in Y$ are

<u>conjugate relatively to</u> (f,g) <u>if and only if</u>

$$\phi\ (a(x)) + \gamma\ (b(y)) = a(x)\ b(y) = \langle x,y \rangle \quad .$$

The first equality may be interpreted by saying that the real numbers a (x) and b (y) are conjugate points with regard to the pair of Young conjugate functions (ϕ,γ) ; if x and y are different from the respective origins of X and Y, the second one expresses a property of the "rays" (i.e. one-dimensional cones) they generate in X and Y ; such rays may be said conjugate relative to the pair of conjugate gauge functions a and b.

J. J. Moreau

3 SUPERPOTENTIALS AND PERFECT CONSTRAINTS

3. a CONFIGURATIONS AND FORCES

In this Chapter is considered a mechanical system $\mathcal{S}$ whose set of possible configurations, denoted by $\mathcal{U}$, is endowed with a linear space structure. Such is traditionnally the case, due to the use of linear approximation, if the system presents only "small deviations" from a certain reference configuration which constitutes the zero of $\mathcal{U}$.

The bilinear form work places the linear space $\mathcal{U}$ in duality with a linear space $\mathcal{F}$ whose elements constitutes, in a general sense, the possible values of forces experienced by the system. Precisely $\langle u,f\rangle$ denotes the work of the force $f \in \mathcal{F}$ for the displacement $u \in \mathcal{U}$ of the system. For sake of clarity, we shall in some cases comply with the habit of denoting a displacement by such a symbol as δu ; this symbol is meant to recall that the considered displacement equals the difference between two elements of $\mathcal{U}$ representing some configurations ; actually, in the present framework, due to the existence of the privileged configuration "zero", configurations as well as displacements are elements of $\mathcal{U}$, thus have the same algebraic nature.

After replacing, if necessary, the considered spaces by some quotients, it may be supposed that this duality is separating.

EXAMPLE. Take as $\mathcal{S}$ a perfectly rigid body performing only "infinitely small" motions in the neighborhood of the reference configuration. From this reference state, each possible configuration of the body may be described by the corresponding field of displacement vectors, say $u : x \mapsto \vec{u}(x)$. Due to the rigidity of the body and to the fact that displacements are, by approximation, treated as infinitely small this field possesses the property of <u>equiprojectivity</u> ; the totality of equiprojective vector fields is well known to form a linear space of dimension 6 : such is $\mathcal{U}$ in the present case. For sake of brevity let us accept only as acting on $\mathcal{S}$ <u>finite families of forces</u> in the sense of elementary Mechanics. Such a family may be described as a vector field $\phi : x \mapsto \vec{\phi}(x)$ taking the value zero everywhere except on a finite set of points and its work for a displacement field $u \in \mathcal{U}$ is classically defined as the finite sum $w = \Sigma\, \vec{u}(x).\, \vec{\phi}(x)$. For a fixed ϕ the mapping $u \mapsto w$ is clearly a linear form on the space $\mathcal{U}$; on the other hand, the set Φ of the possible ϕ' s is naturally endowed with a linear space structure which makes that, for a fixed u, the work w is a linear form of ϕ. But the space Φ clearly has an infinite dimension, so that this bilinear form cannot place $\mathcal{U}$ and Φ in separating duality. The classical procedure consists in treating as <u>equivalent</u> two families of forces, say ϕ and ϕ', such that

$$\forall\, u \in \mathcal{U} \quad : \quad \Sigma\, \vec{u}(x).\, \vec{\phi}(x) \;=\; \Sigma\, \vec{u}(x).\, \vec{\phi}'(x).$$

The corresponding equivalence classes are called torsors. In other words, if Φ_o denotes the linear subspace of Φ formed by the families of forces which yield a zero work for any $u \in \mathcal{U}$, torsors are the elements of the quotient space Φ / Φ_o, with dimension 6. Such is $\mathcal{F}$ in the present case ; the duality between $\mathcal{U}$ and $\mathcal{F}$ is then separating.

PRODUCT SPACES. Suppose the mechanical system $\mathcal{S}$ consists in the conjunction of n possibly interacting systems $\mathcal{S}_1, \mathcal{S}_2, \ldots, \mathcal{S}_n$ whose respective configuration spaces are the linear spaces $\mathcal{U}_1, \mathcal{U}_2, \ldots, \mathcal{U}_n$. Then the configuration space of $\mathcal{S}$ is the product space $\mathcal{U}_1 \times \mathcal{U}_2 \times \ldots \times \mathcal{U}_n$, naturally endowed with a linear space structure. Denote by $\mathcal{F}_i$ the force space corresponding to the system $\mathcal{S}_i$, a linear space placed in separating duality with $\mathcal{U}_i$ by the bilinear form $\langle .,. \rangle_i$. A force f exerted on the total system $\mathcal{S}$ is a n-tuple $(f_1, f_2, \ldots, f_n)$, $f_i \in \mathcal{F}_i$; this is the generic element of the product space $\mathcal{F} = \mathcal{F}_1 \times \mathcal{F}_2 \times \ldots \times \mathcal{F}_n$. The work of f for a displacement $u = (u_1, u_2, \ldots, u_n)$ of $\mathcal{U}$ is by definition the sum

$$\langle u,f \rangle = \sum_i \langle u_i, f_i \rangle_i$$

in which we recognize the natural bilinear form placing the product spaces $\mathcal{U}$ an $\mathcal{F}$ in separating duality (cf. § 2. c).

This construction of $\mathcal{U}$ and $\mathcal{F}$ as the products of the respective spaces corresponding to subsystems of $\mathcal{S}$ is a customary procedure in computation. It prepares also for the application of our general pattern to continuous media, as developed in B. Nayroles's lectures : then $\mathcal{U}$ and $\mathcal{F}$ are some linear spaces of measurable functions, with regard to a certain non-negative measure. The sum which above defines the work is replaced by an integral.

3. b STATICAL LAWS

A <u>statical law</u> is a relation, denote it by $\mathcal{R}$, between the configuration $u \in \mathcal{U}$ that the system $\mathcal{S}$ may occupy and some, say $f \in \mathcal{F}$, among the forces it may experience when it comes through this configuration. Such a relation arises from the study of some of the physical processes in which the system is involved.

Instead of relations as $\mathcal{R}$, one may equivalently speak of <u>multi-mappings</u> from one of the two spaces into the other ; for instance, to every u in $\mathcal{U}$ corresponds the (possibly empty) set, denote it by $R(u)$, of the elements f of $\mathcal{F}$ which are related to u by $\mathcal{R}$.

In particular it may happen that the set $R(u)$ consists, for each u, of a single element ; then the statical law is described as a single-valued mapping $u \mapsto f$ from $\mathcal{U}$ into $\mathcal{F}$. If, in addition, there exists a numerical function $W : \mathcal{U} \to \mathbf{R}$ such that this mapping is expressed by

$$f = - \operatorname{grad} W(u)$$

(weak gradient or "Gâteaux differential" relative to the duality defined above) it is classically said that the considered statical law admits W as potential.

The simplest statical law imposes the value $f_o \in \mathcal{F}$ of a certain force acting on the system, independently of the configuration u. Such a constant mapping from $\mathcal{U}$ into $\mathcal{F}$ evidently admits the potential W expressed by

$$W(u) = -\langle u, f_o \rangle \ .$$

EQUILIBRIUM. Suppose that all the physical processes in which the system $\mathcal{S}$ takes part imply forces, acting on it, which either vanish in any expected equilibrium or are n forces $f_1, f_2, \ldots, f_n$ respectively related to the configuration u by n statical laws independent of time, denoted by $\mathcal{R}_1, \mathcal{R}_2, \ldots, \mathcal{R}_n$. Then the equilibrium problem consists in determining the values of u in $\mathcal{U}$ possessing the following property : there exist $f_1, f_2, \ldots, f_n$ in $\mathcal{F}$ respectively related to u by the relations $\mathcal{R}_1, \mathcal{R}_2, \ldots, \mathcal{R}_n$ and such that $f_1 + f_2 + \ldots + f_n = 0$. According to the "principle of virtual work" and due to the way in which $\mathcal{F}$ has been constructed as a quotient space placed in separating duality with $\mathcal{U}$, these values of u correspond in fact to the equilibrium configurations of $\mathcal{S}$, i.e. the configurations in which immobility is a motion compatible with our physical information about this system.

Equivalently, if $R_1, R_2, \ldots, R_n$ denote the multimappings corresponding as above to the n statical laws, the equilibrium configurations are characterized by

$$0 \in R_1(u) + R_2(u) + \ldots + R_n(u) .$$

Let us stress at last that the concept of statical law, as we just defined it, is not restricted to the study of equilibrium problems. In evolution problems also, statical laws will be considered, possibly depending on time.

3. c FRICTIONLESS BILATERAL CONSTRAINTS

The description of a constraint in Mechanics requires fondamentally more information than merely defining a set of permitted configurations. This description always includes some indication concerning the forces of constraint or reactions experienced by the system and implied by the material process which restricts its freedom. Let us emphasize that perfect, i.e. frictionless, constraints are a special type of statical law.

Consider for instance the situation described in the language of elementary Mechanics as follows : a certain particle s of the system $\mathcal{S}$ is maintained bilaterally, without friction, on a given regular material surface S. Let

$$h(\vec{x}) = 0 \tag{3.1}$$

be the equation of S, where $\vec{x}$ denotes the generic element of a three-dimensional frame of reference E_3, treated as a three-dimensional linear space, and h a smooth numerical function defined on E_3, with nonzero gradient. Let $\vec{p}_o$ denote the position of the particle s in E_3 when the system $\mathcal{S}$ presents the configuration corresponding to the zero of $\mathcal{U}$. For the configuration corresponding to some element u of $\mathcal{U}$, this position is $\vec{p}$ and, due to our framework of small deviations and linearization, the mapping $\vec{\ell} : u \mapsto \vec{p} - \vec{p}_o$ is treated as linear from $\mathcal{U}$ into E_3 ; in all the following, this linear mapping is supposed continuous with regard to some locally convex topology compatible with the duality $(\mathcal{U}, \mathcal{F})$, thus continuous for all such topologies. Similarly, the linearization procedure replaces the function h by its first order expansion in the neighborhood of $\vec{p}_o$ so that the condition $\vec{p} \in S$ takes the form

$$h(\vec{p}_o) + (\vec{p} - \vec{p}_o) \cdot \overrightarrow{\mathrm{grad}}\, h(\vec{p}_o) = 0$$

(scalar product and gradient are understood here in the sense of the three-dimensional Euclidien space E_3) i.e.

$$h(\vec{p}_o) + \vec{\ell}(u) \cdot \overrightarrow{\mathrm{grad}}\, h(\vec{p}_o) = 0 \quad . \tag{3.2}$$

Here arises the need of an additional hypothesis concerning $\vec{\ell}$ for the continuous linear form $u \mapsto \vec{\ell}(u) \cdot \overrightarrow{\mathrm{grad}}\, h(\vec{p}_o)$ not to be identically zero ; as the vector $\overrightarrow{\mathrm{grad}}\, h(\vec{p}_o)$ has been supposed different from zero, the sufficient assumption we shall make in all the following is : the linear mapping $\vec{\ell}$ from $\mathcal{U}$ into the three-dimensional space of the "physical" vectors is <u>surjective.</u> One may express this by saying that the particle s of the system is regular regarding the use of $\mathcal{U}$ as the configuration space of the system. Then the values of u satisfying (3.2) constitute a closed hyperplane

$$\mathcal{L} = U + a \quad , \tag{3.3}$$

where a represents some known element of $\mathcal{U}$ and U denotes the linear subspace with codimension 1

$$U = \{u \in \mathcal{U} : \vec{\ell}(u).\ \overrightarrow{\text{grad}}\, h(\vec{p}_o) = 0\} \ .$$

For the particle s to be maintained in S it must experience in addition to other possible actions, the force of constraint $\vec{R}$, or reaction, arising from this material surface. In the language of the pair of spaces $(\mathcal{U}, \mathcal{F})$ the representation of this force consists, by definition, in the element $r \in \mathcal{F}$ possessing the following property : for any $\delta\, u \in \mathcal{U}$, to which corresponds in the "physical" space E_3 the displacement $\delta\, \vec{p} = \vec{\ell}(\delta\, u)$ of the particle s, the work of $\vec{R}$ equals $\langle \delta\, u, r \rangle$, i.e.

$$\langle \delta u, r \rangle = \vec{\ell}(\delta\, u).\ \vec{R} \quad . \tag{3.4}$$

Let us make use now of the hypothesis that the constraint is frictionless. By definition this means $\vec{R}$ is normal to the surface S at the point p ; equivalently $\vec{R}$ yields a zero work for any displacement vector $\delta\, \vec{p}$ which is tangent to S at this point. Due to the linearization procedure which replaces the equation of S by (3.2), this amounts to

(3.5) $\forall \delta u \in U \; : \; \langle \delta u, r \rangle = 0$.

In other words r belongs to V, the subspace of $\mathcal{F}$ orthogonal to U.

It will be supposed that conversely any value of r, i.e. of $\vec{R}$, satisfying this condition can be produced by the device enforcing the constraint. Physically, this means first the constraint is bilateral : the particle s should more exactly be visualized as guided without friction between two parallel surfaces infinitely close to each other ; secondly these surfaces are strong enough to exert arbitrarily large normal reactions. We propose to summarize these facts by saying that the considered perfect constraint is firm (cf. MOREAU [14], vol. 2, § 9. 2) Except otherwise stated, firmness will always be implicitely assumed in the following.

In short, all our information about the constraint is contained in the two conditions $u \in \mathcal{L}$, $r \in V$; equivalently it may be said that the pair (u,r) belongs to the subset $\mathcal{L} \times V$ of $\mathcal{U} \times \mathcal{F}$ and this indeed constitutes a statical law in the sense defined by § 3. b, i.e. a relation between the possible configuration u of the system and some of the forces it undergoes.

This relation is subdifferential.

In fact consider the indicator function $\psi_{\mathcal{L}}$ of the affine manifold described by (3.3) ; the subdifferential of this closed convex function is easily found to be

$$\partial\, \psi_{\mathcal{L}}(u) = \begin{cases} V & \text{if } u \in \mathcal{L} \\ \emptyset & \text{if } u \notin \mathcal{L} \end{cases} .$$

Therefore the relation $(u,r) \in \mathcal{L} \times V$ is equivalent to

$$- r \in \partial\, \psi_{\mathcal{L}}(u) \quad , \tag{3.6}$$

which is another way of conveying the whole of our information about the considered constraint. The minus sign in the left member is immaterial as the right member is a linear space : this is only for sake of consistency with further developments.

More generally, the system $\mathcal{S}$ may be submitted at the same time to several constraints of the preceding sort, respectively defined by n closed hyperplanes $\mathcal{L}_i = U_i + a_i$, $i = 1, 2, \ldots, n$. The set of the permitted configurations is then $\bigcap_i \mathcal{L}_i$; if this intersection is not empty let us use again the notation $\mathcal{L} = U + a$ to represent it, where U is now the intersection of the closed linear subspaces U_i, each with codimension 1. As the reaction r_i implied by the i-th constraint belongs to V_i, the one-dimensional subspace orthogonal to U_i in $\mathcal{F}$, the sum r of the n reactions belongs to V, the subspace orthogonal to U Conversely, any element of V possesses at least one decompostion into a sum $\sum_i r_i$, $r_i \in V_i$ (this is merely the classical theorem of Lagrange multipliers : the duality between $\mathcal{U}$ and $\mathcal{F}$ being separating, the biorthogonal of a finitely generated subspace equals this subspace itself). Therefore, each of the n perfect constraints being assumed firm, the joint effect of them is fully represented by the same writing as (3.6) and this is also trivially true in the case $\mathcal{L}$ is empty.

Thereby we are induced to consider, in general, statical laws expressed under the form (3.6), where $\mathcal{L}$ represents a closed affine manifold whose codimension is not necessarily finite : we shall refer to such statical laws as (firm) perfect affine constraints.

Note at last that, when studying evolution problems, a perfect constraint described as above may be moving : i.e. the affine manifold $\mathcal{L}$ may depend on time in a given way. Just keep in mind at such event that the so-called displacements, labelled in the preceding by the symbol δ, merely express the comparison between possible configurations at a definite instant ; traditionnally they are qualified as virtual in contrast with the real displacements which occur as a consequence of the actual motion. In most practical cases the subspace U which defines the dimension and the direction of $\mathcal{L}$ is independent of time ; only the element a of $\mathcal{U}$ is moving ; we shall meet such a situation in Chapter 6.

3. d PERFECT UNILATERAL CONSTRAINTS

With the same notations as in the preceding, suppose now that the particle s of the system $\mathcal{S}$, instead of being bilaterally maintained in the surface S, is only confined by some impenetrable block whose S constitutes the boundary. Suppose the function h chosen in such a way that the region of E_3 permitted thereby to the position $\vec{p}$ of s is defined by the inequality

$$h(\vec{p}) \geqslant 0 .$$

Then, using the same linearization procedure as before, the set of the permitted values of u is characterized by the inequality

$$(3.7) \qquad h(\vec{p}_o) + \vec{\ell}(u). \overrightarrow{\text{grad}}\, h(p_o) \geqslant 0$$

which defines in $\mathcal{U}$ a closed half-space $\mathfrak{D}$ with the affine manifold $\mathfrak{L}$ as boundary.

Here again, the description of the mechanical situation requires some information about the force of constraint $\vec{R}$ that the block must exert on s to prevent penetration ; this information will rather be formulated by means of the element $r \in \mathcal{F}$ which represents the force according to (3.4).

First, this reaction vanishes when s does not touch the block, i.e. when (3.7) holds as a strict inequality ; in other words one has the implication

$$(3.8) \qquad u \in \text{int}\, \mathfrak{D} \Rightarrow r = 0 \quad .$$

When, on the contrary, s lies in contact with the boundary S, we still make the no-friction hypothesis, i.e. $\vec{R}$ is normal to S. In addition the unilaterality of the contact imposes that the vector $\vec{R}$ is directed toward the permitted region i.e. directed in concordance with the vector $\overrightarrow{\text{grad}}\, h(\vec{p})$. In terms of work this is expressed as follows : any $\delta\,\vec{p}$ such that $\delta\,\vec{p}\,.\,\overrightarrow{\text{grad}}\, h(\vec{p}) \geqslant 0$ yields $\delta\,\vec{p}\,.\,\vec{R} \geqslant 0$. Now, recalling the regularity assumption made about the mapping $\vec{\ell}$, take as $\delta\,\vec{p}$ the displacement of s in E_3 associated as before with the element δ u of $\mathcal{U}$ by $\delta\,\vec{p} = \vec{\ell}\,(\delta\, u)$. The contact between s and the block means that (3.7) holds as an equality. Besides, due to the linearization procedure, $\overrightarrow{\text{grad}}\, h(p)$ is treated as independent of p. Then $\delta\,\vec{p}.\,\overrightarrow{\text{grad}}\, h(\vec{p}) \geqslant 0$ holds if and only if $\vec{\ell}(\delta\, u).\,\overrightarrow{\text{grad}}\, h(\vec{p}_o) \geqslant 0$; by putting $u' = u + \delta\, u$ the latter is equivalent to $u' \in \mathcal{D}$ so that finally, in view of (3.4), all our information about $\vec{R}$ comes to be equivalent to the following

(3.9) $$\forall\, u' \in \mathcal{D} \quad : \quad \langle u' - u,\, r\rangle \geqslant 0 \quad .$$

This actually implies also (3.8) ; in fact, if $u \in \text{int}\, \mathcal{D}$ the difference $u' - u$, for $u' \in \mathcal{D}$ can be a non zero element of $\mathcal{U}$ with arbitrary direction; hence $r = 0$ for the duality is separating.

In conclusion the geometric condition $u \in \mathcal{D}$ of the constraint is expressed jointly with (3.9) by writing

(3.10) $$- r \in \partial\, \psi_{\mathcal{D}}(u) \quad .$$

Here as in § 3. c let us make conversely the firmness assumption : the block is supposed strong enough to exert any value of $\vec{R}$ agreeing with the preceding requirements ; in other words any value of r satisfying (3.9) is possible. Then relation (3.10) conveys all our information about the considered constraint.

More generally suppose the system $\mathcal{S}$ subjected to n constraints of the preceding sort, corresponding to half-spaces $\mathcal{D}_i$, $i = 1, 2, \ldots, n$. Then the set of the permitted configurations is the closed convex set $C = \bigcap_i \mathcal{D}_i$. As each of the reactions r_i satisfies a relation of the form (3.10), their sum r satisfies

$$- r \in \partial \psi_{\mathcal{D}_1}(u) + \partial \psi_{\mathcal{D}_2}(u) + \ldots + \partial \psi_{\mathcal{D}_n}(u) \quad .$$

The right member is trivially contained in $\partial \psi_C(u)$; actually this sum of sets equals exactly $\partial \psi_C(u)$ because of the "unilateral" counterpart of the multipliers theorem (as the duality $(\mathcal{U}, \mathcal{F})$ is separating, a finitely generated convex cone in $\mathcal{F}$ is closed, thus equal to its bipolar). In conclusion the conjunction of our n unilateral constraints is equivalent to the following statical law

(3.11) $$- r \in \partial \psi_C(u) \quad .$$

Hence we are induced to consider more generally the statical laws defined in the same way by taking as C arbitrary closed convex subsets of $\mathcal{U}$: we call these laws (firm) <u>perfect convex constraints.</u>

Evidently the bilateral constraint studied in § 3. e are a special case of this : take as C a closed affine manifold.

3. e SUPERPOTENTIALS

We shall say that a statical law admits a function

$\phi \in \Gamma_o(\mathcal{U}, \mathcal{F})$ as <u>superpotential</u> if this law consists in the following relation between the configuration u and some force f

$$- f \in \partial \phi (u) \quad .$$

In particular, if a statical law admits some numerical function W as <u>potential</u>, W is also a superpotential if and only if this function is convex. For instance the constant law $f = f_o$ (independent of u) admits as superpotential the linear form $u \mapsto - \langle u, f_o \rangle$.

Another fundamental example is that of a <u>perfect convex constraint</u>, as presented in the preceding paragraph : (3.11) means that the function ψ_C is a superpotential for such a statical law ; by taking as C a closed affine manifold, this includes, according to § 3. c, the traditional bilateral contraints.

Suppose the system subjected at the same time to a finite family of statical laws admitting the respective superpotentials $\phi_1, \phi_2, \ldots, \phi_n$. Then the sum of $f = f_1 + f_2 + \ldots + f_n$ of the corresponding forces is related to u by

$$- f \in \partial \phi_1 (u) + \partial \phi_2 (u) + \ldots + \partial \phi_n (u) \quad .$$

This relation implies

(3.12) $$- f \in \partial (\phi_1 + \phi_2 + \ldots + \phi_n) (u)$$

but is equivalent to it only if some conditions ensuring the additivity of subdifferentials are fulfilled ; according to § 2.f, the usual case where such additivity holds is described as follows : 1° some of the functions ϕ_i are weakly differentiable everywhere in $\mathcal{U}$; 2° there exists a point $u_o \in \mathcal{U}$ at which the others, but possibly one, are finite and continuous for some topology compatible with the duality $(\mathcal{U}, \mathcal{F})$; 3° the last one is finite at u_o .

EQUILIBRIUM. Suppose first that all the mechanical actions to which the system is subjected (except possibly those which vanish in any expected equilibrium) are summarized under the form of a single statical law admitting a superpotential ϕ independent of time. Then, as explained in § 3. b, the equilibrium configurations are characterized by

$$0 \in \partial \phi (u) \quad ;$$

this is a <u>necessary and sufficient condition for</u> u <u>to be one of the points of</u> $\mathcal{U}$ <u>where the numerical function</u> ϕ <u>attains its infimum</u> (cf. § 2. e). Such values of u form a closed convex subset of $\mathcal{U}$, possibly empty.

Suppose more generally that the considered mechanical actions are described by the conjunction of n statical laws admitting as above the respective superpotentials ϕ_i, independent of time. A necessary and sufficient condition for u to be an equilibrium configuration is now

$$0 \in \partial \phi_1 (u) + \partial \phi_2 (u) + \ldots + \partial \phi_n (u) \quad .$$

This implies $0 \in \partial \phi (u)$, with ϕ equal to the sum of the functions ϕ_i ; therefore this sum attains its infimum at the point u. But the converse may not be true, unless the additivity of subdifferentials holds. Actually such a reserve does not seem to be of great pratical importance and B. Nayroles suggests in his lectures a logical attitude which would overcome the difficulty.

EXAMPLE. Make $n = 2$ and suppose that $\phi_1 = \psi_C$, the superpotential of a perfect convex constraint. Then equilibrium is characterizd by

$$0 \in \partial \psi_C(u) + \partial \phi_2(u) \quad .$$

This implies that u is a point in C where the restriction of the function ϕ_1 to this set attains its infimum ; in the vocabulary of mathematical programming, u is one of the solutions of a "constrained" minimization problem. But the converse may not be true, unless the additivity of subdifferentials holds ; particularizing the situation described above, one finds that any of the three following conditions ensures this additivity :

1° The function ϕ_2 is weakly differentiable everywhere in $\mathcal{U}$, i.e. it is a potential in the classical sense.

2° There exists a point in the interior of C where the function ϕ_2 takes a finite value.

3° There exists a point in $\mathcal{U}$ at which the function ϕ_2 is finite and continuous and which belongs to C.

Recall that "interior" or "continuous" may here be understood in the sense of any locally convex topology compatible with the duality $(\mathcal{U}, \mathcal{F})$: the weakest assumption is thus obtained by choosing the finest of these topologies, i.e. the Mackey topology $\tau(\mathcal{U}, \mathcal{F})$; this remark is of course without object in finite dimensional cases.

3. f DUAL MINIMUM PROPERTIES

This paragraph is devoted to the equilibrium problem, in the case where all the mechanical actions exerted on the system $\mathcal{S}$ (except possibly those which vanish at any expected equilibrium) are expressed as the conjunction of two statical laws respectively admitting the superpotentials ϕ_1 and ϕ_2, independent of time. Of course, each of these two superpotentials may in its turn describe the conjunction of several laws ; in practical situations there are usually various possibilities of classifying the mechanical actions into such two groups, so that the statements presented below can generate a great number of different variational properties. It may be imagined that ϕ_1 and ϕ_2 correspond to two different sorts of mechanical action : for instance ϕ_1 is the superpotential of a perfect constraint, while ϕ_2 represents "active forces".

An element u of $\mathcal{U}$ is an equilibrium configuration if and only if there exist $f_1 \in - \partial \phi_1 (u)$ and $f_2 \in - \partial \phi_2 (u)$ such that $f_1 + f_2 = 0$. The determination of such f_1 (or equivalently f_2) prior to that of u, is sometimes called a <u>statical approach of the equilibrium problem</u> (we should prefer to call it <u>sthenic</u>, an adjective meaning "relative to forces"). Privileging ϕ_1, let us agree to call an equilibrium force any value of f_1 associated in this way with some equilibrium configuration.

PROPOSITION 1. Let γ_1 and γ_2 be the respective polar (i.e. dual) functions of ϕ_1 and ϕ_2, relative to the duality $(\mathcal{U}, \mathcal{F})$; denote by $\hat{\gamma}_1$ the function $f \mapsto \gamma_1(-f)$ (it is the polar function of $\hat{\phi}_1 : u \mapsto \phi_1(-u)$). Then any equilibrium force minimizes the function $\hat{\gamma}_1 + \gamma_2$ over $\mathcal{F}$; conversely, if f_1 is a minimizing point of this sum and if $\hat{\gamma}_1$ and γ_2 possess the additivity of subdifferentials at this point, f_1 is an equilibrium force.

In fact if $f_1 \in \mathcal{F}$ corresponds to some $u \in \mathcal{U}$ such that $-f_1 \in \partial \phi_1(u)$ and $f_1 \in \partial \phi_2(u)$ one has equivalently $u \in \partial \gamma_2(f_1)$ and $u \in \partial \gamma_1(-f_1)$; the latter is the same as $-u \in \partial \hat{\gamma}_1(f_1)$; therefore

$$0 \in \partial \hat{\gamma}_1(f_1) + \partial \gamma_2(f_1) \subset \partial(\hat{\gamma}_1 + \gamma_2)(f_1) .$$

Conversely, the assumption that f_1 is a minimizing point of $\hat{\gamma}_1 + \gamma_2$ means that the zero of $\mathcal{U}$ belongs to $\partial(\hat{\gamma}_1 + \gamma_2)(f_1)$; if this set equals $\partial \hat{\gamma}_1(f_1) + \partial \gamma_2(f_1)$, one has

$$0 \in \partial \gamma_2(f_1) - \partial \gamma_1(-f_1)$$

which precisely expresses the existence of some u associated with f_1 in the preceding way.

As far as we can see this Proposition contains as special cases, all the extremal properties of "statical" type in elastostatics. Observe in this connection that if ϕ_2, for instance, is the superpotential of the perfect bilateral constraint defined by the affine manifold $\mathcal{L} = U + a$ (cf. § 3. e), its dual function is defined by

$$\gamma_2 (f) = \psi_V (f) + \langle a,f\rangle \quad .$$

Thus minimizing $\hat{\gamma}_1 + \gamma_2$ over $\mathcal{F}$ is the same as minimizing $\hat{\gamma}_1 + \langle a,.\rangle$ over V, the linear subspace of $\mathcal{F}$ orthogonal to U.

On the other hand, in the usual situations of linear elastostatics, one may take as ϕ_1 the potential of elastic forces, which is a nonnegative quadratic form on $\mathcal{U}$. Calculating its dual γ_1 (equal to $\hat{\gamma}_1$, since quadratic forms are even functions) yields a nonnegative quadratic form defined on some linear subspace of $\mathcal{F}$ and $+\infty$ outside of this subspace ; a special property of the quadratic case is that, if u and $-f$ are conjugate points with regard to ϕ_1, γ_1, one has

$$\phi_1 (u) = \gamma_1 (f) = -\frac{1}{2} \langle u,f\rangle \quad .$$

Thus, γ_1 may be interpreted as "the expression of the elastic energy in terms of the elastic force" and sometimes called the <u>complementary energy</u>. This does not hold anymore in non linear elasticity ; however in the very usual case where the elastic potential ϕ_1 is a <u>quasi-homogeneous</u> convex function, there is still a relation between $\phi_1 (u)$ and $\hat{\gamma}_1 (f)$, if f is the elastic force corresponding to u.

3. g SADDLE - POINT PROPERTY

The notations are the same as in the preceding paragraph. Determining the equilibrium configurations of $\mathcal{S}$ as minimizing points of $\phi_1 + \phi_2$ (cf. § 3. e) and determining the equilibrium forces as minimizing points of $\hat{\gamma}_1 + \gamma_2$ may be considered as dual extremum problems. This is a familiar feature of convex programming and it is habitual to relate such a pair of problems to a saddle-point property for a function called Lagrangian.

PROPOSITION. Define the concave-convex function L on the product space $\mathcal{U} \times \mathcal{F}$ by

$$L(u,f) = \langle u,f \rangle + \hat{\gamma}_1 (f) - \phi_2 (u)$$

with the convention $+\infty - \infty = +\infty$ (or equivalently the convention $+\infty - \infty = -\infty$). A point $u_o \in \mathcal{U}$ is an equilibrium configuration of $\mathcal{S}$, with $f_1 \in \mathcal{F}$ as corresponding equilibrium force, if and only if the element (u_o, f_1) of $\mathcal{U} \times \mathcal{F}$ is a saddle point of L with finite value, i.e. $L(u_o, f_1)$ is finite and for any $u \in \mathcal{U}$ and any $f \in \mathcal{F}$,

(3.13) $$L(u, f_1) \leqslant L(u_o, f_1) \leqslant L(u_o, f) \quad .$$

In fact, suppose first that u_o is an equilibrium configuration with f_1 as equilibrium force, i.e. $-u_o \in \partial \hat{\gamma}_1 (f_1)$ and $f_1 \in \partial \phi_2 (u_o)$; the former of these conditions means

(3.14) $$\forall f \in \mathcal{F} \quad : \quad -\langle u_o, f-f_1 \rangle + \hat{\gamma}_1(f_1) \leqslant \hat{\gamma}_1(f)$$

and the latter

(3.15) $\forall\, u \in \mathcal{U} : \quad \langle u - u_o, f_1\rangle + \phi_2(u_o) \leqslant \phi_2(u)$.

Adding the finite number $-\phi_2(u_o)$ to both members of (3.14) yields the second of inequalities (3.13) ; adding the finite number $\hat{\gamma}_1(f_1)$ to both members of (3.15) yields the first one. The value $L(u_o, f_1)$ is clearly finite.

Conversely, supposing $L(u_o, f_1)$ finite implies that $\hat{\gamma}_1(f_1)$ and $\phi_2(u_o)$ are finite ; then the preceding calculation may be effected backward to deduce (3.14) and (3.15) from (3.13).

REMARK. Exchanging the roles of ϕ_1 and ϕ_2 would yield a quite different function L. Since, in practical situations, there are usually several ways of classifying mechanical action into two groups corresponding to ϕ_1 and ϕ_2, since, on the other hand the $(\mathcal{U}, \mathcal{F})$ pattern may usually be applied in several ways (see § 3. j below), the preceding Proposition generates a pretty great number of saddle point characterisations of the equilibrium in elastostatics.

3. h ONE - DIMENSIONAL EXAMPLES

We consider in this paragraph a system $\mathcal{S}$ whose configuration can be specified by a single numerical variable : it is for instance a rectilinear bar or a string, as far as we are only interested in the distance between its extremities. Denote by $\ell_o + e$ this distance ; in other words, e denotes the <u>elongation</u> of the bar by comparison with some reference state in which the length was ℓ_o. As we are only concerned with static or quasi-static situations, the state of stress of the bar is sufficiently described by the <u>tension</u> s. Classically, for the application of the principle of virtual work to systems comprising the considered bar, the expression of the work of the internal actions must be $-s\,\delta e$. Thus the pattern of the preceding paragraph applies by taking for the linear space $\mathcal{U}$ a copy of the real line R, with e as generic element, and for the linear space $\mathcal{F}$ another copy of R, with s as generic element ; these two one-dimensional linear spaces are placed in separating duality by the bilinear form $\langle .,. \rangle$

$$\langle e,s \rangle = -e\,s \quad . \tag{3.16}$$

This unpleasant minus sign merely comes from our complying with the common habit in solid mechanics of measuring the state of stress by a positive number when it is properly a tension, by a negative number when it is a proper pressure. It has nothing to do with the fact that the considered "actions" are internal : in our formalism, stress is a "force" like any other mechanical action.

This framework permits the formulation of usual behavioral laws of the rectilinear system.

1° <u>Regular elasticity</u>. Suppose that the behavioral law of the bar defines the tension s as a continuous strictly increasing function of the elongation e, namely $s = j(e)$ or equivalently $s = \theta'(e)$, where θ denotes a primitive of j ; observe that θ is then a convex function. Let e_o be some definite value of e and $s_o = \theta'(e_o)$. The affine function

$$e \mapsto (e - e_o)\, s_o + \theta(e_o)$$

is <u>tangent</u> to θ at the point e_o ; now, with regard to the duality defined by (3.16), the <u>slope</u> of this affine function is $- s_o$. In other words the relation $s = \theta'(e)$ may be written as

$$- s = \operatorname{grad} \theta(e) \quad .$$

This means that θ is a potential for the considered statical law (and also, as usual, the expression of the potential energy) ; due to the convexity of θ it is also a superpotential. As we have supposed the function $\theta' = j$ continuous and strictly increasing, it possesses an inverse function j^{-1}, defined on the range of j ; this range is an interval I, possibly unbounded or not closed. The characterization of e and $- s$ as conjugate points

$$\theta(e) + \theta^*(- s) = \langle - s, e \rangle$$

permits the calculation of θ^* by the formula

$$\theta^*(- s) = s\, j^{-1}(s) - \theta\,[j^{-1}(s)]$$

valid for any s in I. The function θ takes the value $+\infty$ outside of the closure of $-I$.

2° Elastic string. We agreed that $\ell_o + e$ represented the distance between the extremities of the considered one-dimensional system. If ℓ_o denotes exactly the length at rest of an elastic string, the corresponding static law has the form $s = j(e)$ where the function j takes now the value zero for $e \leqslant 0$. A primitive of j is a superpotential ; its dual function θ^* with regard to the bilinear form (3.16) takes the value $+\infty$ on $]0,+\infty[$; the values of $\theta^*(-s)$ for s belonging to the range of j are constructed as above if j is continuous and strictly increasing on $[0, +\infty[$.

3° Inelastic string. This may be considered as a boundary case of the preceding. Supposing that ℓ_o is the proper length of the string and that the breaking load is infinite, one finds the following superpotential for the relation between e and s

$$\theta(e) = \begin{cases} +\infty & \text{if } e > 0 \\ 0 & \text{if } e \leqslant 0 \end{cases} .$$

This is the indicator function of the closed convex subset $C =]-\infty, 0]$ of $\mathcal{U}$, so that the present law comes to be a perfect convex constraint. As C is actually a convex cone (see § 2. c) the dual function θ^* is the indicator function of the polar cone, i.e. the subset $]-\infty, 0]$ of $\mathcal{F}$ (it is the set of the possible values of $-s$).

The reader will study other examples such as a cylindrical helix spring, enclosed in a guide tube to prevent buckling ; the length of this spring cannot be less than the length it has when all the spires come into contact. The corresponding behavioral law is equivalent to the conjunction of a law of elasticity and of a perfect convex constraint. This gives a very elementary model of an elastic solid with limited compressibility, a type of material which was studied in generality by W. PRAGER [1] ; the behavior of such a material can be formulated as a statical law admitting a superpotential.

3. i AN EXAMPLE OF COMPOUND SYSTEM

Take as $\mathcal{S}$ a lattice of bars (a truss) whose extremities are articulated with one another through spherical joints. The joints are represented by n points $A_1, A_2, \ldots, A_n$ the nodes of the lattice. To make the description simpler suppose that between each pair of nodes, say A_i and A_j with $i < j$ to avoid repetition, there exists one of the bars denoted by B_{ij}, thus $\frac{1}{2} n (n-1)$ bars in all. The behavior of each bar is treated as one-dimensional ; denote by s_{ij} the tension of the bar B_{ij} and by e_{ij} its elongation with respect to the "zero" state.

Any configuration of the system $\mathcal{S}$ is fully determined by the corresponding positions of the n nodes A_i relative to some three-dimensional Cartesian frame ; these respective positions may be described by the n three-dimensional displacement vectors $\vec{x}_i$ by which they differ from the positions corresponding to the "zero" configuration of the system. Thereby we are induced to consider as the configuration space of $\mathcal{S}$ the 3 n-dimensional linear space X whose generic element x consists in the n-tuple $(\vec{x}_1, \vec{x}_2, \ldots, \vec{x}_n)$.

Here again we restrict ourselves to linearized geometry, by treating the displacements as infinitely small with regard to the lengths of all the bars. Denote by $\vec{\alpha}_{ij}$ (with $i < j$) the unit vector of the oriented line A_iA_j (taken, to fix the ideas, in the zero configuration ; but this precision is immaterial since the bars present only infinitesimal rotations). The elongation of the bar B_{ij} is related to u by

(3.17) $$e_{ij} = \vec{\alpha}_{ij} \cdot (\vec{x}_j - \vec{x}_i)$$

(three-dimensional scalar product).

An <u>external action</u> is a n-tuple of forces $(\vec{y}_1, \vec{y}_2, \ldots, \vec{y}_n)$ respectively exerted on the n nodes ; this n-tuple of three-dimensional vectors, denoted by y, constitutes the generic element of a 3 n-dimensional linear space Y. The bilinear form "work", placing the spaces X and Y in separated duality will be noted $\langle\langle .,. \rangle\rangle$ to prevent confusion in the following, and has the familiar expression

(3.18) $$\langle\langle x,y \rangle\rangle = \sum_{i=1}^{n} \vec{x}_i \cdot \vec{y}_i \quad .$$

In order to formulate the equilibrium problem for the considered system one has to specify the statical laws to which it is subjected. These statical laws are of two sorts:

Some of them concern <u>external actions</u> ; for instance given loads may be applied to some nodes ; or some nodes may be submitted to bilateral or unilateral constraints ; or also some nodes may be subjected to statical laws relating to their positions some of the forces they experience. All this has to be described in the framework of the pair of linear spaces (X,Y).

The other laws, said <u>internal</u>, concern the behavior of the bars and are formulated in terms of the elongations e_{ij} and the tensions s_{ij} : this induces to consider the $\frac{1}{2}n(n-1)$-dimensional linear space E whose generic element, denoted by e, is the $\frac{1}{2}n(n-1)$-tuple of real numbers e_{ij}, $i < j$, and the similar space S whose generic element is s, consisting of the s_{ij}, $i < j$. As explained in § 3. h, the expression of the internal work in the bar B_{ij}, corresponding to a tension measured by the real number s_{ij} and an (increase of) elongation measured by the real number e_{ij} is $-e_{ij}s_{ij}$. Therefore the total internal work in the bars corresponding to given $e = (e_{ij})$ and $s = (s_{ij})$ is

(3.19) $$\langle e,s\rangle = - \sum_{i < j} e_{ij} s_{ij}$$

a bilinear form which places the two linear spaces E and S in separating duality : keep in mind that it differs by the presence of the minus sign from the natural "scalar product" between two spaces whose elements are such $\frac{1}{2} n(n-1)$-tuples of real numbers.

At the present stage, where plasticity is not taken into account, the behavioral laws of the bars are relations between e_{ij} and s_{ij} formulated in the same ways as in § 3. h .This introduces, for each (i,j), i < j, a superpotential θ_{ij} which is a closed convex functions on R and the corresponding statical laws takes the form

(3.20) $$- s_{ij} \in \partial \theta_{ij} (e_{ij}) .$$

By the remarks made in § 2. c about the product of linear spaces, the function θ defined on E by

$$\theta(e) = \sum_{i < j} \theta_{ij} (e_{ij})$$

permits to summarize the $\frac{1}{2} n(n-1)$ relations (3.20) by writing

(3.21) $$- s \in \partial \theta (e) .$$

3. j VARIOUS TREAMENTS OF THE EQUILIBRIUM PROBLEM

Let us pursue the study of the system described above. Continuously distributed external actions, such as gravity, are not taken into account, so that the equilibrium condition of the system consists in the vanishing of the total force experienced by each of the n nodes, i.e. for each value of $i = 1, 2, \ldots, n$ the following three-dimensional vector equation

$$\vec{y}_i + \sum_{i < j} s_{ij}\,\vec{\alpha}_{ij} - \sum_{i > j} s_{ji}\,\vec{\alpha}_{ji} = 0 \quad . \tag{3.22}$$

On the other hand, equalities (3.17) define a linear mapping from X into E which will be denoted by D. By definition the adjoint D^* of D is the linear mapping from S into Y defined by

$$\forall\, x \in X \ ,\ \forall\, s \in S \ :\ \langle D\,x, s\rangle = \langle\langle x, D^*\,s\rangle\rangle \quad .$$

Referring to the definitions of $\langle .,. \rangle$ and $\langle\langle .,. \rangle\rangle$, then identifying the terms of each member yields that the element $D^* s$ of Y consists of the n-tuple of three-dimensional vectors $\overrightarrow{(D^* s)}_i$

$$\overrightarrow{(D^* s)}_i = \sum_{j > i} s_{ij}\,\vec{\alpha}_{ij} - \sum_{j < i} s_{ji}\,\vec{\alpha}_{ji} \quad .$$

Therefore the equilibrium condition (3.22) takes the form

$$y + D^*\,s = 0 \tag{3.23}$$

which of course is equivalent to the principle of virtual work, namely

$$\forall\, x \in X \ :\ \langle\langle x, y\rangle\rangle + \langle D\,x, s\rangle = 0 \ . \tag{3.24}$$

1^0 The method of big spaces.

We give this name to the method which consists in using the pair (x,e), denoted by u as the element which specifies the configuration of our system. Then, with the notations of § 3. a the configuration space is $\mathcal{U} = X \times E$; the corresponding $\mathcal{F}$ is the space $Y \times S$, whose generic element is the pair (y,s) denoted by f. These spaces are placed in separating duality by the expression of the total work $\langle\langle x,y \rangle\rangle + \langle e,s \rangle$, to be denoted by $\langle u,f \rangle$.

Clearly the whole of the space $\mathcal{U}$ is not permitted to u, since the pair (x,e) must belong to the following linear subspace of $\mathcal{U}$

$$U = \{(x,e) \in X \times E \; : \; e = D\,x\}$$

i.e. the graph of D. Let us show that this restriction of freedom may be treated as a perfect constraint.

In fact the equilibrium condition of the system is not the vanishing of the element $f = (y,s)$ but merely equality (3.23). Putting

$$V = \{(y,s) \quad Y \times S \; : \; y + D^{*} s = 0\}$$

we observe that V is precisely the subspace of $\mathcal{F}$ orthogonal to U : this is the same as the equivalence between (3.23) and (3.24). Condition (3.23) is equivalent to asserting the existence of some r in V such that $f + r$ vanishes. Interpreting r as the reaction associated with the considered constraint agrees with our general definition of a perfect affine constraint.

Actually this conception may be related to a physical realization of the constraint : considering $X \times E$ as the configuration space amounts to regarding our system as the conjunction of the following subsystems : the nodes A_i, whose respective configurations are described by the three-dimensional vectors $\vec{x}_i$ and the bars B_{ij}, whose respective states are described by the elongations e_{ij}. The constraint whose geometric effect is expressed by (3.17) merely consists in connecting the bars with the nodes. However, our main motivation in developing the present example is to prepare for the case of continuous media, (cf. B. Nayroles's lectures) ; in this case x is replaced by a field of displacement vectors defined on a region of R^3 and e is replaced by a field of strain tensors ; then $e = D\,x$ is the condition of geometric compatibility between displacements and strains ; this restriction of freedom may be formally considered as a perfect constraint in the same way as above but it does not seem wise to try and visualize a mechanical realization for it.

Suppose that the statical laws concerning the external actions experienced by the system (possibly including constraints acting on the nodes) can be globally described in the framework of the spaces (X,Y) by a superpotential $\zeta \in \Gamma_0\,(X,Y)$; in other words the external force $y \in Y$ is related to the "external" configuration $x \in X$ by

$$(3.25) \qquad -\,y \in \partial\,\zeta\,(\tau) \quad ,$$

where the subdifferential is understood in the sense of the duality (X,Y). Suppose on the other hand that the internal statical laws are expressed by (3.21). By the rules formulated in § 2. c about product spaces, (3.21) and (3.25) are equivalently summarized as

$$- f \in \partial \phi (u)$$

in the sense of the duality between the big spaces with $u = (x,e)$, $f = (y,s)$, and the superpotential ϕ defined by

$$\phi (u) = \zeta (x) + \theta (e) .$$

The equilibrium of the system may then be studied by the methods of §§ 3. e, f, g.

2° The elimination of (E,S)

As the configuration of the system is fully specified when $x \in X$ is given, one may prefer to consider only X as the configuration space, and Y as the force space. Then every mechanical action experienced by the system must be described in terms of elements of Y : precisely it is represented by the element y of Y such that for every displacement δx of the system, the work of the considered action is $\langle\langle \delta x, y \rangle\rangle$. In this way an internal stress $s \in S$ is represented by the element y_s of Y such that

$$\forall \delta x \in X \ : \langle\langle \delta x, y_s \rangle\rangle = \langle D \delta x, s \rangle ,$$

i.e.

$$(3.26) \qquad y_s = D^* s .$$

Thus the statical law (3.21) is transcribed in terms of the pair of spaces (X,Y) as follows

(3.27) $$- y_s \in D^* (\partial \theta (D x)) .$$

If, in particular, there exists a point in the range of D at which θ is finite and continuous (for some topology compatible with the duality (E,S)), the calculation rule (2.15) holds, so that (3.27) amounts to

(3.28) $$- y_s \in D (\theta \circ D) (x)$$

in the sense of the duality (X,Y) ; this constitutes a statical law admitting the function $\theta \circ D$ as superpotential. In this way the techniques of the foregoing paragraphs may be applied with regard to the pair of spaces (X,Y).

3° The elimination of (X,Y)

The mapping $D : X \to E$ is not injective ; this means that the element $e = D x$ does not convey enough information to specify completely the configuration of the system. However one may wish to determine the equilibrium values of e or s prior to that of x or y and in some instances one may be interested in these elements only (in order to discuss strength, for example).

In the principle, the elimination is similar to that of the preceding case. Suppose that all the external laws to which the system is jointly submitted are summarized under the form

(3.29) $$x \in P(y)$$

where P denotes a given multimapping from Y into X. Similarly suppose that all the internal laws are summarized as

(3.30) $$s = R(e)$$

where R denotes a given multimapping from E into S. A system of values of x, y, e, s defines an equilibrium state if and only if it satisfies $e = Dx$ and (3.23), (3.29), (3.30). Thus, as far as e and s only are concerned, the equilibrium condition (i.e. a necessary and sufficient condition for the existence of at least one pair (x,y) associated with (e, s) in such a way that the preceding equilibrium conditions hold) consists in the conjunction of (3.30) with

(3.31) $$e \in D(P(-D^* s)) .$$

In the principle, (3.31) may as well be written under the form

(3.32) $$-s \in Q(e) .$$

Now as far as the interesting unknown is e, the conjunction of (3.30) with (3.32) is equivalently formulated as follows : <u>there exist</u> s_1 <u>and</u> s_2 <u>in</u> S <u>such that</u>

$$s_1 \in R(e)$$

$$s_2 \in Q(e)$$

$$s_1 + s_2 = 0 .$$

Formally we are reduced to the usual pattern of the equilibrium of a system submitted to two statical laws. From this standpoint the relation $s \in Q(e)$ should be considered as <u>the "internal image" of the external</u>

<u>statical law</u> (3.29).

The reader is invited to apply this procedure to an external law of the form $-y \in \partial \zeta(x)$, equivalently written as $x \in \partial \zeta^*(-y)$. Here again the calculation rule (2.15), under some continuity assumption, will yield an image in (E,S) which admits a superpotential. As a first example, take as external statical law a <u>given load</u> $y_o \in Y$ applied to the system ; this may be written under the form (3.29) with

$$P(y) = \begin{cases} X & \text{if } y = y_o \\ \emptyset & \text{if } y \neq y_o \end{cases} .$$

Another primary example is that of a <u>perfect affine constraint</u> formulated relatively to the pair (X,Y).

<u>But it will be more in the spirit of this Chapter to operate with the pair</u> (E,S) <u>in the following way</u> :

Since we choose to deal only with informations formulated in the framework of the paired spaces (E,S), we accept only to speak of the state of the system in terms of e ; on the other hand, a mechanical action experienced by the system will be taken into account only if it can be <u>represented</u> by an element $\sigma \in S$, in such a way that the work of this action for every displacement of the system has the expression $\langle \delta e, \sigma \rangle$. Therefore, if in particular the considered action is an external force $y \in Y$ treated as given, the corresponding σ must be such that

(3.33) $$\forall \delta x \in X \ : \ \langle\langle \delta x, y \rangle\rangle = \langle D \delta x, \sigma \rangle$$

Such a σ does not necessarily exist ; an evident condition for its existence is that y belongs to $D^* S$, the image of S under the linear mapping D^*. The linear subspace $D^* S$ of Y is the orthogonal, in the sense of the duality (X,Y), of the subspace $\operatorname{Ker} D$ of X. Actually the impossibility of representing in the (E,S) framework a load y which would not belong to $D^* S$ does not make any hindrance. In fact suppose, for sake of simplicity, that this load is the only external action exerted on the system ; clearly by (3.23) or by (3.24), $y \in D^* S$ is a <u>necessary condition</u> for the existence of an equilibrium ; this is a familiar fact ; only a family of external forces with zero resultant and zero moment is compatible with equilibrium.

Another fundamental remark about the use of the (E,S) pattern is that <u>all the values of</u> e <u>are not permitted</u>, since necessarily e belongs to the subspace $D X$ (the subspace of E consisting of the "states of strain" which are "geometrically compatible"). On the other hand, if $s \in S$ denotes the sum of all the elements of S representing the mechanical actions exerted on the system, <u>the equilibrium condition is not</u> $s = 0$, but the principle of virtual work, namely

$$\forall \delta x \in X \ : \ \langle D \delta x, s \rangle = 0$$

which means that s belongs to the subspace of S orthogonal to D X (actually the kernel of D^*).

In conclusion the equilibrium problem in (E,S) must be treated by considering the condition $e \in D X$ as a perfect constraint.

The reader will check that given external loads and external perfect affine constraints are transcribed in the (E,S) language by given forces and perfect affine constraints.

It is from this standpoint that the elastoplastic evolution problem will be studied in Chapter 6.

J. J. Moreau

4 LAWS OF RESISTANCE

4. a VELOCITIES AND FORCES

A habitual procedure, when studying a mechanical system, is to associate with each possible configuration of this system a linear space -let us denote it by $\mathcal{V}$- whose elements constitute, in a general sense, the possible values of the velocity of the system if it happens to pass through the considered configuration. Roughly speaking, $\mathcal{V}$ may be interpreted as the tangent space at the corresponding point of the configuration manifold but this need not be made more precise here. This space is of infinite dimension if the system has an infinite degree of freedom.

In the special framework of Chapter 3, where the configuration manifold is treated as a linear space $\mathcal{U}$, a motion of the system is described by a mapping $t \mapsto u(t)$ from some interval of time into $\mathcal{U}$. The velocity is naturally defined in this case as the derivative $\dot{u}(t)$ (taken in the sense of some topology on $\mathcal{U}$) if it exists ; then $\mathcal{V} = \mathcal{U}$, the same for all the configurations.

Let us come back to the general setting. With each configuration is also associated a linear space -denote it by $\mathcal{F}$- whose elements represent in a more or less abstract way, the mechanical actions which may be exerted on the system when it happens to come through the considered configuration : see the construction of the space of torsors in § 3. a. By extension, the elements of $\mathcal{F}$ are called forces. An essential feature in the practice of Mechanics is that several forces are usually applied to the system at the same time. This produces a fondamental dissymmetry between the roles played by $\mathcal{U}$ and $\mathcal{F}$.

To any pair $v \in \mathcal{V}$, $f \in \mathcal{F}$ corresponds the power of the force f if the system possesses the velocity v, a real number denoted by $\langle v,f \rangle$; this defines a bilinear form which places $\mathcal{V}$ and $\mathcal{F}$ in duality.

In the linear framework of Chapter 3 where $\mathcal{V} = \mathcal{U}$, there is no inconsistency in considering the single space $\mathcal{F}$ as the force space associated with any configuration and in using the same bracket as above to denote by $\langle \delta u, f \rangle$ the work of $f \in \mathcal{F}$ corresponding to the displacement $\delta u \in \mathcal{U}$. In fact, suppose this displacement results from a motion $t \mapsto u(t)$ with velocity $\dot{u}$ (derivative understood in the sense of some topology compatible with the duality $(\mathcal{U}, \mathcal{F})$) taking place during a time interval $[t_1, t_2]$, while f is constant in $\mathcal{F}$. The general definition of work as the integral of power yields

$$\int_{t_1}^{t_2} \langle \dot{u}(t), f \rangle \, dt = \int_{t_1}^{t_2} \frac{d}{dt} \langle u(t), f \rangle \, dt =$$
$$= \langle u(t_2), f \rangle - \langle u(t_1), f \rangle = \langle \delta u, f \rangle .$$

4. b PSEUDO - POTENTIALS

Let us agree to call a resistance law a relation, denote it by

$\mathcal{R}$, formulated between the possible velocity $v \in \mathcal{V}$ of the considered system in the considered configuration and one, say $f \in \mathcal{F}$, of the forces it experiences at the same instant. Such a law arises from the study of some of the physical processes in which the system takes part.

It will be said that the law $\mathcal{R}$ is dissipative if the following implication holds

(4.1) $$v \mathcal{R} f \quad \Rightarrow \langle v,f\rangle \leqslant 0 \quad ,$$

which makes it a resistance law in the usual sense.

It will be said that $\mathcal{R}$ admits a function $\phi \in \Gamma_o(\mathcal{V}, \mathcal{F})$ as pseudo-potential if the relation $\mathcal{R}$ is equivalent to

(4.2) $$- f \in \partial \phi (v) \quad .$$

Recall that any subdifferential relation is monotone ; then a law $\mathcal{R}$ of the form (4.2) ensures the implication :

(4.3) $$v \mathcal{R} f \ , \ v' \mathcal{R} f' \ \Rightarrow \langle v-v' \ , \ f-f'\rangle \leqslant 0 \quad .$$

Make in addition the frequently verified hypothesis that zero is among the values that the relation $\mathcal{R}$ permits to f when v is zero, i.e.

(4.4) $$0 \in \partial \phi (0) \quad .$$

Then (4.1) ensues from (4.3) : the corresponding resistance law is dissipative. Observe that (4.4) implies that $\phi(0)$ is finite and constitutes the minimal value of ϕ ; since adding a finite constant to ϕ does not affect the subdifferential, there is no loss of generality in supposing here

(4.5) $$\phi(0) = 0 \quad ;$$

then the function ϕ takes only nonnegative values.

In the following, we shall refer to the situation characterized by (4.2), (4.4), (4.5) by saying that the pseudo-potential ϕ is the resistance function of the considered law.

Recall that, a priori, the pair of linear spaces $\mathcal{V}$, $\mathcal{F}$ is relative to a definite configuration of the system, so that the foregoing concerns only this configuration. However in the usual linear case of Chapter 3, by making $\mathcal{V} = \mathcal{U}$ and considering the single force space $\mathcal{F}$, it will be possible to formulate resistance laws independently of configurations.

REMARK. The example developed in § 3. i, 3. j makes understand also that the pattern of the present Chapter may usually be applied to a definite mechanical situation in several different ways.

A similar example is t! of a continuous medium, occupying in the considered configuration a region Ω of the physical space. A first possibility is to interpret as v the vector field defined on Ω by the velocities of the various particles forming the medium : then the linear space $\mathcal{V}$ will consist of vector fields satisfying some assumptions of integrability, derivability, etc... But in some theories it will be more convenient to consider v as the strain rate tensor field of the medium. Or else, as in § 3. j, one may take for $\mathcal{V}$ a "big space" whose generic element is the couple of a velocity vector field and of a tensor field presumed to be the strain rate field ; then the geometric compatibility between velocity field and strain rate field will be seen as a constraint. To these various standpoints correspond natural choices for the elements f forming the space $\mathcal{F}$: rates of distributed forces, stress tensor fields, etc...

The same pattern will also be applied to formulate <u>local</u> laws : a point of the continuous medium being specified, one considers as $\mathcal{V}$ the linear space of dimension 6 whose elements are the possible values of the local strain rate tensor $\dot{\varepsilon}$ of the medium ; the associated $\mathcal{F}$ is the linear space formed by the possible values of the local stress tensor σ ; the bilinear form which places these two spaces in duality is the classical expression of the density of internal power. A local law, i.e. a relation between the strain rate tensor and the stress tensor at the considered point of the medium, will be formulated by means of a local pseudo-potential, which is a numerical function defined on $\mathcal{V}$. This being done for each point of the medium, it generates a behavioral law of the medium as a whole, i.e. a relation between elements of two function spaces whose generic elements are the strain rate tensor field and the stress tensor field. Under suitable integrability assumptions, these two function spaces are placed in separating duality by the bilinear form defined as the integral of the density of internal power. This permits the description of the considered behavioral law by means of a superpotential which is an integral convex functional. The reader will refer to B. Nayroles's lecture for more details about this mechanical situation and to C. Castaing's lecture for more details about the functional analytic aspect. The basic mathematical material may be found in R. T. ROCKAFELLAR [1], [3], [4].

4. c VISCOUS RESISTANCE

As a first example consider a relation $\mathcal{R}$ of the form

(4.6) $$- f = L v$$

where L denotes a linear mapping from $\mathcal{V}$ into $\mathcal{F}$. In all the phenomena classified as <u>viscosity effects</u> it is always admitted that L is self-adjoint (or "symmetric") with regard to the duality $\langle .,. \rangle$, i.e., for any v and v' in $\mathcal{V}$:

$$\langle v, L v' \rangle = \langle v', L v \rangle \ .$$

From this, one easily deduces that $L v$ is the weak gradient at the point v of the <u>quadratic form</u> ϕ defined on $\mathcal{V}$ by

$$\phi (v) = \frac{1}{2} \langle v, L v\rangle$$

This quadratic form is usually called the Rayleigh function of the considered viscosity law.

Making the additional assumption that the viscosity law is dissipative yields that this quadratic form is nonnegative, thus convex. And at any point v the weak gradient $L v$ constitutes the whole of the subdifferential $\partial \phi (v)$. This means that in the present case, the relation (4.6) may equivalently be written as

$$- f \in \partial \phi (v) .$$

Thus ϕ is pseudo-potential and, more precisely, resistance function of the considered law.

The power of the force f associated with v in this way is

$$\langle v,f\rangle = - \langle v, L v\rangle = - 2 \phi (v) \quad ;$$

the negative of it is frequently called the dissipated power corresponding to v ; hence the name of dissipation function which is given in the present case to the quadratic form $v \mapsto 2 \phi (v)$.

REMARK. Gyroscopic forces give an example of a law of the form (4.6) with a linear mapping L which is not self-adjoint ; on the contrary

$$\langle v, L v'\rangle = - \langle v', L v\rangle \ .$$

Such a law admits no pseudo-potential unless L is the zero mapping ; the dissipated power is essentially zero, so that (4.1) is satisfied :

this law may be said dissipative.

4. d VELOCITY CONSTRAINT

Take back the framework of § 3. e, i.e. the example of the firm perfect constraint whose geometric condition is $u \in \mathcal{L}$, with $\mathcal{L} = U + a$, a possibly moving affine manifold. The linear subspace U is supposed independent of time thus also V which is the subspace of $\mathcal{F}$ orthogonal to U. This geometric condition may equivalently by written, for every t,

$$\forall w \in V \quad : \quad \langle u - a, w \rangle = 0 \; .$$

Supposing that the known function $t \mapsto a$ possesses a weak derivative $\dot{a}$, this yields, by choosing w independent of t, that the velocity $v = \dot{u}$ satisfies

$$\forall w \in V \quad : \quad \langle \dot{u} - \dot{a}, w \rangle = 0$$

i.e.

(4.7) $$v \in U + \dot{a} \; .$$

Recall on the other hand that, by the definition of a firm perfect constraint, the reaction $r \in \mathcal{F}$ exerted on the system by the enforcing device may be an arbitrary element of V. Exactly like in § 3. b, this fact may be expressed jointly with (4.7) by writing :

(4.8) $$- r \in \partial \, \psi_{\dot{\mathcal{L}}} (v)$$

where $\dot{\mathcal{L}}$ denotes the affine manifold $U + \dot{a}$.

<u>This constitutes a resistance law admitting the function</u> $\psi_{\dot{\mathcal{L}}}$ <u>as pseudo-potential</u>. Let us call it a <u>velocity constraint</u>.

It is no place to explain how, in the general setting of a configuration-depending pair of spaces $(\mathcal{V}, \mathcal{F})$, the usual differentiability assumptions let any firm perfect bilateral smooth constraint be expressed under the form (4.8). This form includes more generally the relations between reaction and velocity classically known as <u>non-holonomic perfect constraints</u> ; the standard example of it consists in the <u>perfect rolling without sliding</u> of solid bodies, actually an extreme case of friction.

4. e FRICTION AND PLASTICITY

Suppose given a weakly closed non empty convex subset C of $\mathcal{F}$. Let us formulate a relation $\mathcal{R}$ between v and f by the <u>principle of maximal dissipation</u> namely : the values of $f \in \mathcal{F}$ which this relation associates with a given $v \in \mathcal{V}$ are the elements of C which minimize the power, i.e. minimize the function $\langle v,.\rangle$. In other words $v \mathcal{R} f$ means

$$\begin{cases} f \in C \\ \forall f' \in C \quad : \quad \langle v, f'\rangle \geqslant \langle v, f\rangle \end{cases}$$

which is immediately found equivalent to

$$\forall f' \in \mathcal{F}: \quad -\langle v, f'-f\rangle + \psi_C(f) \leqslant \psi_C(f')$$

i.e.

$$(4.9) \qquad - v \in \partial \psi_C (f)$$

which in turn is equivalent to

$$(4.10) \qquad f \in \partial \psi^*_C (- v)$$

(cf. § 2. e) and also to

$$(4.11) \qquad \psi^*_C (- v) + \psi_C (f) + \langle v, f\rangle = 0 .$$

Denote by ϕ the function $v \mapsto \psi^*_C (- v)$, i.e.

$$\phi(v) = \sup_{f \in C} \langle -v, f\rangle = \sup_{g \in -C} \langle v, g\rangle \quad ;$$

it is the <u>support function of the set</u>. $-C$.

Then (4.10) is transcribed as

$$- f \in \partial \phi (v) \quad ;$$

this means that <u>the considered resistance law admits</u> ϕ <u>as pseudo-potential</u> (or <u>resistance function</u> in the usual case where C contains the origin of $\mathcal{F}$; such is the condition for the present law to be <u>dissipative</u>).

Relation (4.11) may equivalently be written as

$$(4.12) \qquad \begin{cases} f \in C \\ -\langle v, f\rangle = \phi (v) \end{cases} \quad ;$$

in other words the values of f that the considered relation associates with a given v are those elements of C for which the dissipated power $-\langle v, f\rangle$ equals exactly $\phi(v)$.

The reader will check that all the preceding pattern applies to Coulomb's law of friction between two solid bodies $\mathcal{S}_1$ and $\mathcal{S}_2$, when the pressure N, i.e. the normal component of the reaction, is treated as known. Take as v the sliding velocity of $\mathcal{S}_2$ with respect to $\mathcal{S}_1$; then $\mathcal{V}$ is the linear space of dimension 2 consisting of the vectors whose direction is contained in the common tangent plane to the two bodies at the point of contact (this space is not exactly the velocity space for the considered system as a whole, but it is visibly isomorphic to a subspace of it). Take as f the tangential component of the reaction that $\mathcal{S}_2$ undergoes from $\mathcal{S}_1$ so that $\mathcal{F}$ may be considered as the same space as $\mathcal{V}$, the bilinear form $\langle .,.\rangle$ reducing then to the conventional Euclidian scalar product. The customary Coulomb law of isotropic friction consists in taking as C the closed disk centered at the origin, with radius equal to the product of N by the friction coefficient. But anisotropic friction may be described as well, by using convex sets of different shape. See MOREAU [12] about the application of this to discuss the sliding of a vehicle wheel when brake is applied : if the inertia of the wheel is neglected, the resulting effect comes to be equivalent to some anisotropic friction which would take place directly between the vehicle and the ground.

However, the main domain of application of the preceding is plasticity. In its local form the classical law of perfect plasticity (i.e. without strain hardening) is formulated as a relation between the local values of the stress tensor σ and of the plastic strain rate $\dot{\varepsilon}_p$. Giving the yield locus defines a closed convex set C in the six-dimensional space of the variable σ ; among various equivalent formulations, the considered law may be stated as a principle of maximal dissipation which was precisely the starting point of this paragraph. From the local law one obtains the global one by the functional analytic procedure described at the end of § 4. b.

In the study of plasticity as well as in that of friction, an essential feature is the occurence of a relation between the velocity v and the force f which cannot be "solved" to define one of these two elements as a function of the other : to the value zero of v correspond for f all the points of C and to a value of f corresponds as values of v all the elements of the cone $-\partial \psi_C (f)$. This causes much trouble in traditional treatments ; our purpose in Chapter 6, will be to show that such formulations as (4.9), (4.10) or (4.11) permit a very efficient handling in this situation.

4. f DISSIPATION FUNCTION

The relation $\mathcal{R}$ between v and f may equivalently be written under the form

$$f \in R\,(v)$$

where R denotes a multimapping from $\mathcal{V}$ into $\mathcal{F}$. Given v in $\mathcal{F}$, there is a priori no reason for all the values of f in the set R (v) to yield the same value for the dissipated power $-\langle v, f\rangle$. However this precisely happens in many practical instances : in such cases, the dissipated power appears as a single-valued numerical function of the variable v, defined on dom $R = \{v \in \mathcal{V} : R\,(v) \neq \emptyset\}$. Let us denote by D this function, usually called the dissipation function of the considered law.

In the case of viscous resistance presented in § 4. c, the set R (v) reduces to a single element for each v in $\mathcal{V}$; hence the existence of a dissipation function is trivial. In fact we found

$$D\,(v) = 2\,\phi\,(v) \quad .$$

In the case of friction or plasticity presented in § 4. e, (4.12) proves the existence of a dissipation function expressed now, for every v in dom $\partial\,\phi$, as

$$D\,(v) = \phi\,(v) \quad .$$

Both preceding examples exhibit a close connection between the superpotential, or resistance function, ϕ and the dissipation function D. Actually in both cases, the resistance function ϕ happens to be <u>positively homogeneous</u>, with degree m ; this implies $-\langle v, f\rangle = m\,\phi(v)$, which may be considered as a generalization of Euler's identity to "subdifferential calculus". Many practical resistance functions possess such a homogeneity (e.g. usual laws of <u>creep</u>). More generally :

PROPOSITION. <u>Let ϕ be a resistance function (i.e. ϕ is the pseudo-potential of a resistance law, with $0 \in \partial\phi(0)$ and $\phi(0) = 0$) ; suppose $\partial\phi(v) \neq \emptyset$ whichever is v in $\mathcal{V}$. For the existence of a function $h : R \to R$ ensuring the implication</u>

$$-f \in \partial\phi(v) \Rightarrow -\langle v, f\rangle = h(\phi(v))$$

<u>(in other words, for the function $h \circ \phi$ to be dissipation function) it is necessary and sufficient that ϕ has the quasi-homogeneous form $\phi = \alpha \circ j$, where j is an everywhere subdifferentiable gauge function on $\mathcal{V}$ and α a convex differentiable mapping from $[0, +\infty[$ into itself, with $\alpha(0) = 0$.</u>

A sketched proof is given in MOREAU [13], and for more details [16]. It may be remarked that the function h is then strictly increasing. The dissipation function $D = h \circ \phi$ is not convex in general, but only <u>quasi-convex</u> i.e. its "slices" $\{v \in \mathcal{V} : D(v) \leqslant \rho\}$ for $\rho \in R$, are (closed) convex sets ; all these sets are homothetic of

$J = \{v \in \mathcal{V} : j(v) \leqslant 1\}$, the set whose j is the gauge.

By the facts indicated in § 2. h , the dual function of $\phi = \alpha \circ j$ is also a quasi-homogeneous function, namely $\phi^* = \beta \circ k$, where β is the Young conjugate of α and k the gauge function of the polar set K of J.

In the case of plasticity or friction the function α is identity , so that β is the indicator function of the subset $[0,1]$ of $[0, +\infty[$ and $K = -C$.

4. g SUPERPOSITION OF RESISTANCE LAWS

It is usual to take into account at the same time several resistance laws in the same pair $(\mathcal{V}, \mathcal{F})$ of linear spaces. Let ϕ_1 and ϕ_2 the respective pseudo-potentials of two such resistance laws. For every v in $\mathcal{V}$, the set of the possible values of the sum of the two forces is $\partial \phi_1 (v) + \partial \phi_2 (v)$. This is contained in $\partial (\phi_1 + \phi_2) (v)$ and, in particular, if the functions ϕ_1 and ϕ_2 possess the additivity of the subdifferentials, the conjunction of the two resistance laws amounts exactly to the single following one

$$(4.13) \qquad - f \in \partial (\phi_1 + \phi_2) (v) .$$

Suppose for instance $\phi_1(v) = \psi^*_C(- v)$, i.e. the resistance

denote the resistance function of some <u>viscosity</u> law (cf. § 4. c : it is a nonnegative l.s.c. quadratic form on the space $\mathcal{V}$) ; choose a strictly positive constant λ and take more generally

$$\phi_2(v) = \lambda\, q(v) = \frac{1}{\lambda}\, q(\lambda\, v) \quad ,$$

so that λ may be interpreted as a <u>viscosity coefficient</u>. As a condition ensuring the additivity of subdifferentials make, for instance, the following assumption (cf. § 2. f) : <u>the function</u> ϕ_1 <u>is continuous at the origin</u>, at least for the Mackey topology $\tau(\mathcal{V}, \mathcal{F})$; by § 2. c, 5°, this means the convex set C is <u>compact</u> for the weak topology $\sigma(\mathcal{F}, \mathcal{V})$. Then the resulting <u>viscoplastic</u> law may be expressed under the form (4.13).

Now the assumptions made imply, by § 2. d, that the polar function of $\phi_1 + \phi_2$ is the infimal convolute $\phi_1^* \nabla \phi_2^*$. As already mentioned in Chapter 3, the dual q^* of the quadratic form q consists in a positive definite quadratic form, defined on some subspace of $\mathcal{F}$, and extended with the value $+\infty$ outside of this subspace. By § 2.c , 2°, the dual of ϕ_2 is $\frac{1}{\lambda} q^*$. On the other hand, the dual ϕ_1^* of ϕ_1 is the indicator function of the set - C. Thus using the equivalence between (2.6) and (2.7) (§ 2. e) the viscoplastic resistance law (4.13) amounts to

$$v \in \partial\,(\psi_{-C} \nabla \frac{1}{\lambda} q^*)\,(-\,f) \quad ,$$

while the corresponding purely plastic resistance law would be written as

$$v \in \partial \psi_{-C}(-f) \quad .$$

By definition, for every $y \in \mathcal{F}$,

$$(\psi_{-C} \nabla \frac{1}{\lambda} q^*)(y) = \inf_{z \in \mathcal{F}} [\psi_{-C}(z) + \frac{1}{\lambda} q^*(y-z)]$$
$$= \inf_{z \in -C} \frac{1}{\lambda} q^*(y-z)$$

and, due to the assumed compactness of C, the infimum is a minimum. Clearly this expression takes the value 0 for $y \in -C$ and it takes strictly positive values otherwise ; it may be said that $\psi_{-C} \nabla \frac{1}{\lambda} q^*$ is a penalty function for the set - C and the penalty coefficient $\frac{1}{\lambda}$ is the reciprocal of the viscosity coefficient (other remarks about penalty functions will be given, for the special case of Hilbert space, in § 5. d).

Due to quadratic forms being even functions, one may equivalently speak of the set C instead of - C ; in short adding some viscosity effects to a plasticity law is equivalent to replacing the indicator function of the "rigidity set" C, by a penalty function of this set ; the smaller is the viscosity coefficient, the larger is the penalty coefficient.

J. J. Moreau

5. MOVING SETS

5. a HAUSDORFF DISTANCE AND VARIATION

Let $t \mapsto A(t)$ denote a multimapping or multifunction (i.e. a set-valued mapping) from the compact interval $[0,T]$ into a metric space (E,d). As in the following the real variable t will be interpreted as the time, we may refer to A as a moving set in E.

A natural way of formulating regularity assumptions about such a multimapping consists in using the Hausdorff distance between subsets of the metric space E.

If A and B are two subsets of E, we call the excess of A over B the expression

(5.1) $$e(A,B) = \sup_{a\in A} d(a,B) = \sup_{a\in A} \inf_{b\in B} d(a,b)$$

The considered sets may be empty ; let us agree that "sup" and "inf" above are understood in the sense of the ordered set $\bar{R}_+ = [0, +\infty]$: the supremum of an empty collection of elements of this ordered set is 0 and the infimum is $+\infty$. Expression (5.1) defines a non symmetric écart ; it satisfies the triangle inequality. Clearly $e(A,B) = 0$ if and only if A is contained in the closure $\bar{B}$ of B.

The Hausdorff (improper) distance of A and B is then defined as the symmetric expression

$$h(A,B) = \max \{e(A,B), e(B,A)\}$$

with value in $\bar{R}_+$. This is zero if and only if A and B have the same closure.

By means of Hausdorff distance, the classical concept of <u>variation</u> may be applied to moving sets. Let $[s,t]$ be a compact subinterval of $[0,T]$; for any finite subdivision of this interval, namely

$$S \quad : \quad s = \tau_0 \leqslant \tau_1 \leqslant \ldots \leqslant \tau_n = t$$

put

$$V(S) = \sum_{i=1}^{n} h(A(\tau_{i-1}), A(\tau_i)) \quad \in \bar{R}_+ \quad .$$

The supremum of $V(S)$ for S ranging over all the finite subdivisions of $[s,t]$ is called the variation of A on this interval ; notation var (A ; s,t). From h satisfying the triangle inequality one easily deduces that

(5.2) $\quad s \leqslant t \leqslant u \Rightarrow \mathrm{var}(A \; ; \; s,u) = \mathrm{var}(A \; ; \; s,t) + \mathrm{var}(A \; ; \; t,u)$.

In particular if $\mathrm{var}(A \; ; \; 0,T) < +\infty$, the variation is also finite on any subinterval of $[0,T]$; in this case, introducing the non decreasing function v from $[0,T]$ into R_+

(5.3) $\quad v(t) = \mathrm{var}(A \; ; \; 0,t)$

yields

(5.4) $\quad s \leqslant t \Rightarrow \mathrm{var}(A \; ; \; s,t) = v(t) - v(s)$.

The numerical function v is Lipschitz with ratio λ if and only if the multimapping A satisfies itself the Lipschitz condition, with ratio λ, i.e., for any s and t in $[0,T]$,

$$h(A(s), A(t)) \leqslant \lambda \; |t-s|$$

The numerical function v is absolutely continuous on $[0,T]$ if and only if the multimapping A possesses itself the absolute continuity, as formulated by means of Hausdorff distance, i.e. : for any $\varepsilon > 0$, there exists $\eta > 0$ such that the implication

$$\sum_i |\tau_i - \sigma_i| < \eta \Rightarrow \sum_i h(A(\sigma_i), A(t_i)) < \varepsilon$$

holds for any finite family $]\sigma_i,\tau_i[$ of non overlapping subintervals of $[0,T]$. In this case the numerical function v is almost everywhere differentiable ; the derivative, denoted by $\dot{v}$, is a nonnegative element of $L^1(0,T ; R)$ which may be called the <u>speed function of the moving set</u> A. Clearly

$$s \leqslant t \Rightarrow h(A(s), A(t)) \leqslant \int_s^t \dot{v}(\tau)\, d\tau \quad . \tag{5.5}$$

Let us restrict ourselves now to the case where, for any t, the set $A(t)$ is closed ; then the non decreasing function v is constant over some subinterval of $[0,T]$ if and only if the multimapping A is also constant over this subinterval. This implies the existence of a multimapping $\mathcal{A}$ from $[0,v(T)]$ into E yielding the <u>factorization</u>

$$A(t) = \mathcal{A}(v(t)) \quad .$$

Evidently, for $\sigma \leqslant \tau$ in $[0,v(T)]$, one has

$$\mathrm{var}(\mathcal{A} ; s,\tau) = \tau - \sigma$$

so that $\mathcal{A}$ is Lipschitz with ratio 1.

5. b THE CASE OF CONVEX SETS IN A NORMED SPACE

Let E denote a real normed linear space and F its

topological dual endowed with the usual norm. This constitutes a dual pair as considered in Chapter 2 (keep in mind that the norm topology on E is compatible with the duality, but not the norm topology on F unless E is a reflexive Banach space).

Let C and C' be two <u>non empty</u> convex subsets of E ; as we are to deal with distances, it is immaterial to suppose these sets closed or not. Let γ and γ' be the respective support functions of C and C' which are positively homogeneous elements of $\Gamma_o(F,E)$, vanishing at the origin of F.

Denoting by B the closed unit ball of F , one finds

$$e(C,C') = \sup_{y \in B} (\gamma(y) - \gamma'(y)) \tag{5.6}$$

(with the convention $\infty - \infty = -\infty$).

This is easily proved by observing that, for $\rho \in R$, the inequality $\rho \geqslant e(C,C')$ means that, if $\beta(\rho)$ denotes the closed ball centered at the origin with radius ρ , the set $\overline{C'} + \beta(\rho)$ contains C ; express then this inclusion in terms of support functions. Another way of proof would start from the following formula giving the distance of a point a of E to the set C'

$$d(a,C') = \sup_{y \in B} [\langle a,y\rangle - \gamma'(y)] \quad . \tag{5.7}$$

In fact (cf. § 2.)

$$d(a,C') = (\psi_{C'} \nabla |.|)(a) \quad .$$

Since the function $|.|$ is everywhere finite and continuous, since there exists at least one point where $\psi_{C'}$ takes a finite value (namely the value zero), and since both functions are convex, the inf-convolute $\psi_{C'} \nabla |.|$ is convex, everywhere finite and continuous (cf. § 2) thus it equals its bipolar, i.e.

$$(\psi_{C'} \nabla |.|)(a) = \sup_{y \in F} [\langle a,y\rangle - \gamma'(y) - \psi_B(y)]$$
$$= \sup_{y \in B} [\langle a,y\rangle - \gamma'(y)] \quad .$$

which is equality (5.7).

Equality (5.6) implies that the Hausdorff distance between the non empty convex sets C and C' is finite only if dom γ and dom γ' (i.e. the sets of the points of F where γ and γ' take finite values) consist in the same set denoted by D and then

$$(5.8) \qquad h(C,C') = \sup_{y \in B \cap D} |\gamma(y) - \gamma'(y)| \quad .$$

Note that D is a conic convex subset of F ; its polar cone in E is the recession cone of C and C'. Recall that D equals the whole of F if and only if C and C' are bounded.

The expression (5.8) of the Hausdorff distance yields the following :

Let $t \mapsto C(t)$ be a multimapping from $[0,T]$ into the normed space E, with non empty convex values ; denote by $y \mapsto \gamma(t,y)$ the support function of C(t). This multimapping is absolutely continuous (resp. Lipschitz with ratio λ) if and only if the set $D = \text{dom}\,\gamma(t,.)$ is independant of t, with the existence of a finite non decreasing numerical function $\rho : [0,T] \to \mathbb{R}$, absolutely continuous (resp. Lipschitz with ratio λ), such that for any $y \in D$ and any subinterval $[s,t]$ of $[0,T]$ one has

$$|\gamma(t,y) - \gamma(s,y)| \leqslant |y| \, (\rho(t) - \rho(s))$$

($|.|$ denotes here the norm in F).

Equivalently there exists $\dot{\rho}$, a nonnegative element of $L^1(0,T\,;\mathbb{R})$ such that for any y in D, the numerical function $t \mapsto \gamma(t,y)$ is absolutely continuous and its derivative $\dot{\gamma}$ satisfies for almost every t

$$|\dot{\gamma}(t,y)| \leqslant |y| \, \dot{\rho}(t) \tag{5.9}$$

(resp. the same inequality with $\dot{\rho} = \lambda$). If such is the case one may take as $\dot{\rho}$ the speed function of the moving set C.

Characterizing the regularity of the motion of a (closed) convex set $t \mapsto C(t)$ by means of its support function $\gamma(t,.)$ is quite a natural procedure. In fact an essential feature in locally convex topological linear spaces is that a closed convex set equals the intersection of all the closed half-spaces containing it, or equivalently the intersection of the minimal ones among these half-spaces, i.e. the half-spaces which have in the present case the form $\{x \in E : \langle x,y\rangle \leqslant \gamma(t,y)\}$, with $|y| = 1$. Fixing here y yields a moving half space whose boundary hyperplane keeps a constant direction ; the derivative $\dot{\gamma}(t, y)$ may be interpreted as the speed of this moving hyperplane, or as the speed of the moving half-space itself. Then (5.9) expresses a uniform majoration of the speeds for the minimal half-spaces of all directions.

Example. Take as $C(t)$ a convex set moving by translation, i.e.

$$C(t) = C_o + w(t)$$

where C_o denotes a fixed convex set and w a fonction defined on $[0,T]$ with values in E. Then, if γ_o is the support function of C_o

$$\gamma(t,y) = \gamma_o(y) + \langle w(t),y \rangle \quad .$$

One conclues that the multimapping is absolutely continuous if (and only if, in the case where C_o is bounded) the function $t \mapsto w(t)$ is absolutely continuous. When E is a reflexive Banach space, the absolute continuity of w is known to imply for almost every t the existence of the strong derivative $\dot{w}$ (cf. KOMURA [1]) and this yields for the speed $\dot{v}$ of C the majoration

(5.10) $$\dot{v} \leqslant |\dot{w}|$$

(equality when C_o is bounded).

5. c INTERSECTION OF TWO MOVING CONVEX SETS

The practical use of the preceding concepts requires some criteria of absolute continuity for multimappings. The object of this paragraph is to establish the following one (already published in MOREAU [22] or, for more details, [19]) :

PROPOSITION. Let $t \mapsto A_t$ and $t \mapsto B_t$ denote two multimappings from the compact interval $[0,T]$ into the normed space E, with convex values.

Suppose that for any $t \in [0,T]$ *the set* A_t *has a nonempty intersection with the interior of* B_t *and that the diameter of* $A_t \cap B_t$ *is finite. Then if the two multimappings are absolutely continuous (resp. Lipschitz) such is also the multimapping* $t \mapsto A_t \cap B_t$.

We shall decompose the proof into several lemmas which may be of use by themselves.

LEMMA 1. *Let* B_1, B_2 *denote two convex subsets of the normed space* E *and* A_1, A_2 *two arbitrary subsets of* E ; *then* (e *denoting the "excess" as in* § 5. a)

$$e(A_1, E\setminus B_1) \leqslant e(A_2, E\setminus B_2) + e(A_1, A_2) + e(B_1, B_2) . \tag{5.11}$$

Let us prove first that for any $a \in E$

$$d(a,E\setminus B_1) \leqslant d(a,E\setminus B_2) + e(B_1,B_2) . \tag{5.12}$$

One makes calculation easier by performing a translation reducing to the case where a is the origin of E. Let g_1, g_2 be the support functions of B_1 and B_2 , defined on the dual F of E. Let ρ be an arbitrary positive number satisfying the inequality $\rho \leqslant d(0,E\setminus B_1)$, which means that the open ball with center 0 and radius ρ is contained in B_1 ; in terms of support functions this inclusion is equivalent to $\rho \leqslant g_1(y)$ for any y belonging to Σ , the unit sphere of F. Now (5.6) implies

$$\forall\ y \in \Sigma\ :\ g_1(y) \leqslant g_2(y) + e(B_1,B_2) \ ;$$

therefore $\rho - e(B_1,B_2) \leqslant g_2(y)$; inequality (5.12) (trivial if $e(B_1,B_2) = +\infty$) follows. From it one obtains (5.11) by taking suprema for a ranging over A_1, then using the fact that the écart e satisfies the triangle inequality.

LEMMA 2. Let A and B denote two convex subsets of the normed space E ; suppose that B contains an open ball with radius $\rho > 0$, with center a belonging to A. Then

$$(5.13) \qquad \forall\, x \in E \ : \ d(x, A \cap B) \leqslant \left(1 + \frac{|x-a|}{\rho}\right)(d(x,A) + d(x,B)) .$$

Proof : Denote indifferently by $|.|$ the norm in E or the dual norm in F ; let f and g be the support functions of A and B. Similarly to (5.7) we have

$$d(x,A) = \sup \{\langle x,u\rangle - f(u) : u \in F \ , \ |u| \leqslant 1\}$$

and the corresponding expression for $d(x,B)$. Define a positively homogeneous function ϕ on $F \times F$ by

$$\phi(u,v) = \langle x, u + v\rangle - f(u) - g(v) \ .$$

For an arbitrarily chosen constant $k > 0$ this yields

$$(5.14) \qquad k(d(x,A) + d(x,B)) = \sup \{\phi(u,v) : |u| \leqslant k \ , \ |v| \leqslant k\} \ .$$

The hypotheses in the Lemma to be proved imply, by elementary arguments, that the closure $\overline{A \cap B}$ of $A \cap B$ equals the intersection of the closures $\bar{A}$ and $\bar{B}$ of A and B. Then, the support function of $A \cap B$ is the dual function of $\psi_{\bar{A}} + \psi_{\bar{B}}$, i.e. the Γ-hull of $f \nabla g$; by the facts summarized in § 2. d, this Γ-hull is the function $f \nabla g$ itself, i.e.

$$(f \nabla g)(w) = \inf \{f(u) + g(v) \ : \ u + v = w\} \ .$$

Using again the expression (5.7) for the distance from a point to a convex set, this yields

$$(5.15) \qquad d(x, A \cap B) = \sup \{\langle x,w\rangle - (f \nabla g)(w) : |w| \leqslant 1\}$$

$$= \sup \{\phi(u,v) \ : \ |u + v| \leqslant 1\} \ .$$

Let us make calculation easier by supposing that a translation has been performed in E such that $a = 0$; then the hypotheses made about A and B are expressed by $f \geqslant 0$ and $g \geqslant \rho\ |.|$, hence

$$|u + v| \leqslant 1 \ \Rightarrow \ \phi(u,v) \leqslant |x| - \rho\ |v| \ .$$

As $\phi(0,0) = 0$ and in view of (5.15) this implies

$$d(x,A \cap B) \leqslant \sup \{\phi(u,v) : |v| \leqslant \frac{|x|}{\rho} \ , \ |u| \leqslant 1 + \frac{|x|}{\rho}\} \ .$$

After putting $k = 1 + \frac{|x|}{\rho}$ in (5.14), the comparison of the sets over which the suprema are taken yields (5.13).

REMARK. In the case where E is a Hilbert space one may use trigonometry to establish a slightly better inequality ; see MOREAU [19] .

LEMMA 3. <u>Let</u> A <u>and</u> B <u>denote two convex subsets of</u> E ; <u>take</u> α <u>and</u> ρ <u>in</u> $]0, +\infty[$ <u>such that</u> $\alpha < \rho < e(A, E\ B)$. <u>Then, for any</u> x <u>in</u> E <u>such that</u> $d(x,A) + d(x,B) \leqslant \alpha$, <u>one has</u>

$$d(x,A \cap B) \leqslant \frac{\rho + \operatorname{diam}(A \cap B)}{\rho - \alpha} \ (d(x,A) + d(x,B)) \ .$$

This results from (5.13) and from the inequality

$$|x - a| \leqslant \operatorname{diam}(A \cap B) + d(x,A \cap B) \ .$$

Bringing together these lemmas one obtains easily :

LEMMA 4. <u>Let</u> T <u>denote a topological space</u> ; <u>let</u> $t \mapsto A_t$ <u>and</u> $t \mapsto B_t$

be two multimappings from T *into the normed space* E, *with convex values. Let* $s \in T$ *such that*

$$\operatorname{diam}(A_s \cap B_s) < +\infty \quad ,$$

$$A_s \cap \operatorname{int} B_s \neq \emptyset \quad ,$$

$$\lim_{t \to s} e(A_t, A_s) = 0 \qquad (\text{resp. } \lim_{t \to s} e(A_s, A_t) = 0) \quad ,$$

$$\lim_{t \to s} e(B_t, B_s) = 0 \qquad (\text{resp. } \lim_{t \to s} e(B_s, B_t) = 0) \quad .$$

Then

$$\lim_{t \to s} e(A_t \cap B_t, A_s \cap B_s) = 0 \quad (\text{resp. } \lim_{t \to s} e(A_s \cap B_s, A_t \cap B_t) = 0)$$

and the two numerical functions $t \mapsto \operatorname{diam}(A_t \cap B_t)$ *and* $t \mapsto e(A_t, E \setminus B_t)$ *are upper semicontinuous (resp. lower semicontinuous) at the point* s.

Let us now complete the proof of the Proposition :

The hypotheses imply that the two multimappings $t \mapsto A_t$ and $t \mapsto B_t$ are continuous in the sense of Hausdorff distance. The finite numerical function $t \mapsto \operatorname{diam}(A_t \cap B_t)$ is continuous by Lemma 4 on the compact interval $[0,T]$, thus majorized by some constant $R < +\infty$. By the same lemma the numerical function $t \mapsto e(A_t, E \setminus B_t)$ is continuous on $[0,T]$, with strictly positive values since $A_t \cap \operatorname{int} B_t \neq \emptyset$, thus minorized by some constant $\rho > 0$. Choose $\alpha \in]0,\rho[$; the functions $t \mapsto \operatorname{var}(A\, ;\, 0,t)$ and $t \mapsto \operatorname{var}(B\, ;\, 0,t)$ being finite and continuous, there exists $\delta > 0$ such that for σ and τ in $[0,T]$, the condition $|\sigma - \tau| < \delta$ ensures that $h(A_\sigma, A_\tau)$ and $h(B_\sigma, B_\tau)$ are less than $\frac{\alpha}{2}$. Then Lemma 3 implies

$$h(A_\sigma \cap B_\sigma, A_\tau \cap B_\tau) \leqslant \frac{\rho + R}{\rho - \alpha} (h(A_\sigma, A_\tau) + h(B_\sigma, B_\tau))$$

which yields the expected majorations.

5. d DISTANCE AND PENALTY FUNCTION IN A HILBERT SPACE

Let H be a real Hilbert space ; denote by $(.|.)$ the scalar product in it and by $|.|$ the norm. By means of this scalar product, H may be identified by its dual ; in other words $(.|.)$ is a bilinear form on $H \times H$ which places H in duality with itself and the norm-topology is compatible with this duality.

Easy computation yields that the function

$$Q \ : \ x \mapsto \frac{1}{2} |x|^2$$

which clearly belongs to $\Gamma_o(H,H)$ equals its own dual (actually it can be proved that Q is the only fonction equal to its dual).

Let C be a non empty closed convex subset of H ; denote by q the numerical function defined on H by

$$q(x) = \frac{1}{2} [d(x,C)]^2 = (\psi_C \nabla Q)(x) .$$

Elementarily this function is convex, everywhere finite, continuous, Fréchet-differentiable with gradient

(5.16) $$\text{grad } q(x) = x - \text{proj}_C \, x \ ,$$

where $\text{proj}_C \, x$ denotes the nearest point to x in C. (All this is a special case of a theory in which the indicator function ψ_C is replaced by an arbitrary element of $\Gamma_o(H,H)$; see MOREAU [6].)

Choose a strictly positive constant λ ; then $x \mapsto \frac{1}{\lambda} q(x)$ defines what is commonly called a penalty function of the set C, i.e. a finite function which takes the value 0 when $x \in C$ and rapidly growing positive values when the distance from x to C increases. So to speak, the smaller is the constant λ , the greater is the penalty for x of lying at a distance from C. The penalty function may be considered as an approximation of ψ_C in a sense which concerns also the subdifferentials as follows : Denote by A the multimapping $x \mapsto \partial \psi_C(x)$ from H into itself, which constitutes a special case of maximal monotone operator. In general, for a chosen $\lambda > 0$, the single valued, everywhere defined mapping

$$A_\lambda = \frac{I - (I + \lambda A)^{-1}}{\lambda} , \tag{5.17}$$

where I denotes identity, is classically called a Yosida approximation, or Yosida regularization, of A ; it is Lipschitz with ratio $\frac{1}{\lambda}$. Here A_λ may easily be explicited ; by definition the equality $y = (I + \lambda A)^{-1}(x)$ means $x \in (I + \lambda A)(y)$ or equivalently $x-y \in \partial \psi(y)$ for $\partial \psi(y)$ is a cone so that the factor λ may be omitted. This is well known to characterize y as equal to $\text{proj}_C\, x$; hence (5.17) becomes

$$A_\lambda (x) = \frac{1}{\lambda} (x - \text{proj}_C\, x) = \frac{1}{\lambda} \text{grad}\, q(x) \quad . \tag{5.18}$$

5. e MOVING CONVEX SET IN A HILBERT SPACE

With the same notations as in the preceding paragraph, suppose $t \mapsto C(t)$ is an absolutely continuous multimapping from $[0,T]$ into H, with non empty closed convex values ; put

$$q(t,x) = \frac{1}{2}[d(x, C(t))]^2 \quad .$$

Let $t \mapsto z(t)$ be an absolutely continuous mapping from $[0,T]$ into H.

Classically the continuity of $t \mapsto C(t)$ in the sense of Hausdorff distance and the continuity of $t \mapsto z(t)$ imply the continuity of the mapping

$$t \mapsto \text{proj}\,(z(t), C(t)) \quad .$$

The proof of it is based on some majoration of the square of the displacement of the projection which implies nothing about the absolute continuity of this mapping ; however :

LEMMA 1. <u>If</u> $t \mapsto C(t)$ <u>and</u> $t \mapsto z(t)$ <u>are absolutely continuous, on</u> $[0,T]$ <u>so is the numerical function</u> $k : t \mapsto d(z(t), C(t))$.

In fact, with the notation e of § 5. a, one has

$$d(z,C) = e(\{z\}, C)$$

so that, using the triangle inequality concerning the écart e, one obtains finally, for arbitrary σ and τ in $[0,T]$,

$$(5.19) \qquad |d(z(\sigma), C(\sigma)) - d(z(\tau), C(\tau))| \\ \leqslant d(z(\sigma), z(\tau)) + h(C(\sigma), C(\tau))$$

It just remains to apply the definition of absolute continuity.

This lemma implies that the function k possesses for almost every t a derivative denoted by $\dot{k}(t)$; thus the function

$$t \mapsto \frac{1}{2}(k(t))^2 = q(t,\ z(t))$$

possesses, for the same values of t, a derivative equal to $k(t)\,\dot{k}(t)$.

The absolute continuity of the multimapping C means that its <u>variation function</u> $v : t \mapsto \mathrm{var}\,(C\ ;\ 0,t)$ is absolutely continuous, thus possesses a derivative $\dot{v}(t)$ for almost every t. Similarly the absolute continuity of the vector function $t \mapsto z(t)$ implies the existence of its strong derivative $\dot{z}(t)$ for almost every t (by virtue of H being a reflexive Banach space ; see KOMURA [1]).

Let us prove now the following, which will be of use in next paragraph :

LEMMA 2. <u>For any</u> t <u>in</u> $[0,T]$ <u>such that the derivatives</u> $\dot{z}(t)$, $\dot{v}(t)$, $\dot{k}(t)$ <u>exist</u>, <u>one has</u>

(5.20) $$|k(t)\,\dot{k}(t) - (\dot{z}(t)|\ \mathrm{grad}\ q(t,\ z(t))| \leqslant k(t)\,\dot{v}(t)\ .$$

In fact for such a value of t the hypotheses imply the existence of

$$\lim_{s \to t} \frac{q(s,\ z(s)) - q(t,\ z(t))}{s - t} = k(t)\,\dot{k}(t)\ .$$

Now

(5.21) $$\frac{q(s,z(s)) - q(t,z(t))}{s - t} = \frac{q(s,z(s)) - q(s,z(t))}{s - t} + \\ + \frac{q(s,z(t)) - q(t,z(t))}{s - t}\ .$$

As the numerical function $x \mapsto q(s,x)$ is convex on H, its gradient at some point is also a subgradient ; this yields

$$(z(s) - z(t) | \operatorname{grad} q(s,z(t))) \leqslant q(s,z(s)) - q(s,z(t))$$
$$\leqslant (z(s) - z(t) | \operatorname{grad} q(s,z(s))) .$$

The mapping $s \mapsto \operatorname{proj}(x, C(s))$ is continuous, the mapping $x \mapsto \operatorname{proj}(x, C(s))$ is nonexpanding, thus the mapping

$$(s,x) \mapsto \operatorname{grad} q(s,x) = x - \operatorname{proj}(x, C(s))$$

from $[0,T] \times H$ into H is continuous ; hence one obtains the existence of

$$\lim_{s \to t} \frac{q(s,z(s)) - q(s,z(t))}{s - t} = (\dot{z}(t) \mid \operatorname{grad} q(t, z(t))) .$$

Therefore the last term in (5.21) possesses also a limit which may be interpreted as the derivative at the point t for the function

(5.22) $$s \mapsto q(s, z(t)) = \frac{1}{2}[d(z(t), C(s)]^2 .$$

Writing the same inequality as in (5.19), but with constant z , yields

$$|d(z(t), C(s)) - d(z(t), C(t))| \leqslant h(C(s), C(t))$$
$$\leqslant |v(s) - v(t)|$$

so that the derivative of the function (5.22) has its absolute value majorized by $k(t)\,\dot{v}(t)$; this completes the proof of (5.20).

5. f THE SWEEPING PROCESS

Suppose given an absolutely continuous multimapping $t \mapsto C(t)$ from $[0,T]$ into the real Hilbert space H, with nonempty closed convex values ; denote by $x \mapsto \psi(t,x)$ the indicator function of $C(t)$.

We put the problem of finding an absolutely continuous (single valued) mapping $u : [0,T] \to H$ agreeing with some initial condition

(5.23) $$u(0) = a \quad , \text{ given in } C(0)$$

and whose derivative $\dot{u}$ satisfies for almost every t in $[0,T]$

(5.24) $$-\dot{u}(t) \in \partial\, \psi(t, u(t)) \quad .$$

Interpreting u as a moving point in H, we call it a solution of the sweeping process by the moving convex set C. The reason of this name lies in the following mechanical image of condition (5.24) :

As $\partial\,\psi(t,x)$ is empty when $x \notin C$, this condition implies $u(t) \in C(t)$ for almost every t, thus for every t, by virtue of our continuity assumptions. Suppose, to make things clearer, that the moving convex set C possesses a nonempty interior. As long as the point $u(t)$ lies in this interior, the subdifferential $\partial\,\psi(t,u(t))$, i.e. the cone of normal outward vectors at the point $u(t)$ of the convex set (cf. § 2. e) reduces to the single element 0 ; then (5.24) implies that the moving point u remains at rest. It is only when u is caught up with by the boundary of C that it may take a nonzero velocity, so as to go on belonging to C, and by (5.24) this velocity possesses an inward normal direction with regard to C. In other words, condition (5.24) governs the quasistatic evolution of a material point u subject to the following mechanical actions :

1° some resistance acting along the line of its velocity and opposite in direction ;

2° the moving <u>perfect constraint</u> whose geometric condition is $u \in C(t)$ (cf. § 3. d).

Elementarily the initial value problem formulated above possesses at most one solution. Such uniqueness property holds more generally with "evolution equations" of the form

$$(5.25) \qquad -\dot{u}(t) \in A(t, u(t)) \quad ,$$

where $A(t,.)$ denotes, for each $t \in [0,T]$, a <u>monotone multimapping</u> (or multivalued operator) from H into itself. In fact, monotonicity immediately implies that if u_1, u_2, absolutely continuous, are solution of (5.25), <u>the function</u>

$$t \mapsto |u_1(t) - u_2(t)|$$

<u>is non increasing</u> ; therefore these two solutions are equal if they agree with the same initial value.

Equations such as (5.25) have already been studied, but mainly under hypotheses involving that the set

$$\text{dom}\, A(t,.) = \{x \in H \ : \ A(t,x) \neq \emptyset\}$$

is independant of t ; see references in BREZIS [1]. Here, on the contrary, the problem becomes trivial if $\text{dom}\, \partial\psi(t,.)$, namely $C(t)$, is constant ; thus the simple equation (5.24) furnishes the occasion of focusing upon the difficulties which arise from the variation of the domain. In the same line must be quoted :

1° H. BREZIS [2] who studied by a "double regularization" technique the case

$$A(t,.) = \partial \phi(.) + \partial \psi(t,.)$$

with $\phi \in \Gamma_0(H,H)$ independent of t and under some hypotheses involving the projection mapping $x \mapsto \text{proj}\,(x, C(t))$; they do no seem directly comparable with our absolute continuity assumption.

2° C. PERALBA [1], [2] who succeeded in generalizing to the case $A = \partial \phi$, with $\phi \in \Gamma_0(H,H)$ depending on time in a suitable way, the author's regularization method (see MOREAU [17]).

Because of its insertion in this context we also choose a regularization technique, i.e. the use of penalty functions, to prove, in next paragraph, an existence theorem. Another advantage of doing so refers to the application of equation (5.24) to elastoplastic mechanical systems, developed in Chapter 6 below : as explained in § 4. g; when the considered convex is the rigidity set defining a law of plasticity, the replacement of its indicator function by some penalty function comes to take into account some additional viscosity. The reasoning used below could then be adapted to prove that the solution of an elasto-visco-plastic problem tends to the solution of the elastoplastic problem when viscosity tends to zero. From the physical standpoint this may be as important as the existence question itself.

The existence theorem obtained will supply the needs of Chapter 6. Actually a deeper insight into the sweeping process can be gained from a <u>discretization method</u> (published as multigraph in MOREAU [18]) which consists in proving first the convergence of the "catching up algorithm" (cf. § 5. h below) ; this method permits weaker hypotheses, by replacing the concept of the variation of a multimapping by that of <u>retraction</u> : use instead of Hausdorff distance the "unilateral" écart e. On the other hand, a generalization of the process can be defined in this line for the case of a possibly <u>discontinuous</u> moving convex set C, provided its variation (resp. retraction) is finite.

On the application of the discretization method to equations of the form (5.25), with $A(t,.) = A_o(.) - f(t)$ see J. NECAS [1] .

5. g EXISTENCE THEOREM

The study of equation (5.24) is made greatly easier by the following remark : the sweeping process associates the chain of the positions of the moving point u to the chain of the positions of the moving set C in a way which does not depend on the timing. More precisely, the change of variable in Lebesgue integral, along with the fact that the set $\partial \psi$ is a cone, i.e. the multiplication by a nonnegative scalar sends it into itself, implies : <u>Let</u> π <u>denote a non decreasing absolutely continuous mapping from</u> $[0,T]$ <u>onto an interval</u> $[0,T']$;

suppose $C = C' \circ \pi$, i. e.

(5.26) $$\forall\, t \in [0,T] \;:\; C(t) = C'(\pi(t))$$

where C' is an absolutely continuous multimapping from $[0,T']$ into H, with nonempty closed convex values ; let $u' : [0,T'] \to H$ be a solution of the sweeping process for C' ; then the mapping $u = u' \circ \pi$ is a solution of the sweeping process for C.

As explained in § 5. a, taking for π the variation fonction v of the given multimapping C yields a factorization of the form (5.26), with C' Lipschitz with ratio 1. This reduces the existential study of the sweeping problem to the Lipschitz case, i.e. the case where the speed function of the moving convex set belong to $L^\infty(0,T\,;\,\mathbb{R})$, or even is merely a constant.

Let us now proceed to establish :

PROPOSITION. For any a in $C(0)$ the sweeping problem, as formulated in the preceding paragraph, possesses a (unique) solution.

Let n be positive integer. Denote by $u_n : [0,T] \to H$ the solution of the differential equation

(5.27) $$-\dot{u}_n = n \operatorname{grad} q(t, u_n(t))$$

for the initial condition

(5.28) $$u_n(0) = a \;.$$

In fact the expression (5.16) of grad q implies, under the hypotheses made concerning $t \mapsto C(t)$, that the mapping $(t,x) \mapsto n \operatorname{grad} q(t,x)$ is continuous relatively to t and is Lipschitz with ratio n relatively to x ; hence classically the existence and the uniqueness of u_n which is a continuously differentiable function from $[0,T]$ into H.

Observe that the construction of the ordinary differential equation (5.27) consists in replacing the right member $A = \partial \psi$ of (5.24) by its Yosida regularization (5.18), with $\lambda = \frac{1}{n}$; equivalently, the indicator function of C is replaced by the penalty function $n\,q$: thus the moving point $u_n(t)$ is allowed to not belong to $C(t)$ but then, in view of the expression (5.14) of grad q , it must have a velocity directed toward its projection on $C(t)$; the magnitude of this velocity is proportional to the distance from $u_n(t)$ to $C(t)$ and proportional to the penalty coefficient n .

LEMMA 1. <u>If the speed function</u> $\dot{v}$ <u>of the moving set</u> C <u>belongs to</u> $L^2(0,T\,;\mathbb{R})$, <u>the sequence of the derivatives</u> $\dot{u}_n$ <u>is bounded in</u> $L^2(0,T\,;H)$.

Denote by $h_n(t)$ the common value of

$$\frac{1}{n}\,|\dot{u}_n(t)| = |\text{grad } q(t, u_n(t))| = d(u_n(t), C(t)) .$$

Inequality (5.20) (§ 5. e, Lemma 2) yields, for almost every t ,

$$|h_n(t)\,\dot{h}_n(t) - (\dot{u}_n(t)\,|\,\text{grad } q(t, u_n(t)))| \leqslant h_n(t)\,\dot{v}(t)$$

hence, due to (5.27),

$$h_n(t)\,\dot{h}_n(t) + n\,(h_n(t))^2 \leqslant h_n(t)\,\dot{v}(t) \tag{5.29}$$

As $a \in C(0)$, one has $h_n(0) = 0$, thus, by integration over $[0,T]$,

$$\frac{1}{2}(h_n(T))^2 + n\int_0^T (\dot{h}_n(t))^2\,dt \leqslant \int_0^T \dot{h}_n(t)\,\dot{v}(t)\,dt$$

Denoting by $\|.\|$ the norm in $L^2(0,T\ ;\ \mathbb{R})$ as well as the norm in $L^2(0,T\ ;\ H)$, this yields

(5.30) $$\|\dot{u}_n\| \leqslant \|\dot{v}\|$$

which proves the lemma.

REMARK. More may be obtained from inequality (5.29). Suppose only the absolute continuity of $t \mapsto C(t)$ so that the derivatives $\dot{v}(t)$ and $\dot{h}_n(t)$ exist for almost every t. For the values of t such that $h_n(t) \neq 0$, inequality (5.29) implies

$$\dot{h}_n(t) + n\,h_n(t) \leqslant \dot{v}(t)$$

and this is also true when $h_n(t) = 0$ (then $\dot{h}_n(t) = 0$ since zero is the minimal value of h_n). The elementary treatment of this differential inequality, with the initial condition $h_n(0) = 0$, yields :

(5.31) $$|\dot{u}_n(t)| \leqslant \dot{v}(t)$$

for almost every t.

In particular, if $v \in L^p(0,T\ ;\ \mathbb{R})$, with $1 \leqslant p \leqslant +\infty$, the same inequality as (5.30) holds for L^p norms.

From such majorations, there are many ways of establishing the convergence of the sequence u_n to a function which is a solution of the sweeping process. In view of our L^2 framework, the most efficient seems to make use of the following elementary property of Hilbert spaces, due to M. CRANDALL and A. PAZY [1] :

J. J. Moreau

Consider a real Hilbert space with scalar product noted $\langle . | . \rangle$ and norm noted $\| . \|$. Let (r_n) be a sequence of positive real numbers ; let (z_n) be a sequence of elements of this Hilbert space such that

$$\forall n \;, \; \forall m : \; \langle z_n - z_m \mid r_n z_n - r_m z_m \rangle \leqslant 0$$

Then :

If r_n is strictly increasing in n, $\|z_n\|$ is decreasing and $\lim_{n \to \infty} z_n$ exists.

If r_n is strictly decreasing, $\|z_n\|$ is increasing ; if in addition $\|z_n\|$ is bounded, $\lim_{n \to \infty} z_n$ exists.

From this we are to prove :

LEMMA 2. If $\dot{v} \in L^2(0,T ; \mathbb{R})$ the sequence $\dot{u}_n$ is strongly convergent in $L^2(0,T ; H)$.

In fact, let m and n be two positive integers ; for any t in $[0,T]$, the values of the functions u_m, $\dot{u}_m$, u_n, $\dot{u}_n$ satisfy

$$\frac{d}{dt} |u_m - u_n|^2 = 2(u_m - u_n \mid \dot{u}_m - \dot{u}_n) \quad . \tag{5.32}$$

Denote by p_m, p_n the respective projections of $u_m(t)$ and $u_n(t)$ on $C(t)$; by (5.16) and (5.27) one has

$$-\dot{u}_m = m(u_m - p_m) \in \partial \psi(t, p_m)$$

and the same for n ; due to the monotonicity of $\partial \psi$, this yields by easy calculation

$$(u_m - u_n \mid \dot{u}_m - \dot{u}_n) \leqslant -(\frac{1}{m} \dot{u}_m - \frac{1}{n} \dot{u}_n \mid \dot{u}_m - \dot{u}_n)$$

Recall that $u_m(0) = u_n(0) = a$, integrate (5.32) over $[0,T]$, denote by $\langle . | . \rangle$ the scalar product of the Hilbert space $L^2(0,T ; H)$ and by $\|.\|$ its norm ; this inequality implies

$$0 \leqslant \frac{1}{2} |u_m(T) - u_n(T)|^2 \leqslant - \langle \frac{1}{m} u_m - \frac{1}{n} u_n \mid u_m - u_n \rangle .$$

The sequence $r_n = \frac{1}{n}$ is strictly decreasing ; the sequence $\|u_n\|$ is bounded according to Lemma 1 ; apply CRANDALL and PAZY's result in $L^2(0,T ; H)$.

Next :

LEMMA 3. <u>If</u> $\dot{v} \in L^2(0,T ; \mathbb{R})$ <u>the sequence of functions</u> u_n <u>converges uniformly on</u> $[0,T]$ <u>to an absolutely continuous function</u> u <u>whose derivative is the</u> L^2 - <u>limit of the sequence</u> $\dot{u}_n$; <u>this function is solution of the sweeping process for the initial condition</u> $u(0) = a$.

<u>Furthermore</u>, <u>for almost every</u> t,

(5.33) $$|\dot{u}(t)| \leqslant \dot{v}(t) .$$

In fact, denote by $\dot{u}$ the limit of $\dot{u}_n$ in $L^2(0,T ; H)$ and define $u : [0,T] \to H$ by

$$u(t) = a + \int_0^t \dot{u}(s)\, ds ,$$

so that u is absolutely continuous with a strong derivative equal to $\dot{u}$ almost everywhere. Still denoting by $\|.\|$ the norm in $L^2(0,T ; H)$, the inequality

$$|u(t) - u_n(t)| = \left| \int_0^t (\dot{u}(s) - \dot{u}_n(s))\, ds \right| \leqslant \sqrt{t}\, \|\dot{u} - \dot{u}_n\|$$

shows that u is the uniform limit of u_n.

It remains to prove that u and $\dot{u}$ verify (5.24) almost everywhere. Put

$$p_n(t) = \text{proj}\,(u_n(t),\ C(t))\ .$$

Then, in view of (5.16) and (5.27)

$$u_n(t) - p_n(t) = \text{grad}\ q(t,\ u_n(t)) = -\frac{1}{n}\dot{u}_n\ (t)$$

$$(5.34) \qquad -\dot{u}_n(t) \in \partial\,\psi\ (t,\ p_n(t))$$

and, in view of (5.30), the functions p_n converge to u in $L^2(0,T\ ;\ H)$.

The convergences in $L^2\ (0,T\ ;\ H)$ imply the existence of N', an infinite subset of N, such that for any t which does not belong to a certain subset ω of $[0,T]$ with zero measure, the limit of $p_\upsilon\ (t)$ in H, for υ tending to infinity in N', is $u(t)$ and the limit of $\dot{u}_\upsilon\ (t)$ in H is $\dot{u}(t)$. As the graph of the multimapping $x \mapsto \partial\,\psi(t,x)$ is closed in $H \times H$, (5.34) implies that (5.16) holds for any $t \notin \omega$. On the other hand (5.33) follows from (5.31).

From this lemma, the proof of the formulated Proposition is completed, by performing an absolutely continuous change a variable reducing to the case $\dot{v} \in L^\infty\ (0,T\ ;\ R)$, which a fortiori implies $\dot{v} \in L^2\ (0,T\ ;\ R)$.

REMARK. Inequality (5.33) is clearly preserved by such a change of variable, so that in general for any solution u of the sweeping process

$$|\dot{u}(t)| \leqslant \dot{v}(t)\ .$$

By integration, this yields that <u>the length of the path traveled by the moving point</u> u <u>during an interval of time</u> $[t_1,t_2]$ <u>is majorized by</u>

var (C ; t_1,t_2). This property becomes specially suggestive in the special case where C moves by translation i.e.

$$C(t) = C_o + w(t) ,$$

with w absolutely continuous. Then, in view of § 5. b, example,

(5.35) $$\int_{t_1}^{t_2} |\dot{u}(t)|\, dt \leqslant \int_{t_1}^{t_2} |\dot{w}(t)|\, dt .$$

The association of the function u , a solution of the sweeping process, with the given function w defining the translation imposed to C , may be visualized as a driving affected with play ; (5.35) expresses that such a play makes the driven point travel a path which cannot be longer than the path traveled by the driving device.

5. h DISCRETIZATION ALGORITHM

A method of "time discretization" for the approximate solution of the preceding problem consists in choosing a subdivision of $[0,T]$, namely $0 = t_o < t_1 < \ldots < t_n = T$ and constructing a sequence $x_o, x_1, \ldots, x_n$ of points of H such that x_i constitutes an approximation of $u(t_i)$. Adopting $\frac{1}{t_i - t_{i-1}}(x_i - x_{i-1})$ as an approximation of $\dot{u}(t_i)$ induces to replace (5.24) by

(5.36) $$x_{i-1} - x_i \in (t_i - t_{i-1})\, \partial\, \psi(t_i, x_i)$$

which is a recurrence condition of "implicit" type concerning the desired sequence (an "explicit" method would consist in interpreting the same quotient as an approximation of $\dot{u}(t_{i-1})$; but this yields an unworkable recurrence condition). As $\partial \psi(t_i, x_i)$ is a cone, the strictly positive factor $t_i - t_{i-1}$ in the right member of (5.36) may be omitted and this condition equivalently amounts to

(5.37) $$x_i = \text{proj}\,(x_{i-1}, C(t_i)) .$$

Thus, starting with $x_o = a$, the point sequence (x_i) is constructed by successive projections on the sequence of closed convex sets $C(t_i)$. It is as if the moving point u, instead of being swept along with the moving set C was left behind except that, from time to time, it catches up with this set intantaneously, by the shortest way. We propose to call this the catching up algorithm.

The question is wether the step function $x : [0,T] \to H$ defined from this sequence by

(5.38) $$x(t) = x_i \qquad \text{for} \quad t \in]t_{i-1}, t_i] \quad ,$$

converges to the solution u of the sweeping process, for the same initial value a, when finer and finer subdivisions of $[0,T]$ are considered.

A direct proof of the convergence of this family of step functions may be given, yielding another way to establish the existence of the solution u itself (cf. MOREAU [17], [18]). As this existence has been obtained above by a regularization, or penalty, technique we think it interesting and unusual to study also the discretization algorithm by some extension of the penalty method : the trick consists in making the penalty coefficient vary with t (cf. MOREAU [17]).

PROPOSITION. For any $\varepsilon > 0$ there exists $\eta > 0$ such that the majoration

$$\sup_i (t_i - t_{i-1}) < \eta$$

(resp. there exists $\eta' > 0$ such that the majoration

$$\sup_i \operatorname{var}(C ; t_{i-1}, t_i) < \eta')$$

ensures

$$\forall t \in [0,T] \quad : \quad |u(t) - x(t)| < \varepsilon \quad .$$

Let $\rho : [0,T] \to R_+$ a nonnegative ruled function (actually it will sufficein the following to take as ρ a step function). The classical theory of differential equations ensures the existence of $u_\rho : [0,T] \to H$, solution of

$$-\dot{u}_\rho (t) = \rho(t) \operatorname{grad} q(t, u_\rho (t))$$

agreeing with the initial condition $u_\rho(0) = a$. Denote by h_ρ the absolutely continuous numerical function

$$h_\rho(t) = d(u_\rho(t), C(t)) = |\operatorname{grad} q(t, u_\rho(t))| \ .$$

The same calculation as in § 5. g, proof of Lemma 1, yields the differential inequality

$$\dot{h}_\rho + \rho\, h_\rho \leqslant \dot{v} \ , \tag{5.39}$$

from which elementary techniques leads to :

LEMMA 1. If the speed function $\dot{v}$ of C is majorized by some constant $M \geqslant 0$, the function h_ρ is majorized by the constant $M\, J(\rho)$, where

$J(\rho)$ denotes the supremum over $[0,T]$ of the numerical function k defined on this interval by the differential equation $\dot{k} + \rho\, k = 1$ with the initial condition $k(0) = 0$.

Consider now another function σ similar to ρ and the corresponding u_σ and h_σ. The same inequality as in § 5. b, proof of Lemma 2, yields, for any t in $[0,T]$,

$$\frac{1}{2}\,|u_\rho(t) - u_\sigma(t)|^2 \leqslant \int_0^t -(\operatorname{grad} q(s,u_\rho(s)) - \operatorname{grad} q(s,u_\sigma(s))|$$
$$\rho(s)\operatorname{grad} q(s,u_\rho(s)) - \sigma(s)\operatorname{grad} q(s,u_\sigma(s)))\, ds .$$

The integrand is a scalar product in H, majorized by

$$(h_\rho + h_\sigma)(\rho\, h_\rho + \sigma\, h_\sigma) = \rho\, h_\rho^2 + \sigma\, h_\sigma^2 + (\rho + \sigma)\, h_\rho\, h_\sigma .$$

Now from Lemma 1 and inequality (5.39) one obtains

$$h_\rho\, \dot{h}_\rho + \rho\, h_\rho^2 \leqslant M\, J(\rho)$$
$$h_\sigma\, \dot{h}_\rho + \rho\, h_\sigma\, h_\rho \leqslant M\, J(\sigma)$$

and two symmetrical inequalities. Adding them together and integrating gives the proof of the following :

LEMMA 2. If the function $\dot{v}$ is majorized by some constant $M \geqslant 0$ one has, for every t in $[0,T]$,

$$|u_\rho(t) - u_\sigma(t)|^2 \leqslant 4t\, M^2\, (J(\rho) + J(\sigma)) . \tag{5.40}$$

If, in particular, σ is a constant m

$$J(\sigma) = \frac{1}{m}(1 - e^{-mT}) \leqslant \frac{1}{m} .$$

By § 5. g, the solution u of the sweeping process is the limit of the corresponding u_σ when m (for instance an integer) tends to infinity ;

thus (5.40) implies

(5.41) $$|u_\rho(t) - u(t)|^2 \leqslant 4t\, M^2\, J(\rho)\,.$$

For the continuation take as ρ the step function associated with the subdivision

$$0 = t_o < t_1 < \dots < t_n = T$$

as follows : denoting by m_i the middle point of the interval $[t_i,\ t_{i+1}]$, put

(5.42) $$\rho(t) = \begin{cases} A & \text{if } t_i < t < m_i \\ 0 & \text{if } m_i \leqslant t \leqslant t_{i+1} \end{cases} ,$$

where A is a constant independent of i.

Denote by p the supremum of the $t_{i+1} - t_i$; studying the function k associated with ρ as in Lemma 1 yields :

LEMMA 3. *If* ρ *is defined by* (5.42) *and* $A \geqslant 4$ *one has*

$$J(\rho) \leqslant \frac{1}{\sqrt{A}} + \frac{p}{2} \quad .$$

Hint : the function $K : t \mapsto \max\{\frac{1}{\sqrt{A}},\ k(t)\}$ possesses for almost every t a derivative $\dot{K}(t)$. When $t \in]t_i,\ m_i[$ one has

$$\dot{K}(t) \leqslant -1 \qquad \text{if } K(t) > \frac{1}{\sqrt{A}}$$

$$\dot{K}(t) = 0 \qquad \text{if } K(t) = \frac{1}{\sqrt{A}} \quad .$$

When $t \in [m_i,\ t_{i+1}]$ one has

$$\dot{K}(t) = 1 \qquad \text{if } K(t) > \frac{1}{\sqrt{A}}$$

$$\dot{K}(t) = 0 \qquad \text{if } K(t) = \frac{1}{\sqrt{A}} \quad .$$

From these lemmas we can proceed to the proof of the Proposition.

Observe first that the two alternative statements of this Proposition are equivalent since the variation function v of C is continuous on $[0,T]$, thus uniformly continuous.

The statement concerning variations is visibly indifferent to any (absolutely continuous) non decreasing change of variable ; we take profit of this fact in supposing that a change of variable has been performed reducing to the case where the speed function $\dot{v}$ of C is the constant 1 (see § 5. a).

<u>First step</u>. Denote by π the following absolutely continuous non decreasing mapping from the interval $[0,T]$ onto itself (m_i denotes as before the middle point of $[t_i, t_{i+1}]$)

$$\pi(t) = \begin{cases} t_i & \text{if} \quad t_i \leqslant t \leqslant m_i \\ 2t - t_{i+1} & \text{if} \quad m_i \leqslant t \leqslant t_{i+1} \end{cases}$$

and put

$$C(\pi(t)) = C'(t) \quad .$$

In other words, on each interval of the form $[t_i, m_i]$ the convex set C' remains fixed, equal to $C(t_i)$; on the next interval $[m_i, t_{i+1}]$, it runs through the same chain of configurations as C on $[t_i, t_{i+1}]$, with a timing adjusted in such a way that C' catches up with C at the instant t_{i+1}. Call u' the solution of the sweeping process for the moving convex set C' and the same initial value a as u ; in view of the change of variable one has

$$u'(t) = u(\pi(t)) .$$

By virtue of (5.31), the function u is Lipschitz with ratio 1 ; thus, for any $t \in [0,T]$,

(5.43) $$|u(t) - u'(t)| \leqslant \frac{p}{2} .$$

Second step. Put

$$q'(t,x) = \frac{1}{2}(d(x,C'(t))^2 .$$

Defining ρ by (5.42), denote by u'_ρ the solution of

(5.44) $$-\dot{u}'_\rho(t) = \rho(t) \operatorname{grad} q'(t, u'_\rho(t))$$

agreeing with the initial condition $u'_\rho(0) = a$. The integration of this differential equation may be explicited : On each interval of the form $[m_i, t_{i+1}]$ the function ρ vanishes, so that

(5.45) $$t \in [m_i, t_{i+1}] \Rightarrow u'_\rho(t) = u'_\rho(m_i) .$$

For t ranging over an interval of the form $]t_i, m_i[$, ρ takes the constant value A and the function $x \mapsto q'(t,x)$ is independent of t, with

$$\operatorname{grad} q'(t,x) = x - \operatorname{proj}(x, C(t_i))$$

so that, on this interval

(5.46) $$u'_\rho(t) = u'_\rho(t_i) + [y_i - u'_\rho(t_i)] [1 - \exp A(t_i - t)]$$

where

$$y_i = \operatorname{proj}(u'_\rho(t_i), C(t_i)) .$$

Supposing $A \geqslant 4$, it results from (5.41) and from Lemma 3 that, for any $t \in [0,T]$

(5.47) $$|u'_\rho(t) - u'(t)|^2 \leqslant 16\, t\left(\frac{1}{\sqrt{A}} + \frac{p}{2}\right) .$$

Note that (5.45) and (5.46) yield

$$(5.48) \qquad u'_\rho(t_{i+1}) - u'_\rho(m_i) - u'_\rho(t_i) + [y_i - u'_\rho(t_i)][1-\exp \frac{A(t_i - t_{i+1})}{2}]$$

<u>Third step</u>. Let A tend to $+\infty$; as all the $t_i - t_{i+1}$ are < 0 , (5.48) shows that, for each $i < n$, the difference $u'_\rho(t_{i+1}) - y_i$ tends to zero in H. As the mapping proj (., $C(t_i)$) used in the definition of y_i is continuous, this proves by iteration that, for each $i < n$, the value $u'_\rho(t_{i+1})$ tends to $x(t_{i+1})$ as defined by (5.38). Then (5.46) shows that $u'_\rho(t)$ tends to $x(t)$ for any t in $]t_i, m_i]$ and finally also for any t in $[m_i, t_{i+1}]$ by virtue of (5.46).

In view of (5.47) this pointwise convergence yields, for any $t \in [0,T]$,

$$|x(t) - u'(t)| \leqslant \sqrt{8 \, T \, p}$$

which proves the Proposition, by comparing with (5.43).

J. J. Moreau

6 QUASI-STATIC EVOLUTION OF AN ELASTOPLASTIC SYSTEM

6. a FORMULATION OF THE PROBLEM

The framework in all this Chapter is that of a configuration space $\mathcal{U}$ endowed with a linear space structure ; thus the practical applications of the following mainly concern systems whose displacements are treated as "infinitely small".

According to the usual conception of elastoplasticity, every state of the system is represented by two components which both are elements of $\mathcal{U}$:

The visible (or "exposed") component, denoted by x ; it is the part of the system which undergoes external forces, called loads, and may also be submitted to constraints.

The hidden (or "plastic") component denoted by p.

Strictly speaking, the configuration space of the system is then the product space $\mathcal{U} \times \mathcal{U}$.

The difference $x - p = e \in \mathcal{U}$ will be called the elastic deviation.

Let us denote as before by $\mathcal{F}$ the linear space of forces, placed in separating duality with $\mathcal{U}$; the forces experienced by the component p are :

1° The force $s \in \mathcal{F}$ of "elastic restoring toward x" related to e by

(6.1) $$s = A(e) \quad ,$$

where A denotes a given selfadjoint nonnegative linear mapping from $\mathcal{U}$ into $\mathcal{F}$.

2° The force of "plastic resistance" $f \in \mathcal{F}$ related to the velocity $\dot{p}$ (at any instant where this velocity exists) by the resistance law studied in § 4. e.

(6.2) $$\dot{p} \in \partial \psi_C (-f) \ ,$$

where C denotes a fixed nonempty closed convex subset of $\mathcal{F}$.

The forces experienced by the component x are

1° The reaction $r \in \mathcal{F}$ of a perfect affine constraint (cf. § 3. c) ; this constraints maintains x at every instant in an affine manifold which moves in a given way, say

(6.3) $$\mathcal{L} = U + g(t)$$

where U denotes a fixed closed linear subspace of $\mathcal{U}$ and $t \mapsto g(t)$ is a given function of time, with values in $\mathcal{U}$, which may be called the guiding (or "driving"). Such a constraint constitutes the statical law

(6.4) $$- r \in \partial \psi_{\mathcal{L}} (x) \ .$$

2° The load c(t), a given time-dependent element of $\mathcal{F}$.

3° The force - s of "elastic restoring toward p". Supposing in this way that the elastic force acting on x is the negative of the elastic force acting on p merely means that the total power of the elastic forces vanishes in any evolution which preserves the elastic deviation $x - p$; in other words the elastic energy depends on this deviation only.

The problem is that of determining the evolution of x and p in $\mathcal{U}$, under the hypothesis that the motion is sufficiently slow for inertia to be negligible.

Therefore, the dynamical equations amount to express the quasi-equilibrium of x, namely

(6.5) $$r + c - s = 0$$

and the quasi-equilibrium of p, namely

(6.6) $$s + f = 0 \ .$$

To illustrate the preceding formulation by a practical example, the reader may take back the situation of a lattice of bars, presented in § 3. i, j . If the behavior of each bar is elastoplastic, the $\frac{1}{2}n(n-1)$ - uple of their respective elongations, namely the element $e \in E$, has to be written as a sum, say $e' + p$; here e' denotes the "elastic part" of e, related to the tension $s \in S$ by a linear elasticity law such as (6.1) ; p denotes the "plastic part" of e : its "velocity" $\dot{p}$ is related to s by relations of the form (6.2), (6.6). At this stage one may avoid the explicit consideration of the linear mappings D and D^* by using the third procedure of § 3. j , namely the elimination of (X,Y) :

then the sum $e' + p$ is interpreted as the "visible" configuration, to be denoted here by x ; finally write simply e instead of e'.

The same pattern applies to an elastoplastic continuous medium, occupying a domain Ω of the physical space. Then elements e, e', p, s are some tensor fields defined on Ω ; the spaces E and S are some function spaces. The corresponding quasistatic evolution problem may be treated in the line of the following paragraphs, but with some complications which will not be investigated in this lectures ; the difficulty arises from the fact that, with regard to the Hilbert norm defined by means of the elastic energy (see § 6. b) the convex C possesses an empty interior. Then the theorem on the absolute continuity of intersections (§ 5. c) will be applied relatively to some L^∞-norm ; the absolute continuity of the considered intersection will finally hold with regard to the Hilbert norm too, as this latter is majorized by the L^∞-norm (multiplied by a constant).

Observe that the continuous medium problem is studied by G. DUVAUT and J.L. LIONS, [1], Chap. 5 . Their method is that of vanishing viscosity, basically similar to the regularization technique we used in § 5. g ; but they must restrict themselves to the special case where the "load", denoted here by c, is identically zero ; thus the motion is only caused by the "guiding" g. Paragraph 6. c below explains why this special case is more tractable : it corresponds to a set

$(C-c-g) \cap V$ which moves by translation, so that the intersection theorem is not required for proving its absolute continuity (cf. § 5. b).

We shall not deal in the present lectures with systems governed by behavioral laws of Hencky's type ; the reader will refer to H. Lanchon's lectures on this subject. Hencky's law is also studied in the book of DUVAUT and LIONS, by methods involving the duality of convex functionals.

In order to help the reader to visualize the formulated problem let us finally present a very simple model in which the dimension of $\mathcal{U}$ equals 2. The considered system consists of two particles x and p moving in the plane $\mathcal{U}$. The particle x is guided without friction on the material straight line U + g(t), a line which remains parallel to the fixed line U and moves in a given way. The particle p will be visualized as a plot, whose contact with the plane $\mathcal{U}$ is affected by a given friction. The two particles are connected by a spring whose length in the state of zero tension is zero. In addition, a given force c(t) is applied to x. One studies motions during which the various forces equilibrate each

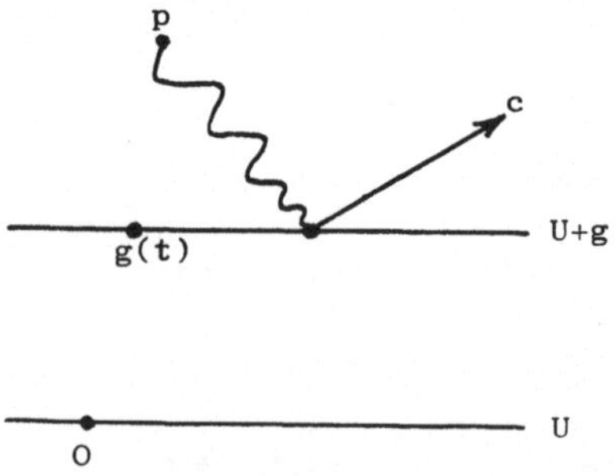

other at any instant ; in particular the friction resistance undergone by p must exactly counterbalance the spring tension.

Investigating this elementary model raises an important observation : though the friction between p and the underlying plane has the characteristics of perfect plasticity, the behavior of the component x exhibits strain hardening. In fact suppose the line U + g is fixed, for instance with g identically zero ; suppose the friction of p is isotropic, i.e. it obeys elementary Coulomb's law. Clearly any motion during which the spring is strained enough for the point p to yield (this imposes a definite value for the distance between x and p) necessarily brings this point closer to the line. Therefore this evolution leaves the system in a state for which the elastic domain, i.e. the set of the values of the load c which may be applied without causing yield, is larger than before.

Such an example suggests that strain hardening can be described, in practical situations, by including in the definition of the hidden component p a sufficient number of internal state variables and postulating that the behavior of such a p is governed by a law similar to that of perfect plasticity. This has been developed, in our framework of convex pseudo-potentials, by Q.S. NGUYEN [1] (see also, for the use of internal state variables without convexity, J. KRATOCHVIL and J. NECAS [1]).

6. b THE HILBERT SPACE NOTATION

Let us restrict ourselves for sake of simplicity to the usual case where the self-adjoint linear mapping $A : \mathcal{U} \to \mathcal{F}$ introduced by the elasticity law (6.1) is one-to-one. Then one makes the treatment of the problem much easier by the notation trick which consists in interpreting the one-to-one mapping A as an identification of the spaces $\mathcal{U}$ and $\mathcal{F}$. Denote by H this single space ; the symmetric bilinear form defined on $\mathcal{U} \times \mathcal{U}$ by

$$(u, u') \mapsto \langle u, A u' \rangle = \langle u', A u \rangle$$

becomes an inner product in H, which will be denoted as $(u' \mid u)$. As the quadratic form

$$u \mapsto \frac{1}{2} \langle u, A u \rangle = \frac{1}{2} (u \mid u)$$

represents the elastic energy, it is nonnegative, thus positive definite due to A being one-to-one. This means that a pre-Hilbert norm $|.|$ is defined on H by

$$|u| = \sqrt{(u \mid u)} .$$

Let us make the assumption that H is complete relatively to this norm, i.e. it is a Hilbert space.

This of course is automatically satisfied in finite dimensional cases. In the case of continuous media also, one is accustomed to formulate the problems in suitable function spaces for this assumption to hold.

Observe that the inner product $(.|.)$ in H and the identification map $A : \mathcal{U} \to \mathcal{F}$ are connected in such a way that subdifferential relations of the form $-f \in \partial \phi(u)$ may equivalently be understood in the sense of the duality $(\mathcal{U}, \mathcal{F} ; \langle .,. \rangle)$, with $u \in \mathcal{U}$ and $f \in \mathcal{F}$, or in the sense of the duality $(H,H ; (.|.))$ with u and f elements of H.

Let us write the formulation of the problem in these notations. Denote by V the subspace of H orthogonal to U ; observe that (6.1) becomes $s = e$; eliminate r by (6.5) and f by (6.6) ; the preceding conditions take the equivalent form

(6.7) $$x \in U + g$$

(6.8) $$s \in V + c$$

(6.9) $$x = p + s$$

(6.10) $$\dot{p} \in \partial \psi_C (s) \quad .$$

Given the compact time interval $[0,T]$, the problem is that of determining the three functions $t \mapsto x$, $t \mapsto p$, $t \mapsto s$, with values in H, absolutely continuous on this interval (this makes the derivative $\dot{p}$ exist for almost every t) satisfying conditions (6.7) to (6.10) for almost every t, and some initial conditions

(6.11) $$x(0) = x_o \quad , \quad s(0) = s_o \quad .$$

Let us make now some assumptions about the data.

ASSUMPTION 1. The given functions $t \mapsto g$ and $t \mapsto c$ are absolutely

<u>continuous on</u> $[0,T]$. In addition, we visibly lose no generality in supposing that c <u>takes its values in</u> U <u>and that</u> g <u>takes its values in</u> V.

ASSUMPTION 2 . <u>The initial data</u> x_o <u>and</u> s_o <u>satisfy the conditions</u>

$$x_o \in U + g(0) \quad , \quad s_o \in V + c(0)$$

<u>evidently required by</u> (6.7) <u>and</u> (6.8), <u>and the condition</u>

(6.12) $$s_o \in C$$

<u>required by</u> (6.10). In fact (6.10) makes that for almost every t, the set $\partial \psi_C(t)$ is non empty, thus $s(t) \in C$, and the latter must also be true for every t in $[0,T]$, by continuity.

Observe also that (6.8) with (6.10) requires the moving affine manifold $V + c$ to meet the convex set C for almost every t, thus for every t by the continuity of c. This may equivalently be written as

(6.13) $$c \in \text{proj}_U C \quad .$$

The mechanical meaning of this necessary condition is clear : a load c (recall that we supposed $c \in U$) which does not satisfy it cannot be counterbalanced by the forces $r \in V$ (the reaction of the affine perfect constraint) and $s \in C$. As the law of plastic resistance (6.10) only permits $s \in C$, this means that if, starting from a configuration defined by some values of x and p, the system experiences a load c which does not verify (6.13), its evolution cannot be quasi-static. Of course, there are in this situation other necessary conditions, namely $\dot{x}-p \in V + c$, a consequence of (6.8) and (6.9).

For mathematical convenience, we shall suppose that the set C possesses a nonempty interior ; then let us agree to replace (6.13) by the stronger following condition.

ASSUMPTION 3. For any t in $[0,T]$ the affine manifold $V + c(t)$ intersects the interior of C.

Without discussing here the physical meaning of this assumption, let us call it the "safe load hypothesis".

In addition, we shall avoid some technical job of covering the interval $[0,T]$ and piecing together local solutions, by making also a last inessential hypothesis :

ASSUMPTION 4. The set C is bounded.

Then :

LEMMA. Assumptions 3 and 4 and the absolute continuity of the function $t \mapsto c$ imply the following : there exists a strictly positive real constant ρ and an absolutely continuous mapping $h : [0,T] \to H$ such that, for every $t \in [0,T]$, one has $h(t) \in C \cap (V + c(t))$ and the closed ball with center $h(t)$ and radius ρ is contained in C.

Outlined proof : Using the notation e of § 5. a, arguments similar to that of § 5. c prove that the numerical function

$$t \mapsto e\,(V + c(t)\,,\,H \setminus C)$$

is continuous on $[0,T]$, with strictly positive values, thus strictly minorized by some constant $\rho > 0$. The set

$$C_\rho = \{x \in H\,:\; d(x, H \setminus C) \geqslant \rho\}$$

is closed and convex, with nonempty interior. For every t in $[0,T]$, the affine manifold $V + c(t)$ intersects the interior of C_ρ. The multi-mapping $t \mapsto V + c(t)$ is absolutely continuous, implying by § 5. c the absolute continuity of the <u>moving convex set</u> $t \mapsto C_\rho \cap (V + c(t))$. Take as h a solution of the <u>sweeping process</u> by this moving non empty closed convex set (cf. § 5. f).

6. c NEW UNKNOWN FUNCTIONS

Conditions (6.7), (6.8), (6.9) may be written as

$$x - g \in U$$

$$c - s \in C$$

$$(x - g) + (c - s) \;=\; p + c - g \;\;;$$

this may equivalently be expressed by means of the orthogonal projectors relative to the complementary orthogonal subspaces U and V

$$x - g \;=\; \mathrm{proj}_U\,(p + c - g)$$

$$c - s \;=\; \mathrm{proj}_V\,(p + c - g)$$

or, as we have supposed $c \in U$ and $g \in V$,

$$\mathrm{proj}_U\, p = x - c - g$$

$$\mathrm{proj}_V\, p = c + g - s \quad .$$

Let us define two new unknowns y and z by

(6.14) $$y = s - c - g = -\mathrm{proj}_V\, p$$

(6.15) $$z = x - c - g = \mathrm{proj}_U\, p$$

which implies

(6.16) $$p = z - y \quad .$$

Due to Assumption 1, the functions $t \mapsto y$ and $t \mapsto z$ are absolutely continuous if only if such are $t \mapsto s$ and $t \mapsto p$.

Under this change of unknowns, conditions (6.7) to (6.10) equivalently amount to

(6.17) $$\begin{cases} \dot{z} - \dot{y} \in \partial \psi_C (y+c+g) \\ z \in U \quad , \quad y \in V \end{cases}$$

to be satisfied for almost every t in $[0,T]$.

Let us first draw a consequence of (6.17).

PROPOSITION. <u>If conditions</u> (6.17) <u>are verified for almost every</u> t, <u>the function</u> $t \mapsto y$ <u>satisfies for these values of</u> t

(6.18) $$-\dot{y} \in \partial \psi_{(C-c-g)\cap V}(y) \quad ;$$

<u>in other words this function is a solution of the sweeping process by the non empty closed convex moving set</u> $t \mapsto (C-c(t)-g(t)) \cap V$.

In fact the second line of (6.17) implies $-\dot{z} \in U$, thus

$-\dot{z} \in \partial \psi_V(y)$. Elementary calculation concerning translation in the space H yields

$$\partial \psi_C(y+c+g) = \partial \psi_{C-c-g}(y) .$$

On the other hand

$$\psi_{(C-c-g) \cap V} = \psi_{C-c-g} + \psi_V ,$$

thus

$$\partial \psi_{C-c-g}(y) + \partial \psi_V(y) \subset \partial \psi_{(C-c-g) \cap V}(y) .$$

Therefore (6.18) follows from the first line of (6.17).

REMARK. As y and $\dot{y}$ essentially belong to V, it is indifferent to understand the subdifferential in (6.18) in the sense of the duality between H and itself or in the sense of the duality between the Hilbert subspaces V and itself.

COROLLARY 1. If two solutions of (6.17) agree with the same initial condition $y(0) = y_o$ they coincide in what concerns the function $t \mapsto y$.

As explained in § 5. f, this uniqueness property follows from the multimapping $\partial \psi_{(C-c-g) \cap V}$ being monotone.

In view of the definition (6.14) of y this Corollary is equivalent to

COROLLARY 2. If two solutions of the system of conditions (6.7) to (6.10) agree with the same initial condition $s(0) = s_o$, these two solutions coincide in what concerns the function $t \mapsto s$.

By the way, (6.18) implies under Assumption 1 that the function $t \mapsto s$ related to y by (6.14) verifies, for almost every t,

$$- \dot{s} \in - \dot{g} + \partial \psi_{C \cap (V + c)} (s) \quad , \tag{6.19}$$

an evolution "equation" analogous to that of the sweeping process. An algorithm of time discretization would also be available for the numerical solution of it.

6. d EXISTENCE THEOREM

Let us proceed to the proof of :

PROPOSITION. Under Assumptions 1, 3, 4, whichever is y_o in $V \cap (C - c(0) - g(0))$, whichever is z_o in U, there exists at least one pair of functions $t \mapsto y$ and $t \mapsto z$, absolutely continuous from $[0,T]$ into H, satisfying (6.17) for almost every t and the initial conditions $y(0) = y_o$, $z(0) = z_o$.

First step. Under the hypotheses made there exists an absolutely continuous function, let us already denote it by $t \mapsto y$, satisfying (6.18) for almost every t and the initial condition $y(0) = y_o$. In fact this function is the solution of the sweeping process, for this initial condition, by the moving convex set $t \mapsto (C - c(t) - g(t)) \cap V$. The existence theorem of § 5. g apply because $t \mapsto C - c(t) - g(t)$ is absolutely continuous (see § 5. b about a translating convex set), thus the considered intersection is also absolutely continuous, by virtue of Assumptions 3 and 4 and the intersection theorem of § 5. c. Defining y in this way, one has $y(t) \in V$ for every t, thus $\partial \psi_V(y) = U$. The additivity of the subdifferentials holds for the functions ψ_{C-c-g} and ψ_V since, by Assumption 3, $V + c + g$ intersects the interior of C (recall that $g \in V$) so that there exists a point at which both functions are finite and the function ψ_{C-c-g} is continuous ; then (6.18) implies for almost every t he existence of at least one element of U, which will be already denoted as $\dot{z}(t)$, such that

(6.20) $$\dot{z}(t) - \dot{y}(t) \in \partial \psi_{C-c-g}(y(t)) = \partial \psi_C(y(t) + c(t) + g(t)).$$

This is the first of conditions (6.17).

Second step. For a value of t such that (6.20) holds the point $\dot{z} - \dot{y}$ is a conjugate of the point $y + c + g$ relatively to the pair of dual functions γ, namely the support function of C, and ψ_C (see § 2. e, Example). This may be written as

(6.21) $$\gamma(\dot{z} - \dot{y}) - (\dot{z} - \dot{y} \mid y + c + g) = 0$$

which implies that for almost every t, the closed convex set

(6.22) $$\Phi(t) = \{w \in H \ : \ \gamma(w) - (w \mid y+c+g) = 0\}$$
$$= \{w \in H \ : \ \gamma(w) - (w \mid y+c+g) \leqslant 0\}$$

possesses a nonempty intersection with the affine manifold $U - \dot{y}(t)$. As γ is a numerical function independent of t and as $t \mapsto y+c+g$ is a continuous mapping from $[0,T]$ into H one observes that $t \mapsto \Phi(t)$ is a <u>measurable multimapping</u> from $[0,T]$ into H (the measurability theory of multimappings is due for a part to C. CASTAING ; see his lectures ; see also, for an exposition of some basic facts in the case of a separable space, R.T. ROCKAFELLAR [4]). Such is also the multimapping $t \mapsto U-\dot{y}(t)$, as the function $t \mapsto \dot{y}$ belongs to $L^1(0,T ; H)$; thus the intersection of the two multimappings is measurable too. Since for almost every t this intersection is nonempty, it possesses a dense collection of measurable <u>selectors</u>. Denote by $t \mapsto \dot{p}(t)$ one of these selectors ; as $\dot{p}(t) \in U - \dot{y}(t)$, by putting $\dot{z}(t) = \dot{p}(t) + \dot{y}(t)$ one has $\dot{z}(t) \in U$ and (6.20) holds for almost every t. If we succeed in proving that $\dot{p}$, thus $\dot{z}$, belong to $L^1(0,T ; H)$, the primitive z of $\dot{z}$ adjusted to the initial value $z(0) = z_0$, will constitute with the function y determined above one of the desired solutions of (6.17).

<u>Third step</u>. As $t \mapsto \dot{p}(t)$ is measurable it just remains to prove that the numerical function $t \mapsto |\dot{p}(t)|$ is majorized by an element of $L^1(0,T;\mathbb{R})$. By the lemma of § 6. b there exists a strictly positive constant ρ and a continuous function $h : [0,T] \to H$ such that for every t one has $h(t) \in V + c(t)$ and the ball with center $h(t)$ and radius ρ is contained in C. This inclusion of convex sets is equivalent to the following inequality between their support functions

(6.23) $$\forall\, w \in H \ : \ \rho\, |w| + (h(t)|w) \leqslant \gamma\,(w) \,.$$

The definition (6.22) of $\Phi\,(t)$ may be transformed by writing

$$(w|y+c+g) = (w|h) + (w+\dot{y}|y+c+g-h) - (\dot{y}|y+c+g-h) \,.$$

Recall that $\dot{p} + \dot{y} \in U$, that $c - h \in V$, that $g \in V$, that $\dot{y} \in V$, that $c \in U$; then

$$(\dot{p}|y+c+g) = (\dot{p}|h) - (\dot{y}|y+g-h) \quad .$$

Therefore, in view of (6.23), $\dot{p} \in \Phi\,(t)$ implies

$$|\dot{p}| \leqslant \frac{1}{\rho}\,(\dot{y}|y+g-h) \leqslant \frac{1}{\rho}\,|\dot{y}|\;|y+g-h| \leqslant \frac{M}{\rho}\,|\dot{y}|$$

where M denotes a majorant of the continuous functions $t \mapsto |y+g-h|$ over the compact interval $[0,T]$. As a solution of the sweeping process, the function $t \mapsto y$ is absolutely continuous, thus the function $t \mapsto \dot{y}$ belongs to $L^1\,(0,T\;;\,H)$; this completes the proof.

By the definitions of y and z, it follows :

COROLLARY. Under Assumptions 1, 2, 3, 4 the evolution problem for the considered elastoplastic system possesses at least one solution ; this solution is unique in what concerns the function $t \mapsto s$.

REFERENCES

BREZIS, H. [1] Operateurs maximaux monotones et semi-groupes de contractions dans les espaces de Hilbert, Cours de 3ème Cycle, Université de Paris 6, 1971 et : Math. Studies, 5, North Holland, 1973.

[2] Un problème d'évolution avec contraintes unilatérales dépendant du temps, C.R. Acad. Sci. Paris, Sér. A 274 (1972), 310-312.

CRANDALL, M. and PAZY, A. [1] Semi-groups of nonlinear contractions and dissipative sets, J. Funct. Anal., 3 (1969), 376-418.

DUVAUT, G. and LIONS, J.L. [1] Les inéquations en mécanique et en physique, Dunod, 1972.

FENCHEL, W. [1] On conjugate convex functions, Canad. J. of Math., 1 (1949), p. 73-77.

GERMAIN, P. [1] Cours de mécanique des milieux continus, Vol. I, Masson, 1973.

KOMURA, Y. [1] Nonlinear semi-groups in Hilbert spaces, J. Math. Soc. Japan, 19 (1967), 493-507.

KRATOCHVIL, J. and NECAS, J. [1] On the solution of the traction boundary-value problem for elastic-inelastic materials, to appear.

LAURENT, P.J. [1] Approximation et optimisation, Hermann, 1972

MINTY, G.J. [1] On the monotonicity of the gradient of a convex function, Pac. J. Math., 14 (1964), p. 243-247.

MOREAU, J.J. [1] Inf-convolution des fonctions numériques sur un espace vectoriel, C. R. Acad. Sci. Paris, 256 (1963), p. 5047-5049.

J. J. Moreau

[2] Fonctionnelles sous-différentiables, C. R. Acad. Sci. Paris, 257 (1963), p. 4117-4119.

[3] Sur la fonction polaire d'une fonction semi-continue supérieurement, C. R. Acad. Sci. Paris, 258 (1964), p. 1128-1131.

[4] Théorèmes "inf-sup", C. R. Acad. Sci. Paris, 258 (1964), p. 2720-2722.

[5] Sur la naissance de la cavitation dans une conduite, C. R. Acad. Sci. Paris, 259 (1965), 3948-3950.

[6] Proximité et dualité dans un espace hilbertien, Bull. Soc. Math. France, 93 (1965) p. 273-299.

[7] One-sided constraints in hydrodynamics, in : ABADIE, J., editor, Non linear programming, North Holland Pub. Co., 1967, p. 261-279.

[8] Quadratic programming in mechanics : Dynamics of one-sided constraints, SIAM J. on Control 4 (1966), 153-158 (Proceedings of the First Int. Conf. on Programming and Control).

[9] Principes extrémaux pour le problème de la naissance de la cavitation, J. de Mécanique, 5 (1966) p. 439-470.

[10] Fonctionnelles convexes, Séminaire sur les Equations aux Dérivées partielles, Collège de France, Paris 1966-67 (multigraph 108 p.).

[11] La notion de sur-potentiel et les liaisons unilatérales en élastostatique, C. R. Acad. Sci. Paris, Sér. A, 267 (1968), p. 954-957.

[12] Convexité et frottement, Université de Montréal, Département d'Informatique, publ. n° 32, 1970 (multigraph 30 p.).

[13] Sur les lois de frottement, de plasticité et de viscosité, C. R. Acad. Sci. Paris, Sér. A, 271 (1970), p. 608-611.

[14] Mécanique classique, Vol. I, 1969 ; Vol. 2, 1971 ; (Masson, Paris).

J. J. Moreau

[15] Sur l'évolution d'un système élasto-visco-plastique, C. R. Acad. Sci. Paris, Sér. A, 273 (1971), p. 118-121.

[16] Fonctions de résistance et fonctions de dissipation, Séminaire d'Analyse convexe, Montpellier 1971, exposé n^0 6 (31 p.).

[17] Rafle par un convexe variable, lère partie, Séminaire d'Analyse Convexe, Montpellier, 1971, exposé n^0 15 (42 p.) ; 2ème partie, ibid, 1972, exposé n^0 3 (36 p.).

[18] Rétraction d'une multiapplication, Séminaire d'Analyse Convexe, Montpellier, 1972, exposé n^0 13 (90 p.).

[19] Intersection de deux convexes mobiles, Séminaire d'Analyse Convexe, Montpellier, 1973, exposé n^0 1 (26 p.).

[20] Sélections de multiapplications à rétraction finie, C. R. Acad. Sci. Paris, Sér. A, 276 (1973), p. 265-268.

[21] Problème d'évolution associé à un convexe mobile d'un espace hilbertien, C. R. Acad. Sci. Paris, Sér. A, 276 (1973), p. 791-794.

[22] Intersection de deux convexes mobiles dans un espace normé, C. R. Acad. Sci. Paris, Sér. A, 276 (1973), p. 1505-1508.

[23] Systèmes élastoplastiques de liberté finie, Séminaire d'Analyse Convexe, Montpellier, 1973, exposé n^0 12 (30 p.).

NAYROLES, B. [1] Essai de théorie fonctionnelle des structures rigides plastiques parfaites, J. de Mécanique, 9 (1970), p. 491-506.

[2] Quelques applications variationnelles de la théorie des fonctions duales à la mécanique des solides, J. de Mécanique, 10 (1971), p. 263-289.

[3] Opérations algébriques en mécanique des structures, C. R. Acad. Sci. Paris, Sér. A, 273 (1971), p. 1075-1078.

NECAS, J. [1] Application of Rothe's method to abstract parabolic equations, to appear.

NGUYEN, Q.S. [1] Matériau élastoplastique écrouissable. Distribution de la contrainte dans une évolution quasi-statique, to appear in : Archives of Mechanics.

PERALBA, J.C. [1] Un problème dévolution relatif à un opérateur sous-différentiel dépendant du temps, C. R. Acad. Sci. Paris, Sér. A, 275 (1972), p. 93-96.

[2] Equations d'évolution dans un espace de Hilbert, associées à des opérateurs sous différentiels, Thèse de 3ème Cycle, Université des Sciences et Techniques du Languedoc, Montpellier 1973.

PRAGER, W. [1] Unilateral constraints in mechanics of continua, Atti del Conv. Lagrangiano, Torino, Accademia delle Scienze, 1964, p. 181-190.

ROCKAFELLAR, R.T. [1] Integrals which are convex functionals, Pacific J. Math. 24 (1968), p. 525-539.

[2] Convex analysis, Princeton University Press, 1970.

[3] Integrals which are convex functionals II, Pacific J. Math., 39 (1971), p. 439-469.

[4] Convex integral functionals and duality, in : E.H. Zarantonello, ed., Contributions to non linear functional analysis, Academic Press, 1971, p. 215-236.

CENTRO INTERNAZIONALE MATEMATICO ESTIVO

(C.I.M.E.)

POINT DE VUE ALGÉBRIQUE. CONVEXITÉ ET INTEGRANDES CONVEXES EN MÉCANIQUE DES SOLIDES

B. NAYROLES

Corso tenuto a Bressanone dal 17 al 26 giugno 1973

B. Nayroles

Introduction

Au cours de ces six exposés on va développer ou seulement évoquer quatre idées principales :

1°) la première est que la structure algébrique d'un problème de Mécanique en constitue la nature profonde, indépendamment des considérations d'analyse fonctionnelle proprement dite : je crois qu'on attache actuellement une importance excessive à celle-ci au détriment de notions vraisemblablement plus importantes.

2°) la seconde est que les structures de dualité sont devenues l'arme la plus puissante du Mécanicien et que la mise en équations d'un problème devrait être toujours effectuée à l'aide du principe des travaux virtuels. Celui-ci, très vieille connaissance des mécaniciens, conduit directement et sans difficultés à des problèmes improprement qualifiés de "faibles". Dans le cas où l'on suppose les fonctions régulières celles-ci satisferont, certes, certaines équations aux dérivées partielles et certaines conditions aux limites, mais nous n'avons aucune raison d'accorder à celles-ci une signification physique plus grande qu'aux équations "variationnelles", c'est à dire de dualité, fournies par le principe du travail virtuel.

3°) la troisième idée est que nous traitons bien souvent des problèmes académiques : nous supposons en effet connues les données nécessaires à la formulation d'un problème bien posé. Les situations pratiques sont fort différentes et présentent le plus souvent un manque d'information considérable en égard aux données des problèmes théoriques. Ainsi en est-il du problème d'évolution élastoplastique où dans la pratique on ne connait ni les conditions initiales ni les efforts exercés au cours du temps, mais seulement certains encadrements des uns et des autres. On peut encore essayer d'obtenir des résultats. D'une façon générale le défaut d'information devrait être systématiquement étudié par la théorie.

4°) La quatrième idée est qu'il y a peut être encore bien des "idées reçues" à reconsidérer dans la conception du rôle de la Mathématique dans les théories physiques. En particulier on peut se demander si la cohérence d'une théorie nécessite vraiment l'"existence des solutions", et si la relation de préordre habituelle entre équations :

"l'équation A est plus forte que l'équation B si toute solution de A est solution de B"

est bien celle qui convient à la mathématique d'une théorie physique.

J'incline à penser que non et pour des raisons qui seront exposées dans le chapitre III : essentiellement parce que toute mesure expérimentale est entachée d'erreur; et aussi parce que, les solutions "exactes" étant inaccessibles au calcul, il me semble plus objectif de les oublier au profit des "solutions approchées", l'approximation étant à définir en liaison avec l'imprécision numérique ou expérimentale.

Ce chapitre III souffre de plusieurs défauts : une rédaction qui pourrait être améliorée, et qui sera sans doute entièrement refondue dans un article ultérieur; le manque d'un résultat sur l'élimination qui m'a obligé, temporairement, à le pallier à l'aide de la proposition 9. On pourra aussi utiliser, plutôt que l'expression pesante : "suite de solutions arbitrairement approchées" (abrégée par s.s.a.b.), l'expression plus légère et plus imagée de "suite approximante".

Quelques exemples seraient, en outre, souhaitables. J'ai cependant voulu saisir l'occasion du C.I.M.E. pour exprimer un certain nombre de réflexions sur ce genre de problème, essentiellement pour attirer l'attention sur leur importance.

Dans le premier chapitre (1 exposé) on expose le thème algébrique couramment rencontré en Mécanique des Solides. Dans le second (2 exposés) on utilise ce thème pour étudier les divers problèmes variationnels qui peuvent être formulés lorsque les lois d'effort sont sous-différentielles. La lecture en suppose une connaissance relativement bonne de la théorie des fonctions convexes (cfr. lectures de J.J. MOREAU et de J. CASTAING. Dans le troisième chapitre (2 exposés) on essaie de construire une algèbre (au sens large de ce terme) des lois sous-différentielles et pour y parvenir on introduit la relation de préordre "approximative" destinée à remplacer la relation usuelle entre équations.

Tout bien considéré on s'apercevra aisément que les deux clés de ce qui va nous intéresser sont

1°) l'égalité éventuelle d'une inf-convolution à une Γ-convolution

2°) l'égalité éventuelle du sous-différentiel d'une somme à la somme des sous-différentiels

Remarque - Le sixième exposé oral traitait de "l'adaptation" des corps élastoplastiques, phénomène essentiel puisqu'il permet le calcul des structures malgré le manque d'information dont nous parlions précédemment.

Comme ce sujet fait l'objet d'une publication dont j'achève la rédaction avec un autre chercheur, il m'était bien évidemment interdit d'y consacrer ici un chapitre particulier. Au demeurant celui-ci aurait nui à l'unité du sujet traité.

B. Nayroles

CHAPITRE I

STRUCTURE ALGEBRIQUE DES PROBLEMES DE MECANIQUE DES SOLIDES

Dans ce chapitre nous définissons le cadre algébrique le plus souvent rencontré dans les problèmes puisqu'il correspond au cas où déplacements et déformations sont supposés très petits et où la géométrie du solide est donnée. Notons que celle-ci est l'inconnue de certains problèmes d'optimisation.

Nous nous proposons d'aboutir à ce cadre algébrique non pas en le déduisant des équations habituelles de la Mécanique, mais en le construisant au fur et à mesure de l'établissement de celles-ci, ce qui sera beaucoup plus instructif. Afin d'alléger l'exposé nous nous référerons à l'exemple des plaques en flexion. Nous prions aussi le lecteur de ne prêter à ce qui suit aucune ambition dogmatique, mais de le considérer comme un rassemblement de quelques idées simples qu'il pourra admettre ou rejeter selon ses propres convictions épistémologiques.

Enfin nous employons souvent les termes "déplacement", "déformation" ou "contrainte" en guise de raccourcis pour les termes "champ de déplacements", de déformation ou de contraintes.

§ 1. DEPLACEMENTS ET EFFORTS

Considérons une plaque plane dont nous voulons étudier ce qu'il est convenu d'appeler les "déformations de flexion"; nous rapportons l'espace euclidien à un repère orthonormé $(O,\vec{x},\vec{y},\vec{z})$ en sorte que, dans son état de référence, la plaque occupe un domaine compact $\overline{\Omega}$ du plan (xOy); Ω désigne l'intérieur de $\overline{\Omega}$. On suppose que le déplacement $\overrightarrow{M_oM}$ d'un point M_o de la plaque est un vecteur "vertical" $u(M_o)\vec{z}$, u étant une fonction réelle définie sur $\overline{\Omega}$.

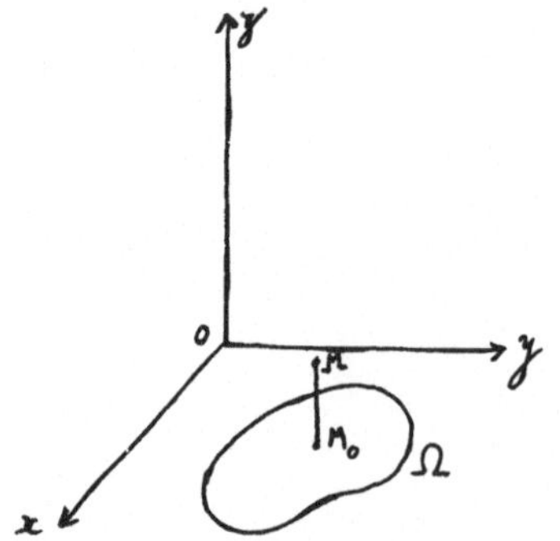

Le choix des fonctions u admissibles n'a pas besoin d'être précisé complètement.

Disons seulement que nous faisons, pour le moment, abstraction des conditions d'appui de la plaque et que nous nous donnons un espace de configuration U, espace vectoriel réel de fonctions réelles définies sur $\overline{\Omega}$. Pour fixer les idées nous pouvons prendre, ce qui se révèlera peut-être maladroit, $U = C^2(\overline{\Omega})$, à titre d'exemple seulement.

Une fois choisi l'espace de configuration U, ou espace des déplacements, on peut définir des efforts exercés sur la plaque. On peut considérer, par exemple, des efforts définis par des densités surfaciques, linéiques ou par des forces concentrées. Ainsi une densité surfacique

$$M \in \Omega \to p(M) \in \mathbb{R}$$

définit un effort ϕ dont le travail virtuel sera la forme linéaire sur U :

$$\text{(1)} \qquad u \in U \to \langle\langle u,\phi \rangle\rangle = \int_{M\in\Omega} u(M)\ \phi(M)dM$$

La Mécanique fondée sur le Principe du travail virtuel ne demande que la connaissance du travail virtuel $\langle\langle .,\phi\rangle\rangle$ développé par un effort ϕ; c'est à dire que la notion d'effort exercé sur un système se confond avec celle de fonctionnelle linéaire sur un espace vectoriel U qui est, dans le cas général, l'espace vectoriel tangent à une certaine variété de configuration (en général l'espace des vitesses), et qui, dans l'hypothèse des déplacements infiniment petits, se confond avec elle.

Ainsi donc l'espace des efforts envisageables est le dual algébrique U^* de l'espace de configuration U. C'est un peu trop grand pour être aisément manipulable, et on se restreint à un sous-espace vectoriel Φ de U^*.

On est alors dans la situation suivante : deux espaces vectoriels l'un de configuration U, l'autre des efforts Φ, mis en dualité par une forme bilinéaire $\langle\langle .,.\rangle\rangle$ qui est le travail virtuel, et coïncide avec (1) lorsque $\phi \in \Phi$ est défini par une densité surfacique de forces.

Bien sûr on peut aussi envisager beaucoup d'autres types d'effort : des couples répartis ou concentrés, par exemple, pourvu qu'on soit capable de définir leur travail virtuel.

D'une façon abstraite nous appellerons "élément mécanique" le triplet des deux espaces et de la forme bilinéaire qui les met en dualité :

$$\mathcal{E} = (U, \langle\langle .,. \rangle\rangle, \Phi)$$

et nous supposerons toujours que cette dualité est séparante.

Soient maintenant $\phi_1, \dots, \phi_n$ des efforts exercés sur la plaque : ils sont en équilibre, par définition, si leur somme est nulle :

(2) $$\phi_1 + \dots + \phi_n = 0$$

puisque, la dualité séparant Φ, ceci équivaut à dire que la somme de leurs travaux virtuels est nulle.

§ 2. LOIS D'EFFORTS

Dans les problèmes de Mécanique les efforts sont quelquefois donnés explicitement, comme les actions de pesanteur, mais plus souvent par des lois. Par exemple :

1°) Loi de résistance élastique : on se donne une application linéaire de U dans Φ :

$$\phi = -k(u)$$

2°) Loi de liaison affine parfaite : on se donne une variété affine $u_o + V$ de U et l'on pose que

$$u \in u_o + V$$

et qu'il y a un effort de liaison associé pouvant prendre n'importe quelle valeur dans l'orthogonal V^o.

3°) Loi fonctionnelle - Une telle loi fait intervenir le temps. Par exemple au "mouvement" $t \to u(t)$ on fait correspondre l'effort

$$\phi(t) = \int_{-\infty}^{t} r(t,\theta)\, u(\theta)\, d\theta$$

où r est une fonction donnée.

Oublions pour l'instant ce dernier exemple pour observer les deux premiers; les généralisant nous poserons la

<u>Définition</u> : on appelle loi d'effort sur l'élément mécanique $\mathcal{E}$ une multi-application A de U dans Φ :

$$\forall u \in U \to A(u) \subset \Phi$$

A(u) étant éventuellement vide.

Lorsque A(u) se réduit, pour tout u, à un seul élément, A est une fonction ordinaire, comme dans le premier exemple. Le second, par opposition, fait bien intervenir une multi-application en posant :

$$A(u) = \emptyset \quad \text{si} \quad u \notin u_o + V$$

$$A(u) = V^o \quad \text{si} \quad u \in u_o + V$$

Nous allons voir en effet que cette définition d'une liaison parfaite correspond bien à l'usage des mécaniciens.

En effet un problème usuel de Mécanique est le suivant : on se donne n lois d'effort $A_1, \ldots, A_n$, et on cherche à résoudre l' "équation d'équilibre" :

(3) $$0 \in A_1(u) + \ldots + A_n(u)$$

c'est à dire le système :

(3') $$\begin{cases} \forall \ i \in \{1, \ldots, n\} \qquad \phi_i \in A_i(n) \\ 0 = \phi_1 + \ldots + \phi_n \end{cases}$$

Si par exemple A_1 définit une liaison affine parfaite l'équation (3) n'admet évidemment de solution que dans la variété affine correspondante.

L'écriture (3) conduit à définir la somme de deux lois d'effort comme somme de deux multi-applications :

$$(A_1 + A_2)(u) = A_1(u) + A_2(u)$$

Remarque

Revenons maintenant à l'exemple 3°). Il ne relève évidemment pas de la définition d'une loi d'effort sur l'élément mécanique $\mathcal{E}$. Mais considérons un élément mécanique $\tilde{\mathcal{E}} = (\tilde{U}, \langle\langle .,. \rangle\rangle, \tilde{\phi})$

où $\tilde{U}$ est un espace de fonctions du temps à valeurs dans U

$\tilde{\phi}$ est un espace de fonctions du temps à valeurs dans Φ

et $$\langle\langle \tilde{u},\tilde{\phi} \rangle\rangle = \int_{t\in T} << \tilde{u}(t), \tilde{\phi}(t) >> dt$$

T désignant l'intervalle de variation du temps. Alors une loi fonctionnelle sur $\mathcal{E}$ pourra être définie comme une loi d'effort sur $\tilde{\mathcal{E}}$.

Ainsi, moyennant le remplacement de l'espace des configurations par celui des mouvements et les remplacements correspondants, nous pourrons faire entrer les lois de comportement dans le cadre des lois d'effort.

§ 3. DEPLACEMENTS SOLIDIFIANTS ET TORSEURS. PREMIER EXEMPLE D'ELEMENT MECANIQUE QUOTIENT

Les déplacements $u \in U$ qui sont des "déplacements" au sens des transformations géométriques sont les fonctions affines du couple $(x,y) = M$ On les nomme encore "déplacements solidifiants". Ils constituent un sous-espace vectoriel U_s de U. Pour le mécanicien il est équivalent de supposer la plaque indéformable ou d'imposer la liaison parfaite :

(4) $$u \in U_S$$

à laquelle on associe donc l'effort de liaison ϕ_S, arbitraire dans U_s°.

Supposons donc que la plaque soit soumise à la liaison d'indéformabilité précédente, donc à l'effort de liaison ϕ_S, et aussi à d'autres efforts $\phi_1, \ldots, \phi_n$ de somme ϕ. Le système composé de

$$\begin{cases} \phi_1 + \ldots + \phi_n + \phi_S = 0 \\ \phi_S \in U_S^{\circ} \end{cases}$$

est équivalent à l'équation

(5) $$\phi_1 + \ldots + \phi_n \in U_S^{\circ}$$

Considérons l'espace quotient $\Phi/U_S^{\circ} = \mathcal{C}$, et notons $\mathcal{L}$ l'application canonique de Φ sur $\mathcal{C}$. L'élément $t_i = \mathcal{L}(\phi_i) \in \mathcal{C}$ est dit "torseur associé à l'effort ϕ". Comme $\mathcal{L}$ est linéaire, l'équation d'équilibre (5) équivaut à

(6) $$t_1 + \ldots + t_n = 0$$

qui est du type (2). Cette ressemblance conduit à considérer la dualité définie entre U_s et $\mathcal{C}$ par

$$(u,t) \in U_S \times \mathcal{C} \longmapsto (u/t) = \langle\langle u,\phi\rangle\rangle$$

où ϕ est un élément arbitraire de $\mathcal{L}^{-1}(t)$. (6) est alors l'équation d'équilibre pour l'"élément mécanique quotient" (U_S, (./.), $\mathcal{C}$), par lequel nous pouvons remplacer l'élément mécanique initial.

Nous reviendrons, d'une façon plus systématique, sur cette notion de quotient et en particulier nous verrons ce qu'il advient d'une loi d'effort par passage au quotient.

§ 4. DEFORMATIONS ET CONTRAINTES

Soit $u \in U$ un champ de déplacements quelconque pour la plaque. On sait dire s'il est ou non solidifiant mais on ne sait pas encore définir à quelle déformation il correspond. D'un point de vue global nous pourrons dire que deux champs u et u' de U correspondent à la même déformation de la plaque s'ils diffèrent d'un déplacement solidifiant; autrement dit le point de vue global permet de définir un état de déformation de la plaque comme un élément de l'espace quotient U/U_s.

Etant donné un système matériel quelconque la Mécanique élémentaire considère d'habitude des efforts "extérieurs" ou "intérieurs" au système. Ces derniers ont pour propriété essentielle de ne pas travailler dans les déplacements solidifiants, c'est-à-dire d'être de torseur nul, et d'être donnés par des lois indépendantes de ceux-ci. Pour ménager la signification usuelle de l'expression "efforts intérieurs", nous emploierons l'adjectif "interne" et nous poserons la

Définition : Une loi d'effort A est dite interne à l'élément mécanique $\mathcal{E}$ si

1°) A est à valeurs dans U_S° (torseur nul)

2°) A est invariante par déplacement solidifiant :

$$\forall u \in U \qquad \forall v \in U_S \qquad A(u+v) = A(u)$$

Ceci signifie qu'une telle loi ne dépend que de la déformation. Cette définition constitue la tentative de caractérisation des efforts intérieurs la plus poussée qu'on puisse effectuer sous le seul point de vue global; elle ne permet pas de rendre compte de la notion de matériau et de loi constitutive : cette notion est essentiellement locale. C'est pourquoi il faut introduire une définition locale de la déformation et des efforts intérieurs.

Nous avons donc à identifier tout d'abord l'espace quotient U/U_S, parfaitement abstrait, à un espace vectoriel de champs définis sur Ω. Considérons l'opérateur différentiel

$$\text{grad grad } u = \begin{vmatrix} \dfrac{\partial^2 u}{\partial x^2} & \dfrac{\partial^2 u}{\partial x \partial y} \\ \dfrac{\partial^2 u}{\partial x \partial y} & \dfrac{\partial^2 u}{\partial y^2} \end{vmatrix} = u_{,i,j}$$

Il associe à tout $u \in U$ (on a supposé par exemple $U = C^2(\bar{\Omega})$) un champ de tenseurs symétriques :

$$M \in \Omega \mapsto \text{grad grad } u(M) \in \mathcal{E}_3 \quad \text{(espace vectoriel de dimension 3)}$$

et

(7) $$u \in U_S \iff \text{grad grad } u = 0$$

Une des représentations possibles du quotient U/U_S est donc l'espace vectoriel des champs grad grad u. C'est évidemment la plus simple et elle est utilisée dans la théorie des plaques minces; mais nous savons qu'elle peut être insuffisante pour la définition des lois constitutives; admettons-la pour l'instant. Elle signifie que le tenseur de courbure est une information suffisante pour la connaissance de la déformation locale de la plaque.

On admet alors que l'effort ϕ_m développé par le matériau peut être défini à l'aide d'un champ de tenseurs symétriques s par :

(8) $$\langle\langle u, \phi_m \rangle\rangle = - \int_\Omega (\text{grad grad } u)(M).s(M)\, d\Omega$$

où le point signifie le double produit contracté des tenseurs :

$$a:b = \sum_{i,j=1,2} a_{ij}\, b_{ji}$$

s est appelé un champ de contraintes. On note que $\phi_m \in U_S{}^{\circ}$.

Ceci conduit à introduire un espace vectoriel réel E de champs de tenseurs symétriques définis sur Ω et tel que l'opérateur de déformation

$$D = \text{grad grad}$$

soit une application de U dans E (par exemple $E = (C^{\circ}(\Omega))^3$)

à introduire un autre vectoriel S de champs de tenseurs symétriques définis sur Ω, telle que la forme bilinéaire

$$(e,s) \in E \times S \mapsto \langle e,s \rangle = \int_\Omega e(M).s(M)\, d\Omega$$

existe, quitte éventuellement, à éteindre la signification de l'intégrale.

E est l'espace des champs de déformation, $D(U) \subset E$ celui des champs de déformations intégrables. S est l'espace des champs de contraintes.

- $\langle e,s\rangle$ est le travail du champ de contraintes s dans le champ de déformation e; le signe - correspond à la convention de signe qui consiste à considérer comme positives les contraintes de traction. Dans le cas de la plaque s est un champ de "tenseurs de flexion".

L'opérateur déformation D est une application de U dans E : il admet un transposé D^t, application de S dans le dual algébrique U^* de U; par définition

$$s \in S \mapsto D^t s \in U^* : \quad \langle\langle u, D^t s\rangle\rangle = \langle Du, s\rangle \qquad \forall u \in U$$

On note que

$$D^t s \in \Phi \implies D^t s \in U_S{}^\circ$$

On supposera que S a été choisi en sorte que

⑨ $$D^t(S) \supset U_S{}^\circ$$

autrement dit que tout élément $\phi_m \in U_S{}^\circ$ peut être représenté par au moins un champ de contraintes $s \in S$, c'est à dire, d'après ⑧ :

⑩ $$-D^t s = \phi_m$$

Supposons la plaque soumise à un effort quelconque ϕ et à l'effort ϕ_m représenté par s : ils sont en équilibre si

$$\phi + \phi_m = 0$$

c'est à dire, compte-tenu de ⑩, si

⑪ $$\phi = D^t s$$

qui est l'équation d'équilibre.

Notons que

$$s \in \ker(D^t) \iff \forall u \quad \langle\langle u, D^t s\rangle\rangle = 0 \iff \forall u \quad \langle Du, s\rangle = 0$$

ce qui montre que le noyau $\ker(D^t)$ est l'orthogonal de D(U).

Le diagramme ci-dessous résume la situation, parfaitement symétrique si (9) est remplacée par

(12) $$D^t S = U_S{}^\circ$$

$$\begin{array}{ccc}
\ker D = U_S \subset U & \langle\langle .,. \rangle\rangle & \Phi \supset U_S{}^\circ = D^t(S) \\
\Big\downarrow D & & \Big\uparrow D^t \\
D(U) \subset E & \langle .,. \rangle & S \supset \ker D^t = D(U)^\circ
\end{array}$$

Une loi d'effort C sur l'élément mécanique (E, <.,.>, S) est une loi constitutive si elle est définie localement par une multi-application Γ_M, dépendant de M, de $\mathcal{E}_3$ dans lui-même; par définition :

$$s \in C(e) \iff \forall M \in \Omega \qquad s(M) \in \Gamma_M(e(M))$$

La loi constitutive C définit sur l'élément mécanique (U, <<.,.>>, S) la loi d'effort interne A :

$$A = D^t \circ C \circ D$$

Du point de vue global seule la loi A est observable. Pour atteindre C c'est à dire pour effectuer des mesures susceptibles de fournir Γ_M il faudrait découper la plaque en morceaux infiniment petits.

§ 5. LIAISONS; DEPLACEMENTS ET DEFORMATIONS IMPOSES

En général une telle plaque est soumise à des conditions d'appui qui empêchent tout déplacement d'ensemble. Nous supposerons que la plaque est fixée par son bord; soit

$$\partial\Omega = \partial_1\Omega \cup \partial_2\Omega \cup \partial_3\Omega$$

une partition de la frontière $\partial\Omega$.

- Sur $\partial_1\Omega$ le bord est libre
- Sur $\partial_2\Omega$ le bord est simplement appuyé ce qui correspond à l'équation de liaison

(13) $$u_{/\partial_2\Omega} = u^o_{/\partial_2\Omega}$$

où u^o est un élément donné de U

- Sur $\partial_3\Omega$ le bord est encastré, ce qui correspond aux équations de liaison

(14) $$u_{/\partial_3\Omega} = u^o_{/\partial_3\Omega} \qquad \text{et} \qquad \frac{\partial u}{\partial n} = \frac{\partial u^o}{\partial n} \qquad \text{sur } \partial_3\Omega$$

Nous supposons donc que les équations de liaison sont affines et qu'il en existe une solution particulière u^o dans U. Alors les solutions du système (13) + (14) constituent une variété affine $u^o + V$ de U, c'est à dire que ces équations se résument à

(15) $$u \in u^o + V$$

V est l'espace vectoriel des déplacements virtuels compatibles avec les liaisons. Au niveau des déformations (15) entraîne

(16) $$e = Du^o + D(V)$$

et si les liaisons interdisent tout déplacement solidifiant, c'est à dire si, comme nous le supposerons toujours désormais :

$$V \cap U_S = \{0\}$$

alors (15) et (16) sont équivalents. Du^o est une "déformation imposée", u^o un "déplacement imposé". Notons cependant qu'on peut envisager aussi une déformation imposée e^o qui ne soit pas de la forme Du^o, par exemple une déformation d'origine thermique.

§ 6. EQUATIONS D'EQUILIBRE DE LA PLAQUE

Précisons maintenant l'opérateur D^t lorsque l'effort ϕ qui apparaît dans l'équation d'équilibre :

(11) $$D^t s = \phi$$

est défini par une densité surfacique de force p, une densité linéique de force f définie sur $\partial\Omega$, et d'une densité linéique de couple e définie sur $\partial\Omega$, en sorte que

(17) $$<<u,\phi>> = \int_{M\in\Omega} p(M)\ u(M) d\Omega + \int_{M\in\partial\Omega} f(M) u(M) d\xi + \int_{M\in\partial\Omega} C(M) \frac{\partial u}{\partial n}(M) d\xi$$

ξ désigne l'abscisse curviligne de la frontière.

Considérons tout d'abord un champ $s \in S$, et deux fois continûment dérivable sur Ω. Nous pouvons écrire :

$$\text{grad grad } u.s = u \text{ div div } s - \text{div } (u.\text{div } s) + \text{div } (\text{grad } u.s)$$

où grad u.s est le produit contracté du vecteur grad u par le tenseur s

Dans ces conditions nous pouvons écrire

$$<Du,s> = \int_\Omega (\text{grad grad } u).s$$

$$= \int_\Omega u \text{ div div } s + \int_{\partial\Omega} (\text{grad } u.s - u \text{ div } s).n$$

n désignant le vecteur unitaire normal extérieur à la frontière. Notons t le vecteur unitaire tangent à celle-ci en sorte que sur la frontière

$$\text{grad } u = \frac{\partial u}{\partial \xi} t + \frac{\partial u}{\partial n} n$$

Alors

$$<Du,s> = \int_\Omega u \text{ div div } s + \int_{\partial\Omega} \frac{\partial u}{\partial \xi} t.s.n + \int_{\partial\Omega} \frac{\partial u}{\partial n} n.s.n - \int_{\partial\Omega} u \text{ div } s.n$$

On peut transformer le second membre en intégrant par parties :

$$\int_{\partial\Omega} \frac{\partial u}{\partial \xi}\, t.s.n = - \int_{\partial\Omega} u \frac{\partial}{\partial \xi}\,(t.m.n)$$

et introduire le rayon de courbure algébrique ρ de la frontière :

$$\frac{dt}{\partial \xi} = \frac{n}{\rho} \qquad \frac{dn}{\partial \xi} = -\frac{t}{\rho}$$

ce qui donne finalement

(18) $$\langle Du,s\rangle = \int_{\Omega} u \operatorname{div} \operatorname{div} s - \int_{\partial\Omega} u\left\{\operatorname{div} s.n + \frac{1}{\rho}(n.s.n.-t.s.t)+t.\frac{\partial s}{\partial \xi}.n\right\} + \int_{\partial\Omega} \frac{\partial u}{\partial n}\, n.s.n$$

En identifiant 17 et 18 on trouve donc

(19) $$\begin{aligned} &(19.1) \quad \operatorname{div} \operatorname{div} s = p \\ &\operatorname{div} s., + \frac{1}{\rho}\,(n.s.n - t.s.t) + t.\frac{\partial s}{\partial \xi}\,.\,n = f \\ &n.s.n = c \end{aligned}$$

qui sont les équations d'équilibre, équivalentes à (11), d'un champ de contraintes régulier s et d'un effort φ défini par p, f et c.

Si f (nul sur $\partial_1\Omega$) et c (nul sur $\partial_1\Omega \cup \partial_2\Omega$) définissent l'effort de liaison tandis que p définit l'effort donné, alors on peut écrire les équations d'équilibre du champ de contraintes s et de l'effort donné, compte tenu des liaisons :

(20) $$\begin{aligned} &\operatorname{div} \operatorname{div} s = p && \text{sur } \Omega \\ &\operatorname{div} s.n + \frac{1}{\rho}\,(n.s.n - t.s.t) + t\,.\,\frac{\partial s}{\partial \xi}\,.\,n = 0 && \text{sur } \partial_1\Omega \\ &n.s.n = 0 && \text{sur } \partial_1\Omega \quad \partial_2\Omega \end{aligned}$$

L'écriture des conditions aux limites a longtemps posé des problèmes aux mécaniciens, et surtout pour le bord libre $\partial_1\Omega$. L'emploi des méthodes de travail virtuel résoud aisément cette difficulté.

Si s n'est pas un champ régulier l'équation (19.1) conserve un sens en considérant la distribution div div s. Par contre les équations sur la frontière n'ont plus de signification en général : la "trace" de s sur $\partial\Omega$ peut fort bien n'être pas définie. C'est là que réside un des avantages majeurs de la Mécanique du travail virtuel pour laquelle l'équation (11) a un sens précis.

C'est pourquoi le mécanicien envisagera toujours des problèmes posés par la méthode du travail virtuel, qui donne une formulation variationnelle directe. Au contraire l'écriture d'équations différentielles doit être évitée le plus souvent possible, du moins dans une première approche d'un problème de Mécanique des Solides.

§ 7. SITUATION DE REFERENCE POUR UN PROBLEME DE MECANIQUE DES SOLIDES

Les liaisons que nous avons imposées à la plaque sont parfaites : l'effort de liaison défini par f et par c développe un travail virtuel nul dans un déplacement virtuel compatible avec les liaisons; autrement dit cet effort de liaison appartient à V°. On peut remplacer Φ par le quotient $F = \Phi/V^\circ$ et l'élément mécanique $\mathcal{E}$ par l'élément quotient

$$(V, \langle\langle .,. \rangle\rangle, F)$$

avec une définition évidente de $\langle\langle .,. \rangle\rangle$. C'est la même démarche qu'au paragraphe 3. Notons Δ la restriction à V de l'opérateur D. Si $\mathcal{Q}$ désigne l'application canonique de ϕ sur F : $\Delta^t = \mathcal{Q} \circ D^t$, et l'on retrouve un schéma analogue à celui du paragraphe 5 :

$$\begin{array}{ccc} V & \langle\langle .,. \rangle\rangle & F = \Delta^t S \\ \downarrow \Delta & & \uparrow \Delta^t \\ e^\circ + \Delta(V) \subset E & \langle .,. \rangle & S \supset \ker(\Delta^t) = (\Delta(V))^\circ \end{array}$$

On a $\Delta^t S = F$ parce que V ne contient aucun déplacement solidifiant. On posera $\quad I' = \Delta(V) \qquad J = \ker(\Delta^t)$.

Supposons donné un effort $f \in F$, et connue une solution particulière s^o de l'équation d'équilibre

$$\Delta^t s^o = f$$

alors on peut se contenter d'étudier l'élément mécanique :

$$(E, <.,.>, S)$$

soumis

1°) à la liaison parfaite $e = e^o + I'$, l'effort de liaison associé étant un élément quelconque τ de J.

2°) à l'effort donné s^o

3°) à une loi constitutive C

On a alors à résoudre le système

$$(21) \quad \begin{cases} e \in e^o + I' & \text{(équation de liaison)} \\ s \in s^o + J & \text{(équation d'équilibre)} \\ s \in C(e) & \text{(loi constitutive)} \end{cases}$$

Par ailleurs on peut considérer V comme un espace de paramétrage de I' et F comme une représentation du quotient S/J. Autrement dit l'élément (V, <<.,.>>, F) apparaît comme élément mécanique quotient de l'élément (E, <.,.>, S).

B. Nayroles

CHAPITRE II

PROBLEMES D'EQUILIBRE POUR DES LOIS CONSTITUTIVES DE TYPE SOUS-DIFFERENTIEL

§ 1. HYPOTHESES GENERALES - EXEMPLES

Soit $(U, \langle\langle .,. \rangle\rangle, \Phi)$ un élément mécanique. On dit qu'une loi d'effort

$$u \in U \mapsto A(u) \subset \Phi$$

est sous-différentielle, s'il existe une application f de U dans $\bar{R}=[-\infty,+\infty]$ telle que

(1) $$A = - \partial f$$

Nous supposerons toujours que f est un élément de $\Gamma_o(U)$; autrement dit, pour les topologies localement convexes compatibles avec la dualité entre U et Φ :

- f est l'enveloppe supérieure de ses minorantes affines continues ou, ce qui est équivalent :

- f ne prend pas la valeur $-\infty$, et f est convexe et semi-continue inférieurement (cfr. J.J. MOREAU, [1], et son exposé sur les fonctions convexes à cette école d'été).

On a alors l'équivalence des quatre lignes suivantes regroupées dans une même accolade :

(2) $$\begin{cases} -\phi \in \partial f(u) & (2.1) \\ u \in \partial f^*(-\phi) & (2.2) \\ f(u) + f^*(-\phi) + \langle\langle u,\phi \rangle\rangle = 0 & (2.3) \\ f(u) + f^*(-\phi) + \langle\langle u,\phi \rangle\rangle \leq 0 & (2.4) \end{cases}$$

où f^* désigne la fonction conjuguée de f.(2.3) exprime que $(u, -\phi)$ est un couple de points conjugués par rapport au couple de fonctions conjuguées (ou duales) (f, f^*).

Celles-ci satisfont l'inégalité suivante, à la base des théorèmes d'extremum classiques :

$$(3) \qquad \forall (u,\phi) \in U \times \Phi \qquad f(u) + f^*(-\phi) + \langle\langle u,\phi\rangle\rangle \geqslant 0$$

d'où résulte d'ailleurs l'équivalence de (2.3) et (2.4).

Venons-en maintenant à quelques exemples usuels.

1.1 Effort ϕ° donné : $\forall u \in U \;:\; A(u) = \{\phi^\circ\}$

$$f = \langle\langle \cdot, -\phi^\circ\rangle\rangle \qquad f^* = \psi_{\{-\phi^\circ\}}$$

où ψ_B désigne, comme d'habitude la fonction indicatrice d'un ensemble B :

$$\psi_B(x) = \begin{cases} 0 & \text{si} \quad x \in B \\ +\infty & \text{si} \quad x \notin B \end{cases}$$

1.2 Liaison unilatérale convexe parfaite - On se donne un convexe fermé Γ de U (information cinématique), dans lequel on impose à la configuration u de rester. La généralisation naturelle de liaison unilatérale sans frottement est alors donnée par la loi d'effort sous-différentielle suivante (cfr. J.J. MOREAU (2)) :

$$f = \psi_C \qquad f^* = \psi^*_C$$

qui exprime que l'effort est nul si u est intérieur à C, dirigé suivant une normale rentrante si u est sur la frontière.

Comme cas particuliers :

$C = u^\circ + V$ alors $\psi^*_C = \langle\langle u^\circ, \cdot\rangle\rangle + \psi_{V^\circ}$ et $A(u) \equiv V^\circ$ ou $\emptyset$

$C = \{u^\circ\}$ $(V = \{0\})$ alors $\psi^*_C = \langle\langle u^\circ, \cdot\rangle\rangle$ et $A(u) \equiv \Phi$ ou $\emptyset$

1.3 Lois de type rigide plastique - C'est le cas dual du précédent; soit Γ un convexe fermé de l'espace Φ. On prend :

$$f = \psi^*_\Gamma \qquad f^* = \psi_\Gamma$$

Selon que U est un espace de configuration ou de vitesses on obtient la plasticité de HENCKY ou de PRANDTL-REUSS. Dans le second cas ψ^*_Γ est la fonction "puissance de dissipation plastique", et dans le premier l' "énergie dissipée".

1.4 Lois différentielles - f est une fonction faiblement différentiable. Alors :

$$u \in U \longmapsto \partial f(u) = \{\text{grad } f(u)\}$$

C'est en particulier le cas de l'élasticité où l'on suppose précisément l'existence de la fonction "énergie potentielle f" telle que

$$\phi = - \text{grad } f$$

La plupart des "lois de résistance" sont aussi de type sous-différentiel (cfr. J.J. MOREAU (3)).

1.5 Lois constitutives - Revenons à cette notion de loi constitutive; on considère l'élément mécanique (E, <.,.>, S) où E est l'espace des déformations, S celui des contraintes, <e,-s> le travail virtuel. Une loi constitutive de type sous-différentiel s'écrira donc sous la forme

$$f(e) + f^*(s) - \langle e,s \rangle = 0 \qquad (4)$$

où f et f^* sont conjuguées par rapport à la forme bilinéaire <.,.>.

La relation entre la forme globale (4) de la loi et sa forme locale sera étudiée plus loin au paragraphe 3.

§ 2. PROBLEME D'EQUILIBRE ET THEOREMES D'EXTREMUM

A la fin du chapitre I nous avons vu que la situation de référence était la suivante : l'élément mécanique (E, <.,.>, S) est soumis

1°) à la liaison parfaite d'équation $e \in e^o + I$, à laquelle est associé la contrainte inconnue de liaison $-\tau \in J = I^o$.

2°) à une contrainte donnée $-s^o$.

3°) à une loi constitutive que nous supposons ici de la forme 4.

On cherche à équilibrer ces trois lois de contraintes, d'ailleurs toutes de type sous-différentiel si I est fermé pour les topologies compatibles avec la dualité, ce que nous supposerons.

Notons $\mathcal{f}$, $\mathcal{f}_1$ et $\mathcal{f}_2$ les fonctionnelles :

$$(5)\quad \begin{cases} (e,s) \quad E \times S \mapsto \mathcal{f}(e,s) = f(e+e^o) + f^*(s+s^o) - \langle e+e^o, s+s^o\rangle \\ e \in E \to \mathcal{f}_1(e) = f(e+e^o) - \langle e+e^o, s^o\rangle \\ s \quad S \to \mathcal{f}_2(s) = f^*(s+s^o) - \langle e^o, s\rangle \end{cases}$$

$\mathcal{f}_1$ est dite "énergie de déformation", $-\mathcal{f}_2$ "énergie complémentaire". Le problème d'équilibre s'écrit :

Problème I : Trouver $(\varepsilon,\tau) \in I \times J$ tel que $\mathcal{f}(\varepsilon,\tau) = 0$

Mais $\mathcal{f}$ est positive d'après (3). Par conséquent elle est minimum si (ε,τ) est solution. D'autre part la restriction de $\mathcal{f}$ à $I \times J$ est la somme des restrictions de $\mathcal{f}_1$ à I et de $\mathcal{f}_2$ à J, puisque $\langle \varepsilon,\tau\rangle = 0$.

$$(6)\qquad \forall (\varepsilon,\tau) \in I \times J \qquad \mathcal{f}(\varepsilon,\tau) = \mathcal{f}_1(\varepsilon) + \mathcal{f}_2(\tau)$$

Par suite toute solution du problème I est solution du

Problème II - Minimiser la restriction de $\mathcal{J}$ à $I \times J$

et ce problème est équivalent à la réunion des deux problèmes partiels suivants :

Problème III - Minimiser la restriction de $\mathcal{J}_1$ à I (i.e. min. $\mathcal{J}_1 + \psi_I$)

Problème IV - Minimiser la restriction de $\mathcal{J}_2$ à J (i.e. min. $\mathcal{J}_2 + \psi_J$)

Nous allons étudier l'équivalence éventuelle de ces divers problèmes. Notons que $\mathcal{J}_1$ et $\mathcal{J}_2$ appartiennent respectivement à $\Gamma_o(E)$ et à $\Gamma_o(S)$. Plus encore :

Proposition 1 - $\mathcal{J}_1$ et $\mathcal{J}_2$ sont un couple de fonctions conjuguées.

En effet, en utilisant les règles de calcul classiques :

$$\mathcal{J}^*_1 = \{f(e^o+.) - \langle e^o+.,s^o\rangle\}^* = \{f - \langle .,s^o\rangle\}^* - \langle e^o,.\rangle$$
$$= f^*(s^o+.) - \langle e^o,.\rangle = \mathcal{J}_2$$

Si nous considérons maintenant $\mathcal{J}_1 + \psi_I$, que minimise le problème III, sa fonction duale est, par définition

$$(\mathcal{J}_1 + \psi_I)^* = \mathcal{J}_2 \,\underline{\nabla}\, \psi_J$$

et l'on sait que cette "Γ-convolution" est la Γ-régularisée de l'inf-convolution (cfr. J.J. MOREAU (1) ch. 9)

$$(\mathcal{J}_2 \nabla \psi_J)^{**} = \mathcal{J}_2 \,\underline{\nabla}\, \psi_J$$

Nous allons revenir sur les cas classiques d'égalité entre la Γ et l'inf-convolution; supposons qu'elle ait lieu; alors :

Proposition 2 - Les trois assertions suivantes sont équivalentes

(7) $\quad (\mathcal{J}_2 \,\underline{\nabla}\, \psi_J)(0) = (\mathcal{J}_2 \nabla \psi_J)(0)$

(8) $\quad (\mathcal{J}_1 \,\underline{\nabla}\, \psi_I)(0) = (\mathcal{J}_1 \nabla \psi_I)(0)$

$$(9) \qquad \inf_{(\varepsilon,\tau) \in I \times J} \mathcal{J}(\varepsilon,\tau) = 0$$

En effet on a

$$\inf_{e \in E} (\mathcal{J}_1 + \psi_I)(e) = \inf_{e \in E} \left(\mathcal{J}_1(e) + \psi_I(-e)\right) = (\mathcal{J}_1 \nabla \psi_I)(0)$$

et d'autre part

$$\inf_{e \; E} (\mathcal{J}_1 + \psi_I)(e) = -(\mathcal{J}_2 \,\underline{\nabla}\, \psi_J)(0)$$

D'où en utilisant aussi les résultats analogues sur $\mathcal{J}_2 + \psi_J$:

$$\inf_{(\varepsilon,\tau) \in I \times J} \mathcal{J}(\varepsilon,\tau) = (\mathcal{J}_1 \nabla \psi_I)(0) - (\mathcal{J}_1 \,\underline{\nabla}\, \psi_I)(0) = (\mathcal{J}_2 \nabla \psi_J)(0) - (\mathcal{J}_2 \,\underline{\nabla}\, \psi_J)(0)$$

ce qui démontre la proposition.

C'est un résultat très puissant qui permet de considérer comme pratiquement inutile tout résultat d'existence, puisqu'il affirme l'existence, beaucoup plus intéressante, de "solutions arbitrairement approchées" c'est à dire de couples (ε,τ) rendant $\mathcal{J}$ arbitrairement petite.

Rappelons maintenant, dans une même proposition, deux cas classiques d'égalité de l'inf-convolution et de la Γ-convolution (cfr. J.J. MOREAU [1] ch. 9).

Proposition 3 - Soient (f,f^*), (g,g^*) deux couples de fonctions duales.

1°) Si l'ensemble cont. (f) des points où f est finie et continue et l'ensemble dom. (g) des points où g est finie satisfont

$$\text{dom. } (g) + \text{cont. } (f) = E$$

alors $\qquad f \nabla g = f \,\underline{\nabla}\, g$

2°) S'il existe un point a de E où f est finie et continue et où g est finie, alors

$$f^* \nabla g^* = f^* \,\underline{\nabla}\, g^*$$

et cette inf-convolution est exacte c'est à dire que l'inf est un min. :

$$\forall s \in S \quad \exists s' \in S \quad (f^* \nabla g^*)(s) = f^*(s-s') + g^*(s')$$

$$= \min_{t \in S} \{f^*(s-t) + g^*(t)\}$$

On peut évidemment inverser les rôles joués par E et S, f et f^*, g et g^*. Le second cas fournit, outre l'égalité cherchée, un résultat d'existence supplémentaire. Appliquons cette proposition aux couples de fonctions duales $(\mathcal{f}_1, \mathcal{f}_2)$, (ψ_I, ψ_J). Supposons par exemple que la condition 1°) soit satisfaite; alors

$$0 \in \text{dom}\,(\psi_I) + \text{cont}\,\mathcal{f}_1$$

puisqu'en général cont. (ψ_I) est vide (sinon I = E). Alors :

$$\exists \tau \in I \qquad \tau \in \text{cont.}\,(\mathcal{f}_1)$$

de sorte que nous sommes aussi dans le second cas. Pour l'application de la proposition 3 à la recherche des cas de validité des hypothèses de la proposition 2, il suffit donc de s'intéresser au second cas.

Si celui-ci a lieu nous avons :

$$(\mathcal{f}_1 + \psi_I)^* = \mathcal{f}_2 \nabla \psi_J = \min_{\tau \in J} \mathcal{f}_2(\tau)$$

de sorte que, outre le résultat d'équivalence fourni par la proposition 2 nous avons celui de l'existence d'au moins une solution pour le problème IV. En regroupant les résultats qui précèdent :

<u>Théorème 1</u> - S'il existe un point de I où f(e°+.) est finie et continue (resp. un point de J où f^*(s°+.) est finie et continue), alors :

$$\inf_{(\varepsilon,\tau)\ \in\ I\times J} \mathcal{f}(\varepsilon,\tau) = 0$$

et en particulier les problèmes I et II sont équivalents. De plus le problème IV (resp. III) admet au moins une solution.

L'unicité des solutions sera assurée dès que seront strictement convexes les fonctionnelles $\mathcal{J}_1$ et $\mathcal{J}_2$.

Les problèmes III et IV, et le théorème d'équivalence I correspondent aux "principes variationnels" les plus fréquemment rencontrés. On voit qu'ils dérivent tous de la même structure mathématique.

§ 3. FORMULATION DES PROBLEMES DE POINT-SELLE

Nous avons dit que les problèmes III et IV étaient duaux l'un de l'autre : cela signifie qu'ils sont les problèmes d'extremum associés à un problème de point-selle. Rappelons rapidement de quoi il s'agit :

Soient $(E, <.,.>, S)$, $(X, (./.), Y)$ deux couples d'espaces vectoriels en dualité et L un Lagrangien :

$$(e,y) \in E \times Y \to L(e,y)$$

fonction selle du couple (e,y), c'est à dire

$$\forall e \in E \qquad L(e,.) \text{ est concave}$$
$$\forall y \in Y \qquad L(.,y) \text{ est convexe}$$

On considère le problème suivant :

Problème P.S. Trouver (e_1,y_1), point-selle de L, c'est à dire solution de la double inéquation

$$\forall (e,y) \in E \times Y \qquad L(e_1,y) \leq L(e_1,y_1) \leq L(e,y_1)$$

On a $\forall (e,y) \in E \times Y \ : \ L(e,y) \leq \sup_{z\in Y} L(e,z)$

d'où $\forall y \in Y \quad \inf_{e\in E} L(e,y) \leq \inf_{e\in E} \sup_{z\in Y} L(e,z)$

et enfin

$$\sup_{y\in Y} \inf_{e\in E} L(e,y) \leq \inf_{e\in E} \sup_{y\in Y} L(e,y) \tag{9}$$

Supposons maintenant qu'il existe un point-selle; alors

$$\inf_{e\in E} \sup_{y\in Y} L(e,y) \leq \max_{y\in Y} L(e_1,y) = L(e_1,y_1) = \min_{e\in E} L(e,y_1) \leq \sup_{y\in Y} \inf_{e\in E} L(e,y)$$

de sorte que, par comparaison avec (9), on obtient

$$L(e_1,y_1) = \min_{e\in E} \sup_{y\in Y} L(e,y) = \max_{y\in Y} \inf_{e\in E} L(e,y) \tag{10}$$

<u>Théorème</u> - Si le problème de point-selle admet (e_1,y_1) pour solution alors on a l'égalité (10) et e_1 et y_1 sont respectivement solutions des problèmes P (primal) et D (dual) suivants :

<u>Problème P</u> : Trouver $e_1 \in E$ tel que

$$\sup_{y\in Y} L(e_1,y) = \min_{e\in E} \sup_{y\in Y} L(e,y)$$

<u>Problème D</u> : Trouver $y_1 \in Y$ tel que

$$\inf_{e\in E} L(e,y_1) = \max_{y\in Y} \inf_{e\in E} L(e,y)$$

Utilisons maintenant les transformations de Fenchel partielles :

$$\Phi(e,x) = (-L_e)^*(x) = \sup_{y\ Y} \{(x/y) + L(e,y)\}$$

$$\chi(s,y) = (L_y)^*(s) = \sup_{e\ E} \{\langle e,s\rangle - L(e,y)\}$$

On a

$$\sup_{y\in Y} L(e,y) = \Phi(e,0) \tag{11}$$

$$\inf_{e\in E} L(e,y) = - \chi(0,y)$$

Par suite le problème primal consiste à minimiser $\Phi(e,0)$ et le problème dual à maximiser $-\chi(0,y)$. On démontre aisément la :

<u>Proposition 4</u> - Si L appartient à $\Gamma_o(E \times Y)$ les fonctions Φ et χ sont conjuguées pour la dualité :

$$(E \times X\ ,\quad \langle .,. \rangle + (./.)\ ,\quad S \times Y)$$

et les solutions du problème de point-selle sont celles de l'équation de conjugaison :

$$\Phi(e,0) + \chi(0,y) = 0 \tag{12}$$

Revenons maintenant aux problèmes variationnels étudiés précédemment, et prenons

$$E = X \qquad S = Y \qquad (./.) = \langle .,. \rangle$$

$$\Phi_1(e,x) = \mathcal{J}_1(e+x) + \psi_I(e)$$

Le problème III est de minimiser $\Phi_1(e,0)$: c'est donc le problème primal associé au problème de point-selle pour le lagrangien

$$L_1(e,y) = \psi_I(e) - \mathcal{J}_2(y) + \langle e,y \rangle \tag{13}$$

La duale de Φ_1 est

$$\chi_1(s,y) = \mathcal{J}_2(y) + \psi_J(s-y)$$

et comme J est un espace vectoriel

$$\chi_1(0,y) = \mathcal{J}_2(y) + \psi_J(y)$$

de sorte que le problème dual est le problème IV. Par ailleurs le problème de point-selle pour L est équivalent à celui de résoudre l'équation de conjugaison

$$(\mathcal{J}_1 + \psi_I)(e) + (\mathcal{J}_2 + \psi_J)(y) = 0$$

<u>c'est à dire au problème I.</u>

On peut choisir aussi, pour le problème III, la "fonction de perturbation"

$$\Phi_2(e,x) = \mathfrak{f}_1(e) + \psi_I(e + x)$$

d'où le lagrangien L_2 et la fonction duale χ_2

$$(14) \qquad L_2(e,y) = \mathfrak{f}_1(e) - \psi_J(y) + \langle e,y\rangle$$

$$\chi_2(s,y) = \mathfrak{f}_2(s - y) + \psi_J(y)$$

ce qui conduit encore au problème IV comme problème dual, mais à un problème de point-selle différent mais lui aussi équivalent au problème I. D'où le

<u>Théorème II</u> - Le problème I est équivalent à chacun des deux problèmes de point-selle :

Trouver (e_1,s_1) solution de

$$(e,s) \quad E \times S \qquad L_1(e_1,s) \leqslant L_1(e_1,s_1) \leqslant L_1(e,s_1)$$

Trouver (e_1,s_1) solution de

$$(e,s) \quad E \times S \qquad L_2(e_1,-s) \leqslant L_2(e_1,-s_1) \leqslant L_2(e,-s_1)$$

où L_1 et L_2 sont définis aux lignes (13) et (14).

Le problème de point-selle associé à L_1 présente l'avantage d'être en pratique sans contrainte puisqu'il peut être remplacé par le problème de point-selle pour le Lagrangien :

$$(u,s) \in U \times S \rightarrow L'_1(u,s) = - \mathfrak{f}_2(y) + \langle Du,y\rangle$$

La fonctionnelle de REISSNER

Dans ce qui précède nous avons supposé connue une solution particulière s^o de l'équation d'équilibre

$$\Delta^T s = \phi$$

où ϕ est donnée dans F (cfr. chap. I, §7). En pratique il n'est pas toujours commode de trouver une telle solution particulière. On peut remédier à cet inconvénient en considérant tout d'abord le problème primal III' suivant, évidemment équivalent au problème III :

Problème III' - Minimiser sur V la fonctionnelle

$$v \to \mathcal{J}_1(\Delta v) + \langle e^o, s^o \rangle = \mathcal{J}'_1(v)$$

On remarque :

$$\mathcal{J}'_1(v) = f(\Delta v + e^o) - \langle \Delta v, s^o \rangle$$

de sorte que, s^o étant en équilibre avec ϕ

$$\mathcal{J}'_1(v) = f(\Delta v + e^o) - \langle\langle v, \phi \rangle\rangle$$

et s^o n'apparaît plus. Introduisons la fonction de perturbation :

$$\Phi(v,\varepsilon) = f(\Delta v + e^o + \varepsilon) - \langle\langle v, \phi \rangle\rangle \qquad (\varepsilon \in E)$$

Elle conduit au Lagrangien

$$L(v,s) = \langle \Delta v + e^o, s \rangle - \langle\langle v, \phi \rangle\rangle - g(s)$$

et à la fonction duale

$$\chi(\rho,s) = +g(s) - \langle e^o, s \rangle + \psi_{\{\phi - \Delta^T s\}}(\rho) \qquad (\rho \in F)$$

(v_1, s_1) est un point selle de L si et seulement si

$$\Phi(v,0) + \chi(0,s) = 0$$

c'est à dire

$$\begin{cases} \phi - \Delta^T s_1 = 0 \\ f(\Delta v_1 + e^o) + g(s^1) - \langle \Delta v_1 + e^o, s_1 \rangle = 0 \end{cases}$$

c'est à dire que $(\Delta v_1, s_1 - s^o)$ est solution du problème I. Le problème de point selle pour L est donc équivalent au problème I. On note qu'il donne directement les quantités les plus intéressantes à savoir le champ des contraintes et celui des déplacements. L est la "fonctionnelle de REISNNER".

§ 4. INTEGRANDES CONVEXES ET LOIS CONSTITUTIVES

Soit Ω un domaine de R^m; E et S sont des espaces de champs mesurables définis sur Ω, à valeurs dans R^n, et sont mis en dualité par

$$(e,s) \in E \times S \to \langle e,s \rangle = \int_{M \in \Omega} e(M).s(M)d\Omega$$

qu'on suppose définie sur E × S. e(M).s(M) désigne le produit scalaire dans R^n.

Nous considérons le cas où la loi constitutive est donnée localement par une loi sous-différentielle : en tout point M de Ω on se donne un couple de fonctions duales f(M,.) et g(M,.) telles que la loi constitutive se définisse par

$$(15) \qquad \forall M \in \Omega \quad f(M,e(M) + g(M,s(M) - e(M).s(M) = 0$$

On définit alors, formellement pour l'instant, les fonctionnelles

$$e \in E \to F(e) = \int_{M \in \Omega} f(M,e(M))d\Omega$$

$$s \in S \to G(s) = \int_{M \in \Omega} g(M,s(M))d\Omega$$

ce qui pose le problème de mesurabilité pour les intégrandes. Supposons que F et G existent : elles sont alors convexes. Par intégration terme à terme (15) entraîne

$$(16) \qquad F(e) + G(s) - \langle e,s \rangle = 0$$

Inversement (16) entraîne que (15) est vérifiée presque partout sur Ω puisque

$$(17) \qquad \forall (e,s) \in E \times S \quad \forall M \in \Omega \quad f(M,e(M)) + g(M,s(M)) - e(M).s(M) \geq 0$$

D'où la

Proposition 5 - Si pour tout couple $(e,s) \in E \times S$ les fonctions

$$M \to f(M,e(M)) \qquad M \to g(M, s(M))$$

sont duales
alors (16) est équivalente à

$$(15') \qquad f(M,e(M)) + g(M,s(M)) - e(M).s(M) = 0 \quad \text{p.p. sur } \Omega$$

Il reste encore une question en suspens : les fonctions F et G sont-elles duales l'une de l'autre ? Par intégration terme à terme de (17) on a :

$$(18) \qquad \forall (e,s) \in E \times S \qquad F(e) + G(s) - < e,s > \geqslant 0$$

et en particulier, par exemple

$$F(e) \geqslant \sup_{s \, S} \{< e,s > - G(s)\}$$

de sorte que F et G sont des fonctions convexes "sur-duales".

La théorie des intégrandes convexes donne des conditions suffisantes d'existence de F et G, et des conditions suffisantes pour qu'elles soient duales. Le lecteur pourra se reporter à R.T. ROCKAFELLAR ((1), (2), (3)) et aux exposés de C. CASTAING sur cette question. Contentons-nous ici de citer deux résultats d'usage courant en Mécanique, et tirés de (1).

Proposition 6 - (Existence de F et de G) Les fonctionnelles F et G sont bien définies dès que l'une des fonctions f ou g possède les trois propriétés suivantes :

1°) $\forall M \in \Omega \quad \forall a \in R^n \qquad f(M,a) > -\infty$

2°) $\forall M \in \Omega \quad \{a \in R^n / f(M,a) < +\infty\}$ est d'intérieur non vide

3°) $\forall a \in R^n \qquad M \to f(M,a)$ est mesurable

Définition - Soit X un espace de champs mesurables définis sur Ω. On dit que X est décomposable s'il possède les deux propriétés suivantes :

1°) X contient toutes les fonctions mesurables bornées nulles hors d'une partie de mesure finie de Ω.

2°) Quels que soient $x \in X$ et A une partie de mesure finie de Ω, partie dont on note χ_A la fonction caractéristique, le champ $x\,\chi_A$ appartient à X.

Théorème III - (Dualité de F et de G) : Si E et S sont décomposables, si f et g satisfont les hypothèses de la proposition 6, et s'il existe au moins un couple $(e,s) \in E \times S$ tel que

$$F(e) < +\infty \qquad\qquad G(s) < +\infty$$

alors F et G sont duales.

§ 5. UN EXEMPLE D'ELASTICITE NON LINEAIRE COMPORTANT POUR CAS LIMITES LE COMPORTEMENT RIGIDE-PLASTIQUE ET LE MATERIAU A BLOCAGE

Appliquons les résultats précédents à un exemple type, d'ailleurs largement présenté par J.J. MOREAU dans (3), ou par B. NAYROLES dans (1) et (2).

On considère une loi constitutive du type étudié au paragraphe précédent, les fonctions f et g étant cependant un peu particulières.

Tout d'abord on définit en tout point M de Ω une sorte d'intensité de déformation à l'aide d'une jauge

$$a \in R^n \to j(M,a)$$

fonction convexe, positivement homogène de degré 1, et que nous supposons de plus finie et positive, nulle à l'origine seulement. Autrement dit le convexe dont j(M,.) est la jauge, à savoir

$$A(M) = \{a \in R^n \;/\; j(M,a) \leq 1\}$$

est borné et son intérieur contient 0. Rappelons que, inversement :

$$j(M,a) = \inf \left\{ \lambda \in]0, +\infty[\ / \ \frac{1}{\lambda} a \in A(M) \right\}$$

Une fois définie l'intensité de déformation on pose que la densité d'"énergie élastique" est la fonction

(19) $$M \in \Omega \quad a \in R^n \mapsto f(M,a) = \frac{1}{p} \left(j(M,a)\right)^p$$

où

p est une constante, pour l'instant $p \in]1, +\infty[$ ce qui correspond à un comportement élastique au sens propre du terme.

5.1 Calcul de g(M,.)

Le calcul de la fonction duale de f(M,.) (qui est continue puisque finie partout sur R^n) est classique; on peut se reporter à J.J. MOREAU ([1], ch. 14, et [3]) et à R.T. ROCKAFELLAR [4]. On considère l'ensemble polaire de A(M) :

$$B(M) = \{b \in R^n \ / \ \forall a \in A(M) \qquad a.b \leq 1\}$$

dont la jauge k(M,.) est dite jauge conjuguée de j(M,.). B(M) est lui aussi borné et possède 0 comme point intérieur. On va montrer tout d'abord que

$$k(M,b) = \sup_{a.b>0} \frac{a.b}{j(M,a)} = \sup_{\substack{a.b>0 \\ j(M,a)=1}} a.b$$

Or la fonction

$$b \longrightarrow \sup_{\substack{a.b>0 \\ j(M,a)=1}} a.b$$

est positivement homogène et strictement positive hors de l'origine. Il suffit donc d'établir que :

$$\sup_{\substack{a.b>0 \\ j(M,a)=1}} a.b \leqslant 1 \iff k(M,b) \leqslant 1$$

Or

$$\sup_{\substack{a.b>0 \\ j(M,a)=1}} a.b \leqslant 1 \iff \forall a' \in A(M) \quad a'.b < j(M,a') \iff b \in A(M)^{\circ} \iff k(M,b) \leqslant 1$$

Ceci étant acquis calculons

$$g(M,b) = \sup_{a \in R^n} \left(a.b - \frac{1}{p} \left(j(M,a) \right)^p \right)$$

vaut 0 si b = 0. Si b n'est pas nul alors on a :

$$\sup_{a \neq 0} \left(a.b - \frac{1}{p} \left(j(M,a) \right)^p \right) = \sup_{\substack{j(M,a')=1 \\ a'.b>0}} \sup_{\lambda>0} \left(\lambda a'.b - \frac{1}{p} \lambda^p \right)$$

$$= \sup_{\substack{j(m,a')=1 \\ a'.b>0}} \frac{1}{q} (a'.b)^q = \frac{1}{q} \left(k(M,b) \right)^q > 0$$

où : $\frac{1}{p} + \frac{1}{q} = 1$

et par suite on a, pour tout b :

$$g(M,b) = \frac{1}{q} \left(k(M,b) \right)^q \tag{20}$$

$\frac{1}{p}|.|^p$ et $\frac{1}{q}|.|^q$ constituent l'exemple le plus classique de "duales de Young".

5.2 Intégration - Si pour tout a $\in R^n$ la fonction f(.,a) est mesurable les hypothèses de la proposition 6 sont évidemment satisfaites puisque dom f(M,.) = R^n. Les fonctionnelles F et G sont donc définies sur n'importe quels espaces E et S de champs mesurables.

D'autre part $F(0) = 0 = G(0)$

de sorte que F et G sont sommables pour tout couple d'espaces décomposables E et S. Cependant un choix optimal des espaces E et S peut être envisagé :

<u>Proposition 7</u> - Soient E' et S' les ensembles de champs mesurables définis par

$$e \in E \iff F(e) < +\infty \qquad s \in S \iff G(s) < +\infty$$

E' et S' sont des cônes convexes et

$$\forall\ (e,s) \in E \times S \qquad \langle e,s\rangle = \int_\Omega e(M).s(M)d\Omega \in [-\infty, +\infty[$$

La convexité de F et de G et leur quasi-homogénéité entraînent que E' et S' sont des cônes convexes, et d'autre part

$$\forall (e,s) \in E' \times S' \qquad \langle e,s\rangle \leq F(e) + G(s) < +\infty$$

achève la démonstration.

Si l'on peut établir que $\langle e,s\rangle$ ne prend pas la valeur $-\infty$ sur $E' \times S'$ le choix optimal est donc :

$$E = E' + (-E') \qquad S = S' + (-S')$$

D'ailleurs, dans la plupart des cas usuels E' et S' seront des espaces vectoriels, par exemple lorsque, pour tout $M, j(M,.)$ est une norme sur R^n (i.e. $A(M)$ est équilibré).

Le cas le plus simple est celui où il existe deux constantes strictement positives α et β telles que

$$(21) \qquad \forall\ (M,a) \in \Omega \times R^n \qquad \alpha\,|a| < j(M,a) < \beta|a|$$

auquel cas le choix optimal précédent donne

$$E = E' = (L^p(\Omega))^n \qquad S = S' = (L^q(\Omega))^n$$

espaces qui ont le mérite d'être bien connus ..., et surtout d'être des Banach réflexifs ce qui assure l'existence des solutions pour les problèmes précédents lorsque I et J sont fermés.

Dans le cas des plaques, par exemple, n = 3. L'opérateur de déformation est D = grad grad, ce qui conduit à prendre pour U un sous-espace du Sobolev $H^{2,p}(\Omega)$, et pour espace des charges son dual topologique. Si, par exemple la plaque est encastrée sur son contour on prendra

$$U = H_o^{2,p}(\Omega) \qquad \Phi = H^{-2,q}(\Omega)$$

et les problèmes posés précédemment admettront une solution unique pour toute charge $\phi \in H^{-2,q}(\Omega)$, et seulement pour des charges appartenant à cet espace.

5.3 Cas limites (p=1, p= +∞)

Les considérations de l'alinéa 1 sont maintenant fort classiques, et les résultats d'existence et d'unicité sont faciles à obtenir lorsque $p \in]1, +\infty[$ et lorsqu'on dispose des inégalités (21). Sans ces dernières on peut encore espérer construire des espaces fonctionnels adéquats, et obtenir l'inf-compacité faible des fonctionnelles $\mathcal{J}_1$ et $\mathcal{J}_2$, donc l'existence des solutions pour les problèmes posés : ce sera surtout une affaire de technique mathématique.

Nous allons plutôt nous occuper du cas (p = 1, q = +∞), qui est, à un échange près des rôles joués par E et S, le même que celui (p = +∞, q = 1). Le premier représente un comportement rigide-plastique, le second un comportement unilatéral du type "matériau à blocage".

Nous allons fixer le choix des applications

$$M \in \Omega \mapsto A(M) \qquad M \in \Omega \mapsto B(M) = A(M)^o$$

et supposer que (21) a lieu, pour simplifier. Enfin, pour $p \in]1, +\infty[$ nous prenons comme précédemment

$$f_p(M,a) = \frac{1}{p}\left(j(M,a)\right)^p \qquad g_q(M,b) = \frac{1}{q}\left(k(M,b)\right)^q$$

Nous avons évidemment

$$\lim_{p \to 1} f_p(M,a) = j(M,a)$$

et d'autre part la ~~duale~~ de $j(M,.)$ est la fonction indicatrice du convexe $B(M)$:

$$\psi_{B(M)}(b) = \begin{cases} 0 & \text{si } b \in B(M) \\ +\infty & \text{si } b \notin B(M) \end{cases}$$

Or précisément on voit immédiatement que, pour tout b de R^p

$$\lim_{q \to +\infty} \frac{1}{q} \left\{k(M,b)\right\}^q = \psi_{B(M)}(b)$$

Pour p=1 on trouve donc le comportement rigide plastique comme limite du comportement élastique.

Si pour tout $a \in R^p$ la fonction $j(.,a)$ est mesurable alors les fonctionnelles

$$F(e) = \int_\Omega j(M,e(M))d\Omega \; ; \quad G(s) = \int_\Omega \psi_{B(M)}(s(M))\, d\Omega$$

sont définies pour tout champ mesurable e ou s, en vertu de la proposition 6.

D'autre part la double inégalité (21) entraîne l'équivalence

$$F(e) < +\infty \iff e \; \left\{L^1(\Omega)\right\}^n \Rightarrow F(e) < \beta \, ||e||_{1,n}$$

pour tout champ mesurable e. Comme de plus $j(M,.)$ est la fonction d'appui de $B(M)$ les inégalités (21) s'écrivent aussi bien

$$(22) \qquad \forall M \in \Omega \qquad \mathcal{B}(0,\alpha) \subset B(M) \subset \mathcal{B}(0,\beta)$$

où $\mathcal{B}(0,r)$ désigne la boule de centre 0 et de rayon r dans R^n. Par suite on a les équivalences :

$$G(s) < +\infty \iff \text{p.p. sur } \Omega \quad s(M) \in B(M) \iff G(s) = 0$$

et les implications

$$G(s) = 0 \implies s \in \left[L^\infty(\Omega)\right]^n$$

$$s \in \left[L^\infty(\Omega)\right]^n \implies \forall \lambda \in \left[0, \frac{\alpha}{\|s\|_{\infty,n}}\right] \quad G(\lambda s) = 0$$

C'est à dire que l'ensemble convexe des champs mesurables

$$C = G^{\leq}(+\infty)$$

est contenu dans $\left(L^\infty(\Omega)\right)^n$, et l'on a, dans cet espace

$$\mathcal{B}(0,\alpha) \subset C \subset \mathcal{B}(0,\beta)$$

Ceci conduit donc au choix optimum

$$(23) \qquad E = \left[L^1(\Omega)\right]^n \qquad\qquad S = \left[L^\infty(\Omega)\right]^n$$

qui ne sont pas réflexifs, comme il est bien connu.

On sait (cfr. NAYROLES (1), DUVAUT et LIONS (1)) que les problèmes variationnels admettent des solutions en contrainte dans $\left(L^\infty(\Omega)\right)^n$ mais que les solutions en déformation ne sont assurées que dans l'espace bien mal connu $\left(L^\infty(\Omega)'\right)^n$.

Ceci n'est pas gênant car on peut obtenir avec le seul choix de (23) suffisamment de renseignements pour que l'existence des solutions devienne pratiquement sans intérêt mécanique. C'est ce que nous allons voir maintenant.

Tout d'abord E et S sont décomposables, F(0) = 0 = G(0), de sorte que F et G sont duales. D'autre part F est continue* sur $E = \left(L^1(\Omega)\right)^n$ puisque finie partout et majorée par β sur la boule de centre 0 et de rayon 1.

Si donc on se donne $e^\circ \in \left(L^1(\Omega)\right)^n$ et $s^\circ \in \left(L^\infty(\Omega)\right)^n$ le problème I peut ne pas admettre de solution dans le produit de ces espaces mais, d'après le théorème 1 nous avons :

* pour la topologie de la norme, ici compatible avec la dualité

$$\inf_{(e,s)\in I\times J} \mathcal{J}(e,s) = 0$$

c'est à dire que le problème I admet des solutions arbitrairement approchées, ou que le

<u>Problème I'</u> : Trouver $(e,s)\in I \times J$

$$\mathcal{J}(e,s) < \varepsilon$$

admet des solutions pour tout ε strictement positif.

Compte-tenu de l'impossibilité qu'il y a, d'une part de calculer la solution exacte, d'autre part de faire aucune mesure de précision infinie, ce résultat remplace avantageusement tout théorème d'existence pour le problème 1, puisqu'il apporte, en plus, une présomption d'accessibilité numérique des solutions approchées.

B. Nayroles

CHAPITRE III

OPÉRATIONS ET OPERATIONS RÉGULARISÉES SUR LES ELEMENTS MÉCANIQUES ET LES LOIS D'EFFORT SOUS-DIFFÉRENTIELLES

Nous avons déjà parlé, très brièvement, d'une opération sur les lois d'effort, l'addition, et d'une opération sur un élément mécanique, qui est le passage au quotient. Nous allons maintenant définir les opérations d'usage courant et signaler dans quelles circonstances elles interviennent. C'était l'objet d'une note ancienne (B. NAYROLES (3)); ici nous nous occuperons plus particulièrement des lois sous-différentielles; ce sera le second aspect de ce chapitre.

Les opérations que nous effectuerons sur des lois sous-différentielles donneront, sous des hypothèses assez généralement vérifiées, des lois sous-différentielles. Dans le cas contraire une régularisation de ces opérations permettra, en dehors de toute hypothèse, d'obtenir le même résultat. Cette régularisation, qui constitue un changement de la mathématique utilisée habituellement en Mécanique des Solides, est-elle justifiée sur le plan de la Physique ? Ce sera le premier aspect de ce chapitre. En bref la régularisation en question se présente à deux occasions :

1°) Soient $(V, \langle .,. \rangle, W)$ un élément mécanique, f une fonction convexe sur V, à valeurs dans $]-\infty, +\infty]$, f^* sa polaire. On considère la loi d'effort définie par l'équation

$$f(v) + f^*(-w) + \langle v,w \rangle = 0 \tag{1}$$

Peut-on la remplacer, sans dommage pour la Physique, par l'équation

$$f^{**}(v) + f^*(-w) + \langle v,w \rangle = 0 \qquad ? \tag{2}$$

2°) Soient $-w \in \partial f_1(v)$

$-w \in \partial f_2(v)$

deux lois d'effort sur l'élément mécanique précédent. Peut-on remplacer leur somme

(3) $$-w \in \partial f_1(v) + \partial f_2(v)$$

Par (4) $$-w \in \partial(f_1 + f_2)(v)$$

sans dommage pour la Physique lorsqu'aucun critère connu d'additivité des sous-différentiels n'est satisfait ?

Je m'efforcerai de convaincre le lecteur que la réponse peut être positive. Dans un cas comme dans l'autre cette "régularisation" conduit à un "affaiblissement" du problème posé initialement, en ce sens que toute solution de (1) ou de (3) est respectivement solution de (2) ou de (4). On considère généralement qu'un tel affaiblissement d'un problème est justifié s'il facilite la recherche des solutions et si l'on dispose, a-posteriori, de théorèmes permettant d'affirmer que les solutions faibles ainsi trouvées sont aussi des solutions fortes. C'est en général le but des théorèmes de régularité des solutions.

Ce n'est pas ce genre de justification que j'espère apporter mais plutôt la suivante : l'équation (1) n'a pas plus de signification physique que l'équation (2), et de même (3) n'est pas fondamentalement plus justifiée que (4), En sorte que la substitution de (2) à (1) et de (3) à (4) n'apparaît pas comme un changement du problème mécanique. La seconde justification est d'ordre pratique puisque cette substitution permet de construire une algèbre des lois sous-différentielles, indépendante des conditions d'équivalence stricte des équations (1) et (2) et (3) et (4); et que la théorie devient singulièrement plus simple.

§ 1. LOIS MATHEMATIQUEMENT OU PHYSIQUEMENT EQUIVALENTES

1.1 Inadéquation de la mathématique utilisée en Mécanique classique - Le processus de mise en oeuvre d'une théorie physique est en général le suivant : un modèle mathématique ayant été construit le physicien en demande les données à l'expérience; puis il pose un certain nombre de problèmes dont les solutions seront fournies par un travail mathématique comprenant classiquement deux parties; une de mathématique pure qui est la recherche de théorèmes sur l'existence et d'autres propriétés des éventuelles solutions; une autre d'analyse numérique et qui fournit des résultats approchés. Enfin ces résultats sont confrontés à l'expérience.

Or dans ce processus l'imprécision intervient inévitablement à la fois dans les mesures expérimentales et dans le calcul numérique, tandis que le travail de mathématique pure en est exempt. Le résultat en est que la mathématique d'une théorie physique donne, le plus souvent, une description beaucoup trop fine des phénomènes par rapport à ce que l'expérience et le calcul peuvent atteindre. Par exemple la notion de "champ de déplacements" défini en chaque point n'a qu'une signification physique indirecte; aucune mesure ne peut donner la valeur du champ en un point mais fournit une valeur qui peut être interprétée, par exemple, comme une valeur moyenne ou comme une borne supérieure du champ sur un domaine ω dont la petitesse dépend de la finesse du capteur. Introduire un champ de déplacement continu signifie seulement que si l'on augmente indéfiniment la précision de la mesure la suite des résultats expérimentaux converge vers une certaine valeur qui est celle du champ de déplacements en un point précis. Cela signifie aussi que le calcul numérique de cette valeur peut être entrepris sans qu'une petite erreur sur la position du point entraîne une erreur importante sur la valeur calculée. Mais la continuité joue un rôle essentiel en ce qu'elle permet à la valeur locale d'être la limite d'une valeur moyenne, ou d'une borne supérieure, c'est à dire la limite d'un résultat de mesure quand la précision croît indéfiniment. A l'inverse si l'on se donne l'équation de liaison sur la frontière

$$u/_{\partial_1\Omega} = 0$$

la question de savoir si la partie $\partial_1\Omega$ de frontière concernée est fermée ou non est sans signification physique car aucune mesure ni aucun calcul n'a la précision infinie qui permettrait de distinguer $\partial_1\Omega$ de ses points adhérents : la description mathématique est ici trop fine.

En résumé la mathématique employée en Mécanique classique des milieux continus fournit un schéma d'une précision infinie alors que l'expérience ou l'approche numérique, si fines soient-elles, n'ont qu'un "pouvoir séparateur" limité, de sorte que certaines distinctions leur échapperont toujours. J'emploie ce terme de "pouvoir séparateur" par analogie avec celui d'un instrument d'optique.

Certaines difficultés de la théorie, je pense tout particulièrement aux problèmes de traces, sont liées à cette trop grande finesse du schéma mathématique et devraient disparaître si la mathématique utilisée était plus adaptée à la Physique des problèmes; ce sont des "difficultés parasites".

Il n'est pas facile (la Mécanique statistique en donne peut être un exemple ...) de créer une mathématique qui tienne compte de l'imprécision expérimentale et numérique puisque ceci demanderait de connaître une fois pour toutes de quelle nature seront ces imprécisions. Je n'ai pas la prétention de le faire ici et mon but essentiel est d'attirer l'attention du lecteur sur la nécessité de rechercher dans cette direction et de considérer l'imprécision, et d'une façon générale le défaut d'information, comme un aspect essentiel de toute la Physique, aspect totalement ignoré par la Mécanique classique. Je me contenterai donc de deux suggestions qui sont seulement proposées à l'assentiment du lecteur, en attendant d'être remplacées par quelque chose de plus élaboré et de mieux fondé.

1.2 Principe d'adhérence

Nous allons supposer que la limitation du "pouvoir séparateur" de l'expérience ou du calcul numérique peut être traduite en termes topologiques. Comme, de toute façon, il s'agit de confronter les résultats numériques et les résultats expérimentaux, c'est, s'ils sont comparables, le pouvoir séparateur le plus faible qui définit l'imprécision. Notre hypothèse, fort critiquable, est la suivante :

Hypothèse : Il existe, pour l'élément mécanique (V, <.,.>, W) un couple $(\mathcal{T}_V, \mathcal{T}_W)$ de topologies compatibles avec la dualité, possédant la propriété suivante :

quelle que soit la mesure effectuée (de précision finie arbitraire) il existe un voisinage $V \times W$ de l'origine tel que deux couples (v_1,w_1) et (v_2,w_2) satisfaisant

$$(v_1 - v_2, w_1 - w_2) \in V \times W$$

ne peuvent être distingués par cette mesure.

Par raccourci nous dirons que $\mathcal{E}_V \times \mathcal{E}_W$ est plus fine que le pouvoir séparateur de l'expérience.

Dans la formulation de cet énoncé nous n'avons parlé que du pouvoir séparateur de l'expérience, pas de celui du calcul numérique, et pour deux raisons : la première est que l'énoncé est déjà suffisamment lourd et qu'il ne servirait à rien de le compliquer par les considérations analogues évidentes sur l'imprécision numérique. Ensuite le "pouvoir séparateur" du calcul est le plus souvent supérieur à celui de l'expérience.

L'hypothèse est essentiellement discutable sur ce point que $\mathcal{E}_V$ et $\mathcal{E}_W$ sont compatibles avec la dualité ... Le reste est beaucoup plus naturel.

Une conséquence de cette hypothèse est la suivante :

<u>Principe d'adhérence</u> - Deux lois d'efforts sont "physiquement équivalentes" si leurs graphes ont même adhérence pour la topologie produit $\mathcal{E}_V \times \mathcal{E}_W$.

En effet, d'après l'hypothèse précédente, aucune expérience ne pourra mettre ces deux lois en contradiction.

Considérons par exemple la relation de liaison pour une plaque simplement appuyée sur une partie $\partial_1\Omega$ de la frontière :

$$u_{/\partial_1\Omega} = 0$$

et choisissons $V = H^2(\Omega)$ pour espace des déplacements, $W = H^2_{\overline{\Omega}}$ pour espace des efforts. La relation de liaison définit dans V un sous-espace vectoriel V_1 et la loi d'effort est $\psi_{V_1}(v) + \psi_{V_1^\circ}(-w) + \langle v,w\rangle = 0$

Sans nous préoccuper de savoir si $\partial_1\Omega$ est une partie fermée de $\partial\Omega$ (ce qui est essentiel dans la théorie classique), nous pourrons écrire que cette loi d'effort est physiquement équivalente à

$$\psi_{V_1^{oo}}(v) + \psi_{V_1^{o}}(-w) + \langle v,w\rangle = 0$$

On évite ainsi les difficultés liées à la continuité de l'opérateur trace. On peut aussi éviter celles qui sont liées à son existence en posant

$$V_1 = \{ v \in C^{\infty}(\overline{\Omega}) \ / \ v = 0 \quad \text{sur } \partial_1\Omega\}$$

et en prenant l'adhérence $V_1{}^{oo}$ de $\dot{V_1}$ dans V; qui peut raisonnablement affirmer que cette façon de procéder soit moins physique que la mise en équation classique ?

1.3 Solutions approchées

La seconde de mes suggestions est de n'accorder à l'existence d'une solution théorique aucune signification physique dès qu'elle ne peut être exactement calculée. A quelques exceptions près, par exemple si l'inconnue est un nombre entier, on ne sait calculer que des solutions approchées. Ce qui sera confronté à l'expérience étant une solution approchée il est beaucoup moins important d'établir l'existence d'une solution "exacte" que de répondre aux questions suivantes :

1°) Peut-on donner une définition mathématique, physiquement acceptable, des solutions approchées ?

2°) Celles-ci ayant été définies peut-on montrer l'existence de solutions arbitrairement approchées et construire un algorithme permettant leur calcul ?

Le choix de la réponse au 1°) comporte une responsabilité physique considérable; une réponse positive à la seconde constitue le seul théorème d'existence intéressant pour la théorie physique, même si la technique mathématique permet quelquefois de l'obtenir par l'intermédiaire d'un théorème d'existence de solutions exactes. Précisément l'habitude est de définir une solution approchée comme un point d'un voisinage, pour une certaine topologie, d'une solution exacte. Pour fixer les idées considérons l'équation

(5) $$\mathcal{f}(x) = 0$$

où $\mathcal{f}$ est une fonction à valeurs dans R et définie sur un espace vectoriel normé X. Supposons, pour simplifier encore, que cette équation admette l'unique solution x_o. On peut définir d'au moins deux façons différentes une solution x_i approchée à ε_i près, ε_i désignant un réel strictement positif

par (6) $$||x_i - x_o|| < \varepsilon_i$$

ou par (7) $$|\mathcal{f}(x_i)| < \varepsilon_i$$

et ces deux définitions donnent les mêmes suites de s. s. a. b. si, par exemple:

(8) $$\exists \alpha > 0 \quad \exists \beta > 0 \quad \forall x \in X \quad \alpha \, ||x - x_o|| < |\mathcal{f}(x)| < \beta ||x - x_o||$$

ce qui est un cas trivial. Mais si $\mathcal{f}$ est seulement continue alors

$$\lim_{i\to\infty} ||x_i - x_o|| = 0 \implies \lim_{i\to\infty} |\mathcal{f}(x_i)| = 0$$

tandis que nous avons l'implication inverse si la seconde des inégalités (8) a lieu. Les définitions (6) et (7) ne sont pas équivalentes et c'est au physicien de dire quel sens il faut donner à l'approximation des solutions.

Soit d'une façon générale une partie A de X : il existe une infinité d'équations dont A est l'ensemble des solutions, et qui sont donc mathématiquement équivalentes; si la résolution ne peut être qu'approchée il importe que le physicien ait défini la nature de cette approximation pour que le problème physique proposé soit correctement posé. C'est exactement le contraire qui se produit habituellement puisque les mathématiciens s'attaquent aux "équations de la Physique", sans que les critères d'approximation les accompagnent; alors ils ne peuvent travailler que sur les solutions exactes dans une première étape; que proposer, dans une seconde étape, des suites de solutions approchées en un sens que le physicien pourra ou non accepter.

Si au contraire on se donne au départ toute équation avec son critère d'approximation le travail proposé au mathématicien est totalement différent. En particulier la notion d'équivalence de deux équations, plus généralement celle d'équation conséquence d'une autre peut être remplacée par une notion d'équivalence physique, ou de conséquence physique qui devra être exprimée mathématiquement. C'est ce qui va être fait maintenant; la construction que je propose n'est vraisemblablement pas la meilleure et sa mise au point laisse beaucoup à disirer. Mais cette école d'été constituait une très belle occasion d'évoquer cette question, à mon avis essentielle pour un progrès réel de la Mécanique; je n'ai donc pas su résister à l'envie de présenter, même prématurément, mes quelques idées sur le sujet.

1.4 Equivalence approximative

On considère un espace vectoriel topologique X (on notera $\mathcal{C}$ sa topologie) et l'ensemble χ des équations (ou systèmes d'équations) dont l'inconnue est un point de X. Soient E_1 et E_2 deux éléments de χ et A_1 et A_2 leurs ensembles respectifs de solutions. On définit habituellement sur χ la relation de préordre partiel

$$E_1 \to E_2 \overset{\text{déf}}{\Longleftrightarrow} A_1 \subset A_2$$

qui s'énonce ou bien "E_2 est conséquence de E_1" ou bien "E_2 est plus faible que E_1". Lorsque

$$E_2 \leftarrow E_1 \qquad E_2 \to E_1$$

on dit que les équations E_1 et E_2 sont équivalentes; et cette relation d'équivalence est systématiquement employée lorsqu'il s'agit d'étudier les solutions "exactes" c'est à dire les ensembles A_1 et A_2.

Lorsque l'intérêt se porte non sur les solutions exactes mais sur les suites de solutions approchées ces relations de préordre et d'équivalence deviennent sans intérêt.

On va maintenant considérer l'ensemble $\tilde{\chi}$ des équations sur X qui sont données avec la définition de leurs "suites de solutions arbitrairement approchées", que nous écrirons, en abrégé, s.s.a.b.. Pour simplifier l'exposé on identifiera deux éléments de $\tilde{\chi}$ possédant les mêmes s.s.a.b. (cfr Remarque 1). Une première idée est de définir E comme plus faible que E si toute s.s.a.b. de E est s.s.a.b. de E . En fait nous élargirons cette définition en utilisant la topologie de X avec, comme motivation, l'utilisation d'une topologie plus fine que le pouvoir séparateur de l'expérience.

Définition 1 - Soient $\tilde{E}_1$ et $\tilde{E}_2$; nous dirons que $\tilde{E}_1$ est "approximativement plus forte que $\tilde{E}_2$", et nous noterons

$$\tilde{E}_1 \nrightarrow \tilde{E}_2$$

si pour tout voisinage $\mathcal{V}$ de l'origine dans X et à toute s.s.a.b. (x_i^1) de E_1 il existe au moins une s.s.a.b. (x_i^2) de $\tilde{E}_2$ satisfaisant

$$\forall_i \qquad x_i^2 \in x_i^1 + \mathcal{V}$$

On dit que $\tilde{E}_1$ est "approximativement équivalente" à $\tilde{E}_2$ et on note $\tilde{E}_1 \nleftrightarrow \tilde{E}_2$ si

$$\tilde{E}_1 \nleftrightarrow \tilde{E}_2 \overset{\text{déf}}{\Longleftrightarrow} (\tilde{E}_1 \nrightarrow \tilde{E}_2 \text{ et } \tilde{E}_2 \nrightarrow \tilde{E}_1)$$

Proposition 1 - La relation $\nrightarrow$ est une relation de préordre partiel sur $\tilde{\chi}$. En effet $\nrightarrow$ est transitive puisque, pour tout $\mathcal{V}$ il existe $\mathcal{V}_1$ et $\mathcal{V}_2$ voisinages de l'origine, tels que $\mathcal{V}_1 + \mathcal{V}_2 \subset \mathcal{V}$; alors

$$\{x_i^2 \in x_i^1 + \mathcal{V}_1 \text{ et } x_i^3 \in x_i^2 + \mathcal{V}_2\} \Rightarrow x_i^3 \in x_i^1 + \mathcal{V}$$

et d'autre part la relation $\nrightarrow$ est réflexive

$$\forall \tilde{E} \in \tilde{\chi} \qquad \tilde{E} \nrightarrow \tilde{E}$$

Remarque 1 - La définition de $\tilde{\chi}$ donnée précédemment est plus intuitive que mathématique; on peut la préciser en considérant X^N espace des suites d'éléments de X : alors $\tilde{\chi}$ peut s'identifier à l'ensemble Φ des parties ϕ de

X^N. Ainsi $\tilde{E}$ peut être définie par l'ensemble ϕ de ses s.s.a.b..

Sur l'espace vectoriel X^N on peut choisir la topologie T pour lesquels les voisinages de l'origine sont les ensembles de la forme $\mathcal{V}^N$ où $\mathcal{V}$ est un $\mathcal{T}$-voisinage de l'origine sur X, c'est à dire

$$\mathcal{V}^N = \{(x_i) \in X^N \; / \; \forall i \qquad x_i \in \mathcal{V}\}$$

Alors la définition 1 équivaut à

$$\tilde{E}_1 \nrightarrow \tilde{E}_2 \iff \phi_1 \subset \overline{\phi}_2$$

où $\overline{\phi}_2$ est l'adhérence de ϕ_2 dans X^N pour la topologie T. La relation d'"équivalence approximative" correspondante est alors

$$\tilde{E}_1 \nleftrightarrow \tilde{E}_2 \iff \overline{\phi}_1 = \overline{\phi}_2$$

c'est à dire que nous retrouvons, après un long détour, le principe d'adhérence.

Remarque 2 - On aurait pu compliquer la définition 1 en considérant non pas une topologie d'espace vectoriel sur X, mais une structure uniforme quelconque et en particulier la topologie discrète; en prenant pour s.s.a.a. de toute équation les suites de points qui en sont solution on aurait alors la relation d'ordre usuelle comme cas particulier de la relation d'ordre "approximative". C'est peut être une satisfaction pour l'esprit...

On considère encore des "systèmes d'équations" dont les solutions sont, par définition, solutions de chacune des équations. On va construire la notion correspondante pour les éléments de $\tilde{E}$:

Définition 2 - Soient $\tilde{E}_1$ et $\tilde{E}_2$ appartenant à $\tilde{X}$, ϕ_1 et ϕ_2 leurs ensembles de s.s.a.b. respectifs. On appellera système $\tilde{E}_1 \perp \tilde{E}_2$ l'élément de $\tilde{X}$ dont l'ensemble des s.s.a.b. est $\overline{\phi}_1 \cap \overline{\phi}_2$.

Cette opération dépend malheureusement de la topologie choisie sur X, mais, ainsi définie, elle permet d'obtenir la proposition 2. En effet :

De l'implication

$$(\bar{\phi}'_1 \supset \phi_1 \quad \text{et } \bar{\phi}'_2 \supset \phi_2) \Rightarrow \bar{\phi}'_1 \cap \bar{\phi}'_2 = \bar{\phi}_1 \cap \bar{\phi}_2$$

on déduit la

Proposition 2 - **La relation $\longrightarrow$ est compatible avec la loi $\perp$. Autrement dit**

$$(\tilde{E}_1 \rightarrow \tilde{E}'_1 \text{ et } \tilde{E}_2 \rightarrow \tilde{E}'_2) \Rightarrow \tilde{E}_1 \mid \tilde{E}_2 \rightarrow \tilde{E}'_1 \mid \tilde{E}'_2$$

1.5 Suites convergentes et quasi-solutions

Il est utile de considérer les suites convergentes; par ailleurs la notion d'unicité de solution, fondamentale dans la mathématique employée usuellement, doit être remplacée par une notion de convergence des s.s.a.b., que nous étudierons plus loin. En outre l'introduction de la notion de "quasi-solution" permettra d'utiliser les outils mathématiques traditionnels.

Définition 3 - Nous dirons que $\tilde{E}$ est définie numériquement s'il existe une application $\mathcal{J}$ de X dans $[0, +\infty)$ telle que les s.s.a.b. de $\tilde{E}$ soient par définition les suites (x_i) telles que

$$\lim_{i\to\infty} \mathcal{J}(x_i) = 0$$

Ce sera le cas pour les lois d'effort sous-différentielles.

Proposition 3 - Soit $\mathcal{T}$ une topologie d'espace vectoriel métrisable sur X; on considère $\tilde{E}$ définie numériquement par $\mathcal{J}$, et $\tilde{E}'$ approximativement plus forte que $\tilde{E}$. Alors à toute suite (resp. de Cauchy) (x'_i) s.s.a.b. de E' correspond au moins une suite (resp. de Cauchy) (x_i) s.s.a.b. de $\tilde{E}$ et telle que

$$\lim_{i\to\infty} (x_i - x'_i) = 0$$

Considérons en effet une base $\mathcal{V}_k$ de voisinages de 0 avec

$$\mathcal{V}_{k+1} \subset \mathcal{V}_k$$

Pour tout k il existe une s.s.a.b. (y_i^k) de $\tilde{E}$ telle que

$$\forall i \qquad y_i^k - x'_i \in \mathcal{V}_k$$

et comme $\tilde{E}$ est définie numériquement il existe une application de $\mathbb{N}$ dans $\mathbb{N}$ telle que

$$(9) \qquad \begin{cases} \forall k \in \mathbb{N} \quad \forall i \geq I(k) \quad \mathcal{J}(y_i^k) \leq \frac{1}{k} \\ I(k+1) > I(k) \end{cases}$$

Considérons alors l'application N de $\mathbb{N}$ dans $\mathbb{N}$ définie par

$$(10) \qquad j \in N \to N(j) = \max \{k \in \mathbb{N} \ / \ I(k) \leq j\}$$

I et N ont les propriétés suivantes, immédiates

(11) . N est croissante

(12) . $\forall j \in N \qquad (I \circ N)(j) \leq j$

(13) . $\forall k \in N \qquad (N \circ I)(k) = k$

De 10 on déduit que

$$\forall k \in \mathbb{N} \qquad \forall j \geq I(k) \qquad N(j) \geq k$$

de sorte que N(j) tend vers l'infini avec j.

On considère alors la suite

$$j \to x_j = y_j^{N(j)}$$

En vertu de (9) et (12) on a :

$$(y_j^{N(j)}) \leq \frac{1}{N(j)}$$

et comme N(j) tend vers l'infini avec j, (x_j) est une s.s.a.b. de $\tilde{E}$.

D'autre part on a

$$y_j^{N(j)} - x'_j \in N(j)$$

et comme N(j) tend vers l'infini avec j, $\lim_{j\to\infty} (x_j - x'_j) = 0$, ce qui dé-démontre la proposition pour (x'_i) s.s.a.b. de $\tilde{E}$.

Si de plus (x'_i) est une suite de Cauchy il est immédiat de montrer que la suite (x_i) ainsi construite en est une autre. En effet soit $\mathcal{V}$ un voisinage arbitraire de l'origine : il existe trois voisinages de l'origine, $\mathcal{V}_1, \mathcal{V}_2, \mathcal{V}_3$ tels que

$$\mathcal{V}_1 + \mathcal{V}_2 + \mathcal{V}_3 \subset \mathcal{V}$$

et par hypothèse, ou d'après ce qui précède

$$\exists i_1 \quad \forall j \geq i_1 \quad \forall k \geq i_1 \qquad x'_j - x'_k \in \mathcal{V}_1$$
$$\exists i_2 \quad \forall j \geq i_2 \qquad x_j - x'_j \in \mathcal{V}_2$$
$$\exists i_3 \quad \forall k \geq i_3 \qquad x'_k - x_k \in \mathcal{V}_3$$

d'où

$$\forall j \geq \max(i_1, i_2, i_3) \quad \forall k \geq \max(i_1, i_2, i_3)$$
$$x_j - x_k \in \mathcal{V}_1 + \mathcal{V}_2 + \mathcal{V}_3 \subset \mathcal{V}$$

La notion de quasi-solution va nous permettre, dans nombre de cas usuels, de ramener l'étude des s.s.a.a. convergentes à celle des solutions, au sens usuel, d'une équation.

Définition 4 - Soit $\tilde{E}$ numériquement définie par $\mathcal{J}$. On appelle quasi-solution de $\tilde{E}$ tout $\bar{x} \in X$ tel que :

$$\forall \mathcal{V}, \text{ voisinage de } 0 \quad \forall \varepsilon > 0 \quad \exists x \in \bar{x} + \mathcal{V} \quad \mathcal{J}(x) \leq \varepsilon$$

Le rapport entre les quasi-solutions de $\tilde{E}$ et ses s.s.a.b. convergentes est précisé par les trois propositions suivantes, de démonstration immédiate :

Proposition 4 - Soit (x_i) une s.s.a.b. de $\tilde{E}$ convergeant vers $\bar{x}$: $\bar{x}$ est quasi-solution de $\tilde{E}$.

Proposition 5 - Si $\mathcal{X}$ est métrisable toute quasi-solution est limite d'une s.s.a.b.

Proposition 6 - Si $\mathcal{C}$ est métrisable et si $\mathcal{f}$ est continue, les trois assertions suivantes sont équivalentes

. $\bar{x}$ est solution de $\mathcal{f}(x) = 0$

. $\bar{x}$ est quasi-solution de $\tilde{E}$

. $\bar{x}$ est limite d'une suite de s.s.a.b. de $\tilde{E}$

Essayons d'imaginer ce qui peut tenir lieu, dans la mathématique "approximative" que nous proposons, des habituels théorèmes d'existence et d'unicité. On peut rechercher, pour $\tilde{E}$, des résultats du type suivant :

Existence - Selon les besoins on recherchera un résultat plus ou moins fort :

. Il existe une s.s.a.b., c'est à dire que l'ensemble ϕ associé à $\tilde{E}$ n'est pas vide

. Il existe une s.s.a.b. de Cauchy, ce qui revient à l'existence d'une quasi-solution si X est un Banach.

Unicité - l'une des s.s.a.b. est une suite de Cauchy, et quelles que soient deux s.s.a.b. (x_i) et (x'_i) $\lim (x_i - x'_i) = 0$

Il faut faire une remarque à ce sujet : soit $\mathcal{C}$ et $\mathcal{C}'$ deux topologies d'e.v.t. sur X, $\mathcal{C}$ étant à base dénombrable et plus fine que $\mathcal{C}'$, elle même plus fine que le pouvoir séparateur de l'expérience. On pourra souvent transformer des systèmes d'équations en utilisant l'équivalence approximative pour $\mathcal{C}'$ et obtenir une équation

$$\mathcal{f}(x) = 0$$

dont les s.s.a.b. seront définies par

$$\lim \mathcal{f}(x_i) = 0$$

Si l'on ne s'intéresse alors qu'aux s.s.a.b. convergentes pour $\mathcal{C}$, si $\mathcal{f}$ est continue pour $\mathcal{C}$, il suffit, d'après la proposition 6 de considérer les solutions de $\overset{\mathcal{J}(\alpha)=0}{V}$. Les énoncés proposés pour tenir lieu de théorèmes d'existence et d'unicité se ramènent alors à ceux-ci.

§ 2. APPROXIMATION ET REGULARISATION DES LOIS SOUS-DIFFERENTIELLES

Soit (V, <.,.> W) un élément mécanique sur lequel est donnée la loi d'effort

(1) $$f(v) + f^*(-w) + \langle v,w\rangle = 0$$

On pose

$$\mathcal{J}(v,w) = f(v) + f^*(-w) + \langle v,w\rangle \geqslant 0$$

et l'on se propose de justifier, d'un point de vue physique, la définition des solutions approchées à ε près de (1) par

(14) $$\mathcal{J}(v,w) < \varepsilon$$

d'où suit naturellement la définition des s.s.a.b. par

(14') $$\lim_{i\to\infty} \mathcal{J}(v_i,w_i) = 0$$

Eliminons tout de suite le cas où l'une des fonctions est une indicatrice : (14) équivaut alors à (1). Nous ne pouvons évidemment passer tous les autres cas en revue et nous étudierons le plus délicat : celui où f et f^* sont définies par des intégrales. Je pense qu'il suffit d'ailleurs d'étudier le cas de l'élasticité linéaire, tout à fait exemplaire, pour se donner un semblant de conviction. C'est donc, en fait, l'élément mécanique (E, <.,.>, S) des chapitres précédents qui est en cause, ainsi que la théorie des intégrandes convexes.

Considérons tout d'abord l'élément mécanique ($\mathcal{E}_p$,., $\mathcal{E}_p$) de dimension finie p; $\mathcal{E}_p$ est l'espace euclidien de dimension p mis en dualité avec lui-même par le produit scalaire usuel. La loi de contrainte est

(15) $$s = k\ e$$

où k est une application linéaire symétrique définie positive de $\mathcal{E}_p$ sur lui-même. On considère alors les fonctions duales

$$f(e) = \frac{1}{2}\, e.ke \qquad f^*(s) = \frac{1}{2}\, k^{-1}\, s.s$$

et l'identité

(16) $$\forall (e,s) \in \mathcal{E}_p \times \mathcal{E}_p \quad f(e) + f^*(s) - e.s = \frac{1}{2}\,(e-k^{-1}s).k(e-k^{-1}s)$$

de sorte que (15) équivaut à

$$f(e) + f^*(s) - e.s = 0$$

L'inégalité

(17) $$f(e) + f^*(s) - e.s \lesssim \varepsilon$$

équivaut, compte-tenu de (16) et m désignant la plus petite valeur propre de k

$$|e-k^{-1}s| < \sqrt{\frac{2\varepsilon}{m}}$$

ce qui constitue une définition physiquement acceptable des solutions approchées à ε près de (15).

On considère maintenant (E, <.,.>, S) où E et S sont des espaces de champs définis sur un ouvert Ω de R^n, à valeurs dans ε_p. On suppose en outre que la fonction

$$x \in \Omega \to k(x)$$

et les espaces fonctionnels E et S satisfont les hypothèses de ROCKAFELLAR en sorte que les fonctionnelles

$$e \in E \to F(e) = \int_\Omega f(x,e(x))d\Omega$$

$$s \in S \to G(s) = \int_\Omega g(x,s(x)d\Omega$$

existent. Alors l'inégalité

(18) $$F(e) + G(s) - \langle e,s\rangle \lesssim \varepsilon$$

entraîne que sur toute partie ω mesurable de Ω

$$\int_\Omega \left(f(x,e(x)) + g(x,s(x)) - e(x).s(x)\right)d\Omega \lesssim \varepsilon$$

c'est à dire que sur tout ensemble Ω de mesure finie

$$\int_\Omega |e - k^{-1}s|^2 \, d\Omega < \frac{2\,\varepsilon}{\inf\limits_{x\in\omega} K(x)}$$

Or toute mesure de déformation ou de contraintes ne peut se faire que sur un ensemble non réduit à un point; si cette mesure ne peut donner, comme c'est probable, qu'une valeur moyenne sur un domaine ω, alors pour ε suffisamment petit, elle ne pourra distinguer les solutions (1) de celles (14) qui donne donc une définition physiquement acceptable de l'approximation. Par ailleurs la notion même de milieu continu est macroscopique c'est à dire que la déformation et la contrainte n'ont de sens que sous un symbole d'intégration, c'est à dire par leur valeur moyenne.

Revenons maintenant à l'équation (1) posée au début de ce paragraphe. J'admettrai, jusqu'à preuve du contraire, que (14) définit de façon physiquement acceptable ses solutions approchées à ε près, et (14') ses suites de solutions arbitrairement approchées. Nous pouvons alors démontrer la

Proposition 7 - On considère l'équation (1) dans laquelle f est une fonction convexe sur V, f^* sa polaire. Elle est approximativement équivalente à

$$(2) \qquad f^{**}(v) + f^*(-w) + \langle v,w\rangle = 0$$

pour laquelle les s.s.a.b. sont définies par

$$\lim \left(f^{**}(v_i) + f^*(-w_i) + \langle v_i,w_i\rangle\right) = 0$$

et ceci pour toute topologie $\mathcal{E}_V \times \mathcal{E}_W$ produit d'une topologie $\mathcal{E}_V$ localement convexe et compatible avec la dualité sur V par une topologie quelconque sur W.

Eliminons d'abord le cas trivial où f n'admet pas de minorantes affines continues : alors f^{**} vaut $-\infty$ partout, f^{*} vaut + l'infini partout et ni (1) ni (2) n'admettent de s.s.a.b. : elles sont donc approximativement équivalentes et mêmes équivalentes au sens usuel. Si tel n'est pas le cas alors f^{**} est aussi la régularisée s.c.i. de f pour $\mathcal{C}_V$. Comme pour tout $w \in W$ la fonction linéaire $<.,w>$ est continue pour $\mathcal{C}_V$, $f^{**}-<.,w>$ est la régularisée s.c.i. de $f-<.,w>$, c'est à dire

$$(19) \qquad \forall v \in V \qquad f^{**}(v) - <v,w> = \sup_{\mathcal{V}_v} \inf_{v' \in \mathcal{V}_v} \{f(v') - <v',w>\}$$

où $\mathcal{V}_v$ désigne un quelconque voisinage de v pour $\mathcal{C}_V$.

Des inégalités

$$0 \leq f^{**}(v) + f^{*}(-w) + <v,w> \leq f(v) + f^{*}(-w) + <v,w>$$

on déduit que toute s.s.a.a. de (1) est s.s.a.a. de (2) ; donc

$$(1) \leftrightarrow (2)$$

Inversement soit $\mathcal{V}$ un $\mathcal{C}_V$ voisinage de l'origine et (v_i,w_i) une s.s.a.b. de (2). On a, d'après (19)

$$\forall i \in N \quad \forall \mathcal{V} \ \exists v'_i \in v_i + \mathcal{V} \qquad f(v'_i) + <v'_i,w_i> \leq f^{**}(v_i) + <v_i,w_i> + \frac{1}{i}$$

La suite (v'_i,w_i) ainsi construite satisfait donc

$$\forall i \in N \qquad v'_i \in v_i + \mathcal{V} \text{ et } f(v'_i) + f^{*}(-w_i) + <v'_i,w_i> \leq f^{**}(v_i) + f^{*}(-w_i) + <v_i,w_i> + \frac{1}{i}$$

et donc, puisque (v_i,w_i) est s.s.a.b. de 2

$$\lim_{i\to\infty} \{f(v'_i) + f^{*}(-w_i) + <v'_i,w_i>\} = 0$$

ce qui démontre la proposition

Notation - Soit A une loi sous -différentielle (resp. l'inverse d'une loi sous-différentielle) :

$$A = -\,\partial f \qquad (\text{resp. } A^{-} = \partial g(-.))$$

où f(resp. g) est convexe sur V (resp. W) - On appellera loi régularisée de A la loi

$$\hat{A} = -\,\partial f^{**} \qquad (\text{resp. } \hat{A}^{-} = \partial g^{**}(-.))$$

§ 3. ADDITION DES LOIS SOUS-DIFFERENTIELLES

On a défini l'addition de deux lois d'effort A_1 et A_2 sur (V, .,. , W) par

$$(A_1 + A_2)(v) = A_1(v) + A_2(v)$$

Lorsque A_1 et A_2 sont sous-différentielles :

$$A_1 = -\,\partial f_1 \qquad A_2 = -\,\partial f_2$$

leur somme s'écrit

$$(3) \qquad -w \in \partial f_1(v) + \partial f_2(v)$$

équation plus forte que

$$(4) \qquad -w \in \partial(f_1 + f_2)(v)$$

La proposition suivante (cfr. J.J. MOREAU (1) ch. 10) donne une condition suffisante d'équivalence

Proposition 8 - Si f_1 et f_2 sont convexes et s'il existe un point où les deux fonctions sont finies, l'une d'entre elles y étant continue (pour une topologie compatible avec la dualité) alors pour tout v de V $\partial f_1(v) + \partial f_2(v) = \partial(f_1 + f_2)(v)$

Ce résultat très général s'applique à de nombreux cas mécaniques, mais ne couvre pas tous les besoins. Par exemple si f_1 et f_2 sont deux indicatrices de sous-espaces vectoriels de V, disons V_1 et V_2 :

$$\text{pour} \quad v \in V_1 \cap V_2 \quad : \quad \partial\psi_{V_1}(v) + \partial\psi_{V_2}(v) = V_1{}^\circ + V_2{}^\circ$$

$$\partial\psi_{V_1 \cap V_2}(v) = (V_1 \cap V_2)^\circ \supset V_1{}^\circ + V_2{}^\circ$$

et en général la somme $V_1{}^\circ + V_2{}^\circ$ n'est pas fermée. Oubliant l'aspect sthénique des liaisons les mécaniciens ont toujours traité la somme de ces deux liaisons parfaites comme la liaison parfaite :

$$\psi_{V_1 \cap V_2}(v) + \psi_{(V_1 \cap V_2)^\circ}(-w) = 0$$

par application du principe du travail virtuel.

Notre but est de montrer qu'on peut toujours remplacer l'équation (3) par l'équation (4). Nous verrons même que cette dernière est en général approximativement plus forte que la première pour un choix naturel des s.s.a.b. de (3), alors qu'elle est plus faible au sens mathématique usuel.

La difficulté est de définir les s.s.a.b. de (3); en effet dire que (v,w) est solution de (3) signifie qu'il existe un triplet (v,w,w_1) solution du système (3.1), ou encore qu'il existe un quadruplet (v,w,w_1,w_2) solution du système (3.2). Ecrivons ces deux systèmes en posant

$$\text{pour} \quad k = 1,2 \qquad \mathcal{J}_k(v,w) = f_k(v) + f_k^*(-w) + \langle v,w\rangle$$

$$(3.1) \quad \begin{cases} \mathcal{J}_1(v,w_1) = 0 \\ \mathcal{J}_2(v,w-w_1) = 0 \end{cases} \qquad (3.2) \quad \begin{cases} \mathcal{J}_1(v,w_1) = 0 \\ \mathcal{J}_2(v,w_2) = 0 \\ w = w_1 + w_2 \end{cases}$$

On peut définir les s.s.a.b. de l'équation $w = w_1 + w_2$ pour une topologie $\mathcal{T}_W$ sur W par

$$\lim(w_i - w_{1,i} - w_{2,i}) = 0$$

les s.s.a.b. des autres équations sont définies au paragraphe 2.

On peut alors utiliser la définition 2 des s.s.a.b. d'un système pour définir celles de l'équation (3) et ceci nous donne deux définitions a-priori différentes

Deux définitions des s.s.a.b. de 3

On dira que (v_i, w_i) est s.s.a.b. de (3) au sens 1 s'il existe une suite $(w_{1,i})$ telle que $(v_i, w_i, w_{1,i})$ soit s.s.a.b. de (3.1).

On dira que (v_i, w_i) est s.s.a.b. de (3) au sens 2 s'il existe une suite $(w_{1,i}, w_{2,i})$ telle que $(v_i, w_i, w_{1,i}, w_{2,i})$ soit s.s.a.b. de (3.2).

On aurait évidemment pu introduire encore d'autres systèmes équivalents aux précédents. Nous limitant à ceux-ci nous établissons la proposition suivante, passablement rassurante.

Proposition 9 - Les deux définitions précédentes sont équivalentes pour tout couple $(\mathcal{T}_V, \mathcal{T}_W)$ de topologies d'espace vectoriel.

Soit en effet $(v_i, w_i, w_{1,i})$ une s.s.a.b. de (3.1). Pour tout voisinage $\mathcal{V} \times \mathcal{W}$ de l'origine il existe donc des suites (v_i^1), (v_i^2), (w_i^2), $(w_{1,i}^1)$, $(w_{1,i}^2)$, et un voisinage équilibre $\mathcal{W}'$ de 0 dans W tels que :

$$(20)\quad \left\{ \begin{array}{lll} \mathcal{W}' + \mathcal{W}' \subset \mathcal{W} & & \\ \lim \mathcal{f}_1(v_i^1, w_{1,i}^1) = 0 & & \lim \mathcal{f}_2(v_i^2, w_i^2 - w_{1,i}^2) = 0 \\ v_i^1 \in v_i + \mathcal{V} & & v_i^2 \in v_i + \mathcal{V} \\ w_i^2 \in w_i + \mathcal{W}' & w_{1,i}^1 \in w_{1,i} + \mathcal{W}' & w_{1,i}^2 \in w_{1,i} + \mathcal{W}' \end{array} \right.$$

Nous allons montrer que la suite $(v_i, w_i, w_{1,i}, w_{2,i})$, où $w_{2,i}$ est défini par

$$w_{2,i} = w_i - w_{1,i}$$

est s.s.a.b. de (3.2). Posons

$$w_{2,i}^2 = w_i^2 - w_{1,i}^2$$

Alors

$$\lim \mathcal{f}_1(v_i^1, w_{1,i}^1) = 0 = \lim \mathcal{f}_2(v_i^2, w_{2,i}^2)$$

et

$$w_{2,i}^2 - w_{2,i} = (w_i^2 - w_i) + (w_{1,i} - w_{1,i}^2) \in \mathcal{W}' + \mathcal{W}'$$

Enfin

$$\lim (w_i - w_{1,i} - w_{2,i}) = \lim 0 = 0$$

de sorte que $(v_i, w_i, w_{1,i}, w_{2,i})$ est s.s.a.b. de (3.2).

Inversement soit $(v_i, w_i, w_{1,i}, w_{2,i})$ une s.s.a.b. de (3.2). Alors quel que soit $\mathcal{V} \times \mathcal{W}$ il existe $\mathcal{W}'$, voisinage équilibré de 0 dans W, et des suites (v_i^1), (v_i^2), $(w_{1,i}^1)$, $(w_{2,i}^2)$, (w_i^3), $(w_{1,i}^3)$, $(w_{2,i}^3)$ tels que

$$(21) \quad \begin{cases} \mathcal{W}' + \mathcal{W}' + \mathcal{W}' + \mathcal{W}' \subset \mathcal{W} \\ \lim \mathcal{f}_1(v_i^1, w_{1,i}^1) = 0 \qquad \lim \mathcal{f}_2(v_i^2, w_{1,i}^2) = 0 \\ \alpha) \qquad \lim(w_i^3 - w_{1,i}^3 - w_{2,i}^3) = 0 \\ v_i^1 \in v_i + \mathcal{V} \qquad v_i^2 \in v_i + \mathcal{V} \\ w_{1,i}^2 \in w_{1,i} + \mathcal{W} \qquad w_{2,i}^2 \in w_{2,i} + \mathcal{W}' \\ w_i^3 \in w_i + \mathcal{W}' \qquad w_{1,i}^3 \in w_{1,i} + \mathcal{W}' \qquad w_{2,i}^3 \in w_{2,i} + \mathcal{W}' \end{cases}$$

Posons

$$\begin{cases} w_i^2 = w_i^3 \\ w_{1,i}^2 = w_i^3 - w_{2,i}^2 \end{cases}$$

Nous avons

$$(22) \qquad w_i^2 - w_i = w_i^3 - w_i \in \mathcal{W}' \subset \mathcal{W}$$

D'autre part

$$w_{1,i}^{2} - w_{1,i} = (w_i^{3} - w_{2,i}^{3} - w_{1,i}^{3}) + (w_{2,i}^{3} - w_{2,i}) + (w_{2,i} - w_{2,i}^{2})$$
$$+ (w_{1,i}^{3} - w_{1,i})$$

d'où

$$(23) \qquad w_{1,i}^{2} - w_{1,i} \in w_i^{3} - w_{2,i}^{3} - w_{1,i}^{3} + \mathcal{W}' + \mathcal{W}' + \mathcal{W}'$$

Maintenant l'équation (21,α)) entraîne

$$\exists\, i_o \qquad \forall i \geq i_o \qquad w_i^{3} - w_{1,i}^{3} - w_{2,i}^{3} \in \mathcal{W}'$$

En posant donc

$$\text{pour } i < i_o \qquad \bar{w}_{1,i}^{2} = w_{1,i}$$
$$\text{pour } i \geq i_o \qquad \bar{w}_{1,i}^{2} = w_{1,i}^{2}$$

On a, d'après (23)

$$\forall i \qquad \bar{w}_{1,i}^{2} \in w_{1,i} + \mathcal{W}' + \mathcal{W}' + \mathcal{W}' + \mathcal{W}' \subset w_{1,i} + \mathcal{W}$$

et d'après la seconde ligne de (21)

$$\lim \mathcal{f}_1(v_i^{1}, w_{1,i}^{1}) = 0 = \lim \mathcal{f}_2(v_i^{2}, \bar{w}_{1,i}^{2})$$

de sorte que $(v_i, w_i, w_{1,i})$ est s.s.a.b. de (3.1), ce qu'il fallait démontrer.

Proposition 10 - Soient f_1 et f_2 dans $\Gamma_o(V)$. Alors (4) est approximativement plus forte que (3) pour toute topologie $\mathcal{C}_V$ d'espace vectoriel sur V et pour toute topologie $\mathcal{C}_W$ compatible avec la dualité. De plus si $(v_i, w_i, w_{1,i}, w_{2,i})$ satisfait :

$$(24)\quad \left|\begin{array}{l} \lim \mathcal{J}_1(v_i, w_{1,i}) = 0 \\ \lim \mathcal{J}_2(v_i, w_{2,i}) = 0 \\ \lim (w_i - w_{1,i} - w_{2,i}) = 0 \end{array}\right.$$

alors (v_i, w_i) adhérente à l'ensemble des s.s.a.b. de (4).

En effet la proposition 7 nous apprend que, pour les topologies de l'énoncé (4) est approximativement équivalente à

$$(f_1 + f_2)(v) + (f_1^* \nabla f_2^*)(-w) + \langle v, w\rangle = 0 \tag{25}$$

et, par définition

$$(f_1^* \nabla f_2^*)(-w) = \inf_{w_1 + w_2 = w} \{f_1^*(-w_1) + f_2^*(-w_2)\} \tag{26}$$

Il suffit de comparer (25) à (3). Soit (v_i, w_i) une s.s.a.b. de (25). En vertu de (26) il existe une suite $(w_{1,i}, w_{2,i})$ telle que, pour tout i :

$$w_{1,i} + w_{2,i} = w_i$$

$$f_1^*(-w_{1,i}) + f_2^*(-w_{2,i}) < (f_1^* \nabla f_2^*)(-w_i) + \frac{1}{i}$$

d'où

$$0 \leq \mathcal{J}_1(v_i, w_{1,i}) + \mathcal{J}_2(v_i, w_{2,i}) \leq (f_1 + f_2)(v_i) + (f_1^* \nabla f_2^*)(-w_i) + \langle v_i, w_i\rangle + \frac{1}{i}$$

Comme $\mathcal{J}_1$ et $\mathcal{J}_2$ sont positives on en déduit que la suite $(v_i, w_i, w_{1,i}, w_{2,i})$ satisfait (24); elle est donc a fortiori s.s.a.b. de (3.2) et (v_i, w_i) est s.s.a.b. de 3 . Par suite

$$(4) \Longleftrightarrow (25) \Longrightarrow (3)$$

Montrons maintenant la seconde partie de la proposition. On a, ~~d'après (25) et (26)~~, ~~et quels que soient~~ $v_i, w_{1,i}, w_{2,i}$:

$$0 \leq (f_1 + f_2)(v_i) + (f_1^* \nabla f_2^*)(-w_{1,i} - w_{2,i}) + \langle v_i, w_{1,i} + w_{2,i} \rangle \leq \mathcal{L}_i$$

avec $$\mathcal{L}_i = \mathcal{J}_1(v_i, +w_{1,i}) + \mathcal{J}_2(v_i, +w_{2,i})$$

Par suite si $(v_i, w_i, w_{1,i}, w_{2,i})$ satisfait (24) la suite $(v_i, w_{1,i} + w_{2,i})$ est s.s.a.b. de (25); comme

$$\lim (w_i - w_{1,i} - w_{2,i}) = 0$$

pour tout $\mathcal{W}$ voisinage de 0 dans W il existe une valeur i_o telle que

$$i \geq i_o \Rightarrow \quad w_i - w_{1,i} - w_{2,i} \in \mathcal{W}$$

Prenant alors

$$\bar{w}_i = w_i \qquad \text{pour} \quad i < i_o$$

$$\bar{w}_i = w_{1,i} + w_{2,i} \in w_i + \mathcal{W} \qquad \text{pour} \quad i \geq i_o$$

la suite $(v_i, \bar{w}_i)$ ainsi obtenue est bien s.s.a.b. de (25).

Utilisons maintenant les propositions 2 et 7; il vient la :

Proposition 11 - Si f_1 et f_2 sont des fonctions convexes, à valeurs dans $]-\infty, +\infty]$, pour toutes topologies $\mathcal{C}_V, \mathcal{C}_W$ compatibles avec la dualité l'équation

$$-w \in \partial(f_1^{**} + f_2^{**})(v)$$

est approximativement plus forte que l'équation

$$-w \in \partial f_1(v) + \partial f_2(v)$$

Ceci nous permet de remplacer la seconde par la première en encourant deux risques :

1°) d'agrandir l'ensemble des solutions, ce qui nous indiffère dans la mesure où celles-ci ne sont ni physiquement observables ni numériquement calculables.

2°) de restreindre l'ensemble des s.s.a.b.

<u>Notation</u> - Soient $A_1 = -\partial f_1$, $A_2 = -\partial f_2$ deux lois sous-différentielles; on appellera somme régularisée de ces lois la loi sous-différentielle

$$A_1 \hat{+} A_2 = -\partial(f_1^{**} + f_2^{**})$$

On a, d'après ce qui précède

$$A_1 \hat{+} A_2 = \hat{A}_1 \hat{+} \hat{A}_2$$

<u>Proposition 12</u> - L'addition régularisée est commutative et associative.

En effet elle se ramène à l'addition des Γ régularisées de f_1, f_2, f_3 etc...

<u>Application au problème général</u>

Revenons au problème général posé au chapitre II. Sur l'élément mécanique $(V,<.,.>,W)$ on se donne trois lois d'effort sous-différentielles, A_1, A_2, A_3, à savoir

(27)
$$\begin{cases} 1°)\ \text{Effort donné } w^o : & A_1 = -\partial f_1 \quad \text{avec } f_1 = <.,-w_o> \\ 2°)\ \text{Liaison affine parfaite} : & A_2 = -\partial f_2 \quad \text{avec } f_2 = \psi_{v^o+I} \\ 3°)\ \text{Loi constitutive} : & A_3 = -\partial f_3 \end{cases}$$

On suppose que f_2 et f_3 appartiennent à $\Gamma_o(V)$. Le problème d'équilibre s'écrit :

(27') $$0 \in (\partial f_1 + \partial f_2 + \partial f_3)(v)$$

équation approximativement plus faible

(28) $$0 \in \partial(f_1 + f_2 + f_3)(v)$$

Celle-ci admet toujours des s.s.a.b. puisque

$$\inf_{v \in V} \left((f_1 + f_2 + f_3)(v)\right) + (f_1 + f_2 + f_3)^{*}(0) = 0$$

Toute suite $(v_i, w_i = 0)$ telle que

$$\lim\ (f_1 + f_2 + f_3)(v_i) = \inf_{v\ V}\ (f_1 + f_2 + f_3)(v) \quad (\text{d'où } v_i \in v^o + I)$$

est s.s.a.b. de 27 et il lui correspond deux suites $(w_{2,i})$ $(w_{3,i})$ telles que

$$\begin{cases} w_{2,i} \in I^o \\ \lim\ \left(f_3(v_i) + f_3^*(-w_{3,i}) + \langle v_i, w_{3,i}\rangle\right) = 0 \\ \lim\ (w_o + w_{2,i} + w_{3,i}) = 0 \end{cases}$$

pour toute topologie sur W compatible avec la dualité.

Ceci n'ajoute rien à la connaissance pratique de l'équation (28); mais nous avons seulement voulu montrer que (28) possède autant de signification physique que (27') et que le problème ainsi posé est satisfaisant.

§ 4. SOMME GAUCHE DE DEUX LOIS D'EFFORT

C'est l'opération symétrique de l'addition, obtenue en inversant les rôles de V et de W. Nous la noterons par le symbole de la Γ-convolution.

Par définition

$$\overline{(A_1 \,\underline{\nabla}\, A_2)} = \overline{A_1} + \overline{A_2}$$

au second membre le signe + désigne l'addition des multiapplications.

Lorsque $A_1 = -\partial f_1$ $A_2 = -\partial f_2$ avec f_1 et f_2 dans $\Gamma_o(V)$ on a la relation

$$A_1 \,\underline{\nabla}\, A_2 \longleftrightarrow -\partial(f_1 \nabla f_2)$$

ce qui justifie la notation.

Un premier exemple est la loi d'élastoplasticité de Hencky :

$$A = A_1 \,\underline{\nabla}\, A_2$$

où A_1 est une loi d'élasticité linéaire et A_2 une loi du rigide-plastique parfait.

Un second exemple est :

$$\partial\psi_{u^o+V} = \partial\psi_{\{u^o\}} \,\underline{\nabla}\, \partial\psi_V$$

d'emploi courant, comme on vient de le voir.

D'une façon générale la somme de deux lois d'effort correspond à ce que les rhéologistes et les électriciens nomment le "montage en parallèle", tandis que la somme gauche correspond au "montage en série".

§5. QUOTIENT DROIT ET QUOTIENT GAUCHE

Les deux opérations de quotient sont les plus intéressantes de cette algèbre et, bien qu'elles soient d'usage courant depuis le début de la Mécanique, leur formalisation n'est que très récente (cfr. J.J. MOREAU (4) et NAYROLES (3)). Elles correspondent, comme tout passage d'un ensemble à un ensemble quotient, à une perte d'information : les informations ainsi perdues étant jugées ou bien parfaitement inutiles, ou bien trop lourdes à prendre en compte. C'est d'ailleurs une attitude courante, quoique souvent inconsciente, des physiciens de restreindre l'information sur un phénomène à un petit nombre de variables, faute de pouvoir même savoir quelle serait l'information totale.

Commençons par définir les deux quotients : ils sont symétriques l'un de l'autre par échange des rôles de V et de W. Nous nous contenterons donc d'étudier, dans cet alinéa, quelques propriétés du quotient droit qu'il suffira de transposer pour obtenir celles du quotient gauche. Dans le paragraphe suivant nous en verrons quelques applications.

<u>Définition 5</u> - Soient $\mathcal{E} = (V, <.,.>, W)$ un élément mécanique, U un sous-espace vectoriel fermé de V (resp....), $\mathcal{Q}$ l'application canonique de W sur son quotient $\Phi = W/U^\circ$ (resp....). On appelle élément quotient droit (resp. gauche) de $\mathcal{E}$ par U° l'élément mécanique

$$\mathcal{F} = (U, <<.,.>>, \Phi) \qquad \text{(resp...)}$$

où la forme bilinéaire <<.,.>> est définie par :

$$(u,\phi) \in U \times \Phi \to <<u,\phi>> = <u,w> \qquad w \in \mathcal{Q}^{-1}(\phi)$$

(resp...)

Les (resp...) sont, pour abréger le texte, laissés à la diligence du lecteur. On note que la forme bilinéaire est bien définie puisque, u appartenant à U, <u,w> ne dépend que de u et de ϕ.

Le quotient droit intervient si l'on admet que le déplacement u reste dans U et si l'on considère comme équivalents deux efforts qui développent le même travail virtuel dans tout déplacement virtuel $u \in U$. C'est exactement ce qui se produit lorsque l'élément mécanique $\mathcal{E}$ est soumis à deux lois d'effort

$$\left.\begin{array}{ll} \text{l'une de liaison :} & \psi_U(v) + \psi_{U^*}(-w_\ell) = 0 \\ \text{l'autre (éventuellement une somme)} & w \in A(v) \end{array}\right\} \qquad (29)$$

Le système (29) est évidemment équivalent à

(30) $$\begin{cases} \mathcal{Q} \circ A(v) \ni 0 \\ v \in \mathcal{U} \end{cases}$$

Considérons maintenant l'immersion de $\mathcal{U}$ dans V_i c'est l'application transposée de $\mathcal{Q}$, par définition de la forme bilinéaire $\langle\langle .,. \rangle\rangle$; notons-la donc $\mathcal{Q}^T$. Alors (30) équivaut à

(31) $$\mathcal{Q} \circ A \circ \mathcal{Q}^T(u) \ni 0$$

<u>Définition 6</u> - Soit A une loi d'effort sur $\mathcal{E}$. On appelle loi quotient de A la loi $\mathcal{Q} \circ A \circ \mathcal{Q}^T$ qui est une loi d'effort sur $\mathcal{F}$.

<u>Proposition 12</u> - Le quotient droit est distributif par rapport à l'addition :

$$\mathcal{Q} \circ (A_1 + A_2) \circ \mathcal{Q}^T = \mathcal{Q} \circ A_1 \circ \mathcal{Q}^T + \mathcal{Q} \circ A_2 \circ \mathcal{Q}^T$$

La démonstration est immédiate. Venons-en maintenant à l'étude des lois quotients de lois sous-différentielles.

<u>Proposition 13</u> - Soient f une fonction définie sur V, à valeurs dans $]-\infty, +\infty]$, et $\tilde{f}_U$ sa restriction à $\mathcal{U}$. La polaire de celle-ci, pour la dualité entre U et Φ vaut

$$\forall \phi \in \Phi \qquad \forall w \in \mathcal{Q}^{-1}(\phi) \qquad \tilde{f}_U^*(\phi) = (f + \psi_U)^* (w)$$

En effet nous avons

$$\tilde{f}_U^*(\phi) = \sup_{u \in U} \left(\langle\langle u,\phi \rangle\rangle - \tilde{f}_U(u)\right)$$

d'où

$$\forall w \in \mathcal{Q}^{-1}(\phi) \quad \tilde{f}_U^*(\phi) = \sup_{u \in U} \left(\langle u,w \rangle - f(u)\right)$$

$$\forall w \in \mathcal{Q}^{-1}(\phi) \quad \tilde{f}_U^*(\phi) = \sup_{u \in V} \left(\langle u,w \rangle - (f + \psi_U)(u)\right)$$

On en déduit la

Proposition 14 - Sous les hypothèses de la proposition 13 la loi

$$\phi \in -\partial\tilde{f}_U(u)$$

est la loi quotient de la loi sous-différentielle

$$w \in -\partial(f + \psi_U)(u)$$

En particulier si f est convexe, finie et continue en un point de U :

$$\mathcal{Q} \circ -\partial f \circ \mathcal{Q}^T = -\partial\tilde{f}_U$$

La notion d'équivalence approximative nous a permis de remplacer dans tout les cas $\partial f + \partial\psi_U$ par $\partial(f + \psi_U)$. Nous régulariserons donc ce quotient en remplaçant $\mathcal{Q} \circ -\partial f \circ \mathcal{Q}^T$ par la loi plus forte $-\partial\tilde{f}_U$.

§ 6. DEUX UTILISATIONS PRATIQUES DU QUOTIENT DROIT

5.1 Approximations cinématiques

La mécanique des milieux continus s'intéresse tout d'abord au milieu tridimensionnel; puis, comme il est en pratique bien difficile de résoudre numériquement les équations obtenues, on essaye de tenir compte, lorsque l'occasion s'en rencontre, des faibles valeurs de certaines données pour obtenir une théorie simplifiée, par exemple celle des plaques.

Une méthode utilisée bien souvent, et qui peut s'appliquer à de très nombreux cas mathématiques est la suivante. On part de l'équation (ou du système)

$$\mathcal{F}(\varepsilon,x) = 0$$

ou ε est le petit paramètre, puis on fait un développement limite par rapport à ε :

$$\mathcal{F}(\varepsilon,x) = \mathcal{F}_o(x) + \varepsilon\mathcal{F}_1(x) + \varepsilon^2\mathcal{F}_2(x) + \dots$$

et on considère, selon la qualité de l'approximation désirée, les équations

$$\mathcal{J}_0(x) = 0$$

$$\mathcal{J}_0(x) + \varepsilon \mathcal{J}_1(x) = 0$$

ce qui donne des "théories linéarisées au premier ordre, au second ordre etc...". Cette méthode a été particulièrement préconisée par J.M. SOURIAU, mais elle est d'usage constant, et vraisemblablement très ancien.

L'ennui est que les résultats théoriques obtenus sur le problème initial peuvent devenir inapplicables aux problèmes linéarisés, certaines des propriétés de $\mathcal{J}$ étant perdues au cours de la linéarisation.

La Résistance des Matériaux a ceci de particulier que les approximations utilisées conservent la structure algébrique examinée au chapitre I. Ces approximations peuvent être de trois types : cinématique (théorie de Kirschoff pour les plaques et les coques, éléments finis "déplacements etc.) sthénique (théorie de Reissner pour les plaques et les coques, éléments finis "forces") ou mixte (théorie des poutres en flexion, éléments finis mixtes).

Les approximations de type cinématique consistent à supposer (à une approximation près) que le déplacement v reste dans un sous-espace vectoriel U de V : pour ne pas perdre la structure algébrique le mécanicien posera cette hypothèse, non pas en termes d'approximation, mais comme l'introduction d'une loi de liaison parfaite. Alors il travaillera sur l'élément mécanique quotient ($\mathcal{U}$, <<.,.>>, ϕ). Si les efforts dérivent d'un potentiel f il minimisera donc non pas f mais sa restriction à $\mathcal{U}$.

On pourra consulter VALID (1), BREUNEVAL ((1), (2), (3)) à propos de l'application de la technique de linéarisation précédente aux approximations cinématiques dans les coques. Dans SOUCHET (1) on trouvera un exposé clair de la théorie de REISSNER pour les plaques, ainsi évidemment que dans REISSNER lui-même ((1), (2)).

5.2 Problème de Saint-Venant (MAISONNEUVE [1])

Ici c'est la perte d'information qui est elle-même recherchée et qui va nous conduire à un quotient droit. On considère une milieu continu tridimensionnel qui occupe le domaine Ω dont $\partial\Omega$ désigne la frontière. Il est soumis à divers efforts, dont certains sont des efforts extérieurs appliqués sur une partie $\partial_1\Omega$ de la frontière.

Avec SAINT-VENANT et MAISONNEUVE, on souhaite, pour des raisons qui n'ont pas lieu d'être ici précisées, ne s'intéresser qu'au torseur des efforts appliqués. (suite page 72)

§ 7. UNE UTILISATION PRATIQUE DU QUOTIENT GAUCHE : LA SOUS-STRUCTURATION

Le quotient gauche intervient dans les approximations de type sthéniques de la même façon que le quotient droit dans les approximations cinématiques : il n'y a donc pas lieu de développer ce point. Une autre occasion d'utiliser le quotient gauche est la sous-structuration.

Le calcul des grandes structures, c'est à dire dépendant d'un grand nombre de paramètres de liberté, ne peut être entrepris d'un seul coup si l'ordinateur utilisé ne possède pas une mémoire de taille suffisante. Pour calculer ces structures on les découpe en sous-structures que l'on calcule séparément, puis qu'on assemble ensuite. L'intérêt de l'opération est de réduire le nombre de paramètres de liberté dont dépend une sous-structure au nombre de paramètres nécessaires à l'écriture des conditions de liaison avec les autres sous-structures.

Soit par exemple S_1 une sous-structure représentée par l'élément mécanique $(V_1, <.,.>, W_1)$, et constituée par un solide continu. Supposons la

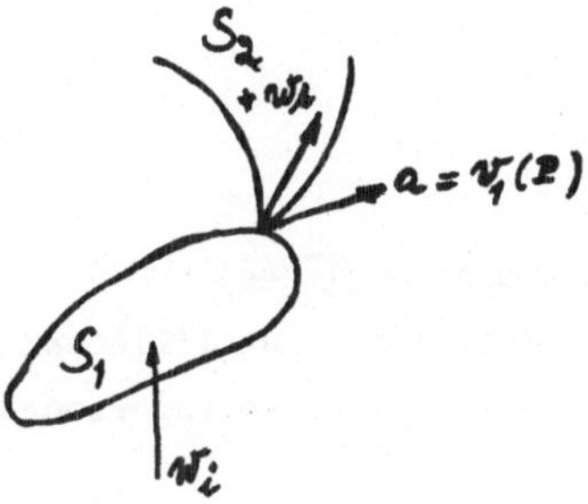

reliée au reste S_2 de la structure en un seul point P : l'équation de liaison égalera le déplacement $a = v_1(P)$ de S_1 au déplacement $v_2(P)$ de S_2.

S_1 est soumise à des efforts $w_1, \ldots, w_n$ et à l'effort $-w_s$ exercé par S_2 sur S_1 en P; w_s est ici une force concentrée en P.

Pour le calcul de la structure complète nous aurons à disposer du renseignement suivant : quel est l'effort w_s qui correspond au déplacement a du point P de S_1, compte tenu des efforts $w_1, \ldots, w_n$?

On note que dans ce problème deux champs de déplacement v et v' de V sont équivalents s'ils prennent la même valeur en P. Parallèlement l'effort de liaison w_s appartient à un sous-espace vectoriel Φ (ici de dimension 3) de W, et que Φ° est précisément l'ensemble des $v \in V$ qui s'annulent en P. Autrement dit le renseignement demandé porte sur l'élément quotient gauche

$$(\mathcal{U} = V_1/\Phi^\circ \; , \; \langle\langle .,. \rangle\rangle, \; \Phi)$$

Soient A la somme des lois d'effort donnant les $w_1, \ldots, w_n$, $\mathcal{Q}$ l'application canonique de V_1 sur $\mathcal{U}$. Donnons-nous $-w_s$ et supposons le en équilibre avec A :

$$0 \in -w_s + A(v)$$

est l'équation qui donne v de fonction de w_s. Elle s'écrit aussi

(32) $$v \in A^{-}(w_s)$$

qui entraîne

(33) $$\mathcal{Q}(v) \in \mathcal{Q} \circ A^{-1}(w_s) = \mathcal{Q} \circ A^{-1} \circ \mathcal{Q}^{T}(w_s)$$

ce qui donne les $a = \mathcal{Q}(v)$ cherchés en fonction de w_s. $\mathcal{Q}^T$ est encore l'immersion de Φ dans W_1.

Inversement

$$a \in \mathcal{Q} \circ A^{-1} \circ \mathcal{Q}^T (w_s)$$

entraîne qu'il existe v solution de (32) et appartenant à $\mathcal{Q}^{-1}(a)$.

Si donc on ne désire utiliser que les informations sur l'élément quotient on devra remplacer la loi A par sa loi quotient, inverse inférieure de $\mathcal{Q} \circ A^{-1} \circ \mathcal{Q}^T$.

On note que la transformation d'une loi d'effort par quotient gauche est distributive par rapport à l'addition gauche (notée $\underline{V}$) mais pas par rapport à l'addition. Dans ce qui précède il faut donc que A soit la somme de toutes les lois d'effort exercés sur la sous-structure et qui ne sont pas des lois normales à Φ^o ; on peut en effet vérifier aisément que si A_1 est normale à Φ^o (c'est à dire une loi d'effort donnant une force appliquée à P en fonction du déplacement de ce point) et si A_2 est une loi d'effort quelconque sur (V , .,. , W) alors

$$(\mathcal{Q} \circ (A_1 + A_2)^- \circ \mathcal{Q}^T) = (\mathcal{Q} \circ A_1 \circ \mathcal{Q}^T)^- + (\mathcal{Q} \circ A_2 \circ \mathcal{Q}^T)^-$$

Additum - (Suite et fin du § 6, ch. IV, oubliées à la suite d'une erreur de mise en page).

Soit $\mathcal{E} = (V, \langle .,. \rangle, W)$ l'élément mécanique constitué par le solide envisagé. L'ensemble U des champs qui solidifient $\partial_1\Omega$ est un sous-espace vectoriel de V, que nous supposerons fermé. U° est l'ensemble des efforts appliqués à $\mathcal{E}$ et qui ne travaillent pas dans les déplacements qui solidifient $\partial_1\Omega$: $W/I^\circ = \Phi$ peut donc être considéré comme l'espace des torseurs des efforts appliqués à $\partial_1\Omega$.

Si l'on se donne

1°) la somme w_1 de tous les efforts appliqués au solide exceptés ceux qui s'exercent sur $\partial_1\Omega$.

2°) le torseur ϕ des efforts exercés sur $\partial_1\Omega$

l'équilibre du solide s'écrit

$$\begin{cases} w_1 + w_2 = 0 \\ w_2 \in \mathcal{D}^{-1}(\phi) \end{cases}$$

Supposons que w_1 soit donné par la loi sous-différentielle

$$-w_1 \in \partial f(v) \qquad\qquad (f \in \Gamma_o(V))$$

le système précédent équivaut à

$$\begin{cases} w_2 \in \partial f(v) \\ w_2 \in \mathcal{D}^{-1}(\phi) \end{cases}$$

dont la donnée est ϕ, les inconnues v et w_2. Il n'est pas déterminé : pour le déterminer on va choisir, avec Maisonneuve, l'élément de $\mathcal{D}^{-1}(\phi)$ qui minimise l'énergie complémentaire f^*, c'est à dire remplacer le système précédent par

$$(34) \quad \begin{cases} w_2 \in \partial f(v) \\ w_2 \in \mathcal{D}^{-1}(\phi) \\ f^*(w_2) = \min\limits_{w \in \mathcal{D}^{-1}(\phi)} f^*(w) \end{cases}$$

Soit $(\bar{v},\bar{w}_2)$ une solution de ce système. On a

$$0 \in \partial\left(f^* + \psi_{\bar{w}_2 + U^o}\right)(\bar{w}_2)$$

c'est à dire, en supposant que les sous-différentiels s'ajoutent

$$\exists\, u \in \mathcal{U} \qquad u \in \partial f^*(\bar{w}_2)$$

de sorte que $(u,\bar{w}_2)$ est solution du système (34). Si f^* est faiblement différentiable on en déduit que $\bar{v} \in \mathcal{U}$ de sorte que résoudre le système

B. Nayroles

BIBLIOGRAPHIE

J. BREUNEVAL (1) "Schéma d'une théorie générale des coques minces élastiques" Journal de Mécanique - vol. 10 n° 2 juin 1971

(2) "Coques minces soumises à des liaisons internes" Journal de Mécanique - vol. 11 n° 2 juin 1972

(3) "Géométrie des déformations des surfaces et équations de la Mécanique des coques" Thèse - Marseille 1972

G. DUVAUT et J.L. LIONS (1) "Les inéquations en Mécanique et en Physique" - DUNOD - 1972

O. MAISONNEUVE (1) "Sur le principe de Saint-Venant" - Thèse - Poitiers 1971

J.J. MOREAU (1) "Fonctionnelles convexes" - Séminaire sur les équations aux dérivées partielles - Collège de France 66-67

(2) "La notion de sur-potentiel et les liaisons unilatérales en élastostatique" C.R.A.S. Paris t. 267 p. 954-957 (16.12.68) - Série A

(3) "Fonctions de résistance et fonctions de dissipation" Séminaire d'Analyse Unilatérale - Montpellier 1971 Exposé n° 6

(4) "Convexité et frottement" Université de Montréal- Département d'Informatique - Pub. n° 32 (1970)

B. NAYROLES (1) "Essai de theorie fonctionnelle des structures rigides plastiques parfaites" - Journal de Mécanique - vol. 9 - n° 2 (1970) p. 491-506

(2) "Quelques applications variationnelles de la théorie des fonctions duales à la Mécanique des Solides" Journal de Mécanique - vol. 10 - n° 2 (1971) p. 263-289

(3) "Opérations algébriques en Mécanique des Structures" C.R. Acad. Sci. Paris t. 273 p. 1075-1078 (1971)

E. REISSNER (1) "On the theory of bending of elastic plates" - J. of Math. Phys. 1944 - 23 (184-191)

(2) "On bending of elastic plates" - Quart. of Appl. Math. 1947 - 5 - 1 (55-68)

R.T. ROCKAFELLAR (1) "Integrals which are convex functionals" - Pacific Journal of Mathematics - vol. 24 n° 3 (1968)

(2) "Integrals which are convex functionals, II" Pacific Journal of Mathematics - vol. 39 n° 2 (1971)

(3) "Convex integral functional and duality" - Contributions to non linear functional analysis (1971) - Academic Press

(4) "On the maximality of sums of non linear monotone operators" Transactions of the American Mathematical Society - vol. 149 - may 1970

R. SOUCHET (1) "Sur la théorie de la flexion des plaques minces élastiques de E. REISSNER" - Thèse - Poitiers 1970

R. VALID (1) "Sur la théorie des coques" - Thèse - Poitiers 1973

CENTRO INTERNAZIONALE MATEMATICO ESTIVO

(C.I.M.E.)

ON CERTAIN CONVEX SETS OF MEASURES AND PHASES OF REACTING MIXTURES

WALTER NOLL

The lecture is not printed here because its content is substantially contained in the paper published in the "Archive for rational mechanics" Volume 38 - 1970, Pag. 1-12.

Corso tenuto a Bressanone dal 17 al 26 giugno 1973

CENTRO INTERNAZIONALE MATEMATICO ESTIVO

(C.I.M.E.)

ON COMPLEMENTARY VARIATIONAL INEQUALITIES

W. VELTE

Corso tenuto a Bressanone dal 17 al 26 giugno 1973

ON COMPLEMENTARY VARIATIONAL INEQUALITIES

W. Velte

1. Introduction

Pairs of complementary variational principles are well known in various fields of applications. In elasticity, for example, the equilibrium state of an elastic medium can be characterized by the principle of minimal potential energy as well as by Castigliano's principle of complementary energy. In electrostatics, the electrostatic field can be characterized by Dirichlet's principle as well as by the complementary principle of Thomson. The method of Trefftz [8] (see also Michlin [5]) is a counterpart of the method of Ritz and uses a complementary variational principle for the underlying self-adjoint elliptic boundary value problem.

A systematical approach to complementary variational principles beginning with Friedrichs [3] (see also Courant and Hilbert [1]) was developed by several authors, considering also inequalities as constraints. Bibliographies are found in Robinson [6] and Sewell [7] .

In the following, we give a simple approach to pairs of complementary variational principles and complementary variational inequalities for a certain class of non-linear elliptic boundary value problems. It is related to problems considered by Fichera [2] and to the class of problems studied by Lions and Stampacchia [4] . The classical examples cited below (involving no inequalities as constraints) are covered as special cases.

2. The variational problem

Let E denote a linear space. In our examples (section 5) E will be a real Hilbert space. But in order to formulate pairs of complementary extremal problems and complementary variational inequalities it is sufficient to have the following situation:

Let E be a real Banach space (or, more general, a linear, locally convex space) , E' the dual of E and $\langle \, , \, \rangle$ the pairing between E and E' . Let be given a symmetric bilinear form $a(\, , \,): E \times E \rightarrow R$, non-negative on E and positiv on some linear subspace $V \subset E$:

(1) $\qquad a(u,v) = a(v,u) \qquad u,v \in E$

(2) $\qquad a(u,u) \geq 0 \qquad u \in E$

(3) $\qquad a(u,u) > 0 \qquad u \in V \, , \; u \neq o$

In our examples $a(\, , \,)$ will be a bilinear form corresponding to a linear elliptic differential operator, and V will be a linear subspace of functions which satisfy certain linear, homogeneous boundary conditions. The linear variety

(4) $\qquad M = \{ u \in E \, / \; u - u_o \in V \}$

with given element u_o then corresponds to the set of functions which satisfy certain inhomogeneous boundary conditions.

We are interested in problems, for which besides of the (homogeneous or inhomogeneous) boundary conditions there are constraints given in the form $u \in K_1$, where $K_1 \subset E$ is a convex set. Then the set K of admissible functions is given by

(5) $\qquad K = M \cap K_1$

Clearly, K is a convex set, too.

Firstly, we shell consider for given $f \in E'$ the following variational problem:

Problem 1. Find the solution $\bar{u}$ of the extremal problem

$$(6) \quad J(u) = a(u,u) - 2\langle f,u\rangle \rightarrow \min \quad , \quad u \in K$$

respectively of the equivalent variational inequality

$$(7) \quad a(u,v-u) - \langle f,v-u\rangle \geq 0 \qquad \forall v \in K .$$

We are not concerned here with the problem of the existence of a solution. Existence holds under additional assumptions (see for example Lions and Stampacchia [4]). The solution is, however, unique by assumption (3) .

Secondly, consider a function $F(\ ): E \rightarrow R$ which is bounded from below and convex. Suppose that $F(\ )$ has for any $u \in E$ a derivative $f(u) \in E'$ in the sense of Gateaux. Then

$$(8) \quad F(v) - F(u) \geq \langle f(u),v-u\rangle \qquad u,v \in E .$$

Problem 2. Find the solution $\bar{u}$ of the extremal problem

$$(9) \quad J(u) = a(u,u) + 2F(u) \rightarrow \min \quad , \quad u \in K$$

respectively of the equivalent variational inequality

$$(10) \quad a(u,v-u) + \langle f(u),v-u\rangle \geq 0 \qquad \forall v \in K .$$

Again, we are not concerned here with sufficient conditions for the existence of a solution. For sufficient conditions see [4], for example. We simply suppose, that a solution exists. Uniqueness follows from (3) and (8) .

3. Error estimate in energy norm

Since Problem 1 is only a special case of Problem 2 with $F(u) = - \langle f,u \rangle$ we can treat both problems in the form (9) .

Let $\bar{u}$ denote the solution of (9) resp. of (10) . In general, it will be not possible to give the solution $\bar{u}$ by explicit terms. Then, if $u \in K$ is any numerical approximation to $\bar{u}$, the question arises how to obtain an error estimate.

In the theory of variational methods for quadratic extremal problems without inequalities as constraints the role of error estimates in energy norm is well known (see, for example, Michlin [5]) . Therefore, it seems quite natural to look for an error estimate in energy norm in the case of variational inequalities, too .

The energy norm is given by

$$(11) \qquad |u| = a(u,u)^{1/2} \quad .$$

This is, by assumption (2) and (3) , a norm in V and a half-norm in E.

Proposition 1. Let $\bar{u} \in K$ denote the solution of (9) resp. of (10) and let $u \in K$ be any numerical approximation to $\bar{u}$. Then

$$(12) \qquad |u-\bar{u}|^2 \leq J(u) - J(\bar{u}) \quad .$$

Proof. The solution $\bar{u} \in K$ satisfies the variational inequality (10) for all $v \in K$. Hence, with $v = u$,

$$|u-\bar{u}|^2 \leq |u-\bar{u}|^2 + 2a(\bar{u},u-\bar{u}) + 2\langle f(\bar{u}),u-\bar{u}\rangle \quad .$$

Using (8) and rearranging the terms on the right hand side, one obtains immediately

$$|u-\bar{u}|^2 \leq |u-\bar{u}|^2 + 2a(\bar{u},u-\bar{u}) + 2F(u) - 2F(\bar{u})$$
$$= J(u) - J(\bar{u}) \quad .$$

Corollary 1. Let d be a lower bound for $J(\bar{u})$, $d \leq J(\bar{u})$, then from (12)

$$(13) \qquad |u-\bar{u}|^2 \leq J(u) - d \quad .$$

Naturally, the question arises how to obtain a lower bound d which is sufficiently close to $J(\bar{u})$. In the next section a systematical approach to a lower bound will be given in terms of complementary extremal problems involving complementary variational inequalities.

4. Complementary variational problems

Since by assumtion the bilinear form $a(\ ,\ )$ is non-negative, one has $a(u-v,u-v) \geq 0$ or

$$(14) \qquad a(u,u) \geq -a(v,v) + 2a(v,u) \qquad u,v \in E ,$$

and (8) may be written in the form

$$(15) \qquad F(u) \geq F(v) + \langle f(v,u-v\rangle \qquad u,v \in E .$$

Consider the functional $J(u) = a(u,u) + 2F(u)$. From (14) and (15) one has for any $u \in K$ and any $v \in E$

$$J(u) \geq -a(v,v) + 2a(v,u) + 2F(v) + 2\langle f(v),u-v\rangle$$
$$= J(v) + 2\left\{ a(v,u-v) + \langle f(v),u-v\rangle \right\} .$$

Hence,

$$(16) \qquad J(u) \geq J(v)$$

provided, that the inequality

$$(17) \qquad a(v,u-v) + \langle f(v),u-v\rangle \geq 0$$

holds. Since we wish to obtain a lower bound for $J(u)$ for any $u \in K$, we require the inequality to hold for any $u \in K$.

Therefore, we introduce the set K_c of all elements $v \in E$ satisfying the <u>complementary variational inequality</u>

$$(18) \qquad a(v,u-v) + \langle f(v),u-v\rangle \geq 0 \qquad \forall\, u \in K .$$

<u>Remark</u>. In the case of Problem 1 the set K_c consists of all elements satisfying the variational inequality

$$(19) \qquad a(v,u-v) \geq \langle f,u-v\rangle \qquad \forall u \in K .$$

Note the similarity between the variational inequality (10) and the complementary variational inequality (18). The roles of u and v are interchanged. The solutions v of (18), however, are not restricted to be in K and are in general not unique.

<u>Proposition 2</u>. Let $\bar{u} \in K$ denote the solution of the variational problem (9) and v any element in K_c. Then

$$(20) \qquad |v-\bar{u}|^2 \leq J(\bar{u}) \dot{-} J(v) . \qquad .$$

<u>Proof</u>. For any two elements $u \in K$, $v \in K_c$ the inequality (18) holds. Hence, with $u = \bar{u}$,

$$\begin{aligned} |v-\bar{u}|^2 &\leq |v-\bar{u}|^2 + 2a(v,\bar{u}-v) + 2\langle f(v),\bar{u}-v\rangle \\ &\leq |v-\bar{u}|^2 + 2a(v,\bar{u}-v) + 2F(\bar{u}) - 2F(v) \\ &= J(\bar{u}) - J(v) \qquad . \end{aligned}$$

<u>Theorem 1.</u> Suppose the variational problem

$$J(u) = a(u,u) + F(u) \rightarrow \min \qquad , \quad u \in K$$

has a solution $\bar{u} \in K$. Then

$$(21) \qquad \{\bar{u}\} = K \cap K_c$$

and

$$(22) \qquad \min_{u \in K} J(u) = J(\bar{u}) = \max_{v \in K_c} J(v) \qquad .$$

Further, for any two elements $u \in K$, $v \in K_c$ one has

$$(23) \qquad 4\left|\frac{u+v}{2} - \bar{u}\right|^2 \leq |u-\bar{u}|^2 + |v-\bar{u}|^2 \leq J(u) - J(v) .$$

<u>Proof</u>. En element $w \in E$ is solution of (10) if and only if $w \in K \cap K_c$. But (10) has only one solution. Hence (21). For any pair $u \in K$, $v \in E$ satisfying (17) one has from (16) $J(u) \geq J(v)$. Since $\bar{u} \in K \cap K_c$ one has $J(u) \geq J(\bar{u})$ for any $u \in K$, and $J(\bar{u}) \geq J(v)$ for any $v \in K_c$. Hence (22).

Inequality (23) follows from

$$4\left|\frac{u+v}{2} - \bar{u}\right|^2 = |(u-\bar{u}) + (v-\bar{u})|^2 \le |u-\bar{u}|^2 + |v-\bar{u}|^2$$

together with (12) and (20) .

One can modify Theorem 1 in different ways, if the functional $J(\ ): E \to R$ is considered as restriction of an other functional. In the following we consider only one extension.

Let $\Psi(\ ,\ ): E \times E \to R$ denote the functional given by

$$(24) \qquad \Psi(v,w) = a(v,v) + 2F(w) \qquad .$$

Clearly, $J(u) = \Psi(u,u)$ for any $u \in E$. We now use (14) and (15), written in the form

$$a(u,u) \ge -a(v,v) + 2a(v,u)$$

and

$$F(u) \ge F(w) + \langle f(w), u-w \rangle \quad .$$

For any $u \in K$ and any $v,w \in E$ follows

$$J(u) = a(u,u) + 2F(u)$$
$$\ge \Psi(v,w) + 2\left\{a(v,u-v) + \langle f(w),u-w\rangle\right\} \quad .$$

Hence,

$$(25) \qquad J(u) \ge \Psi(v,w)$$

provided, that the inequality

$$(26) \qquad a(v,u-v) + \langle f(w),u-w\rangle \ge 0$$

holds. This time, we introduce the set $\tilde{K}_c$ of all pairs $(v,w) \in E \times E$ satisfying the variational inequality

$$(27) \qquad a(v,u-v) + \langle f(w),u-w\rangle \ge 0 \qquad \forall\, u \in K \ .$$

<u>Proposition 3</u>. Let $\bar{u} \in K$ denote the solution of the variational problem (9) and the pair v,w any solution of the variational inequality (27). Then

$$(28) \qquad |v-\bar{u}|^2 \le J(\bar{u}) - \Psi(v,w)$$

Proof. From (27) with $u = \bar{u}$

$$\begin{aligned} |v-\bar{u}|^2 &\le |v-\bar{u}|^2 + 2a(v,\bar{u}-v) + 2\langle f(w),\bar{u}-w\rangle \\ &\le |\bar{u}|^2 - |v|^2 + 2F(\bar{u}) - 2F(w) \\ &= J(\bar{u}) - \Psi(v,w) \quad . \end{aligned}$$

Theorem 2. Suppose the variational problem

$$J(u) = a(u,u) + 2F(u) \rightarrow \min \, , \qquad u \in K$$

has a solution $\bar{u} \in K$. Then

$$\min_{u \in K} J(u) = J(\bar{u}) = \max_{(v,w)\in \tilde{K}_c} \Psi(v,w) \quad .$$

Minimum and maximum are attained for $u = v = w = \bar{u}$. For any $u \in K$ and any pair $(v,w) \in \tilde{K}_c$

$$4\left|\frac{u+v}{2} - \bar{u}\right|^2 \le |u-\bar{u}|^2 + |v-\bar{u}|^2 \le J(u) - \Psi(v,w) \quad .$$

Proof. The theorem follows immediately from (12) and (28) .

Corollary 2. Choose $w \in K$. Then

$$J(w) \ge J(\bar{u}) \ge \Psi(v,w)$$

for any $v \in E$ satisfying the variational inequality

$$(29) \qquad a(v,u-v) + \langle f(w),u-w\rangle \ge 0 \qquad \forall\, u \in K \, .$$

Remark. In general it will be easier to find solutions v of inequality (29) with given $w \in K$ rather than solutions of inequality (18). Therefore, in comparision with Theorem 1, Theorem 2 and Corollary 2 are easier to apply.

5. Applications

Example 1.

(i) Consider a bounded region $\Omega \subset R^N$ with points $x = (x_1, \dots, x_N)$ and smooth boundary Γ consisting of two parts Γ_1 and Γ_2. Consider

$$a(u,v) = \sum a_{jk} \frac{\partial u}{\partial x_j} \frac{\partial u}{\partial x_k} \quad , \quad L(u) = -\sum \frac{\partial}{\partial x_j}(a_{jk} \frac{\partial u}{\partial x_k})$$

$(j,k = 1,\dots,N)$ with sufficiently regular coefficients. Let $E = H^1(\Omega)$, i.e. the completion of $C^1(\bar{\Omega})$ with respect to the norm

$$\|u\|_1 = (\|u\|_o^2 + \sum \|\partial u / \partial k_j\|_o^2)^{1/2} \quad ,$$

$(u,v)_o$ and $\|u\|_o$ denoting inner product and norm in $L_2(\Omega)$.

Further, let

$$V = \{ u \in H^1 \,/\, u = 0 \text{ on } \Gamma_1 \}$$

$$K = \{ u \in H^1 \,/\, u = 0 \text{ on } \Gamma_1 \text{ and } u \geq 0 \text{ on } \Gamma_2 \} .$$

Suppose

$$a(u,v) = a(v,u) \qquad u,v \in H^1$$

$$a(u,u) \geq 0 \qquad u \in H^1$$

$$a(u,u) \geq c\|u\|_1^2 \qquad u \in V \quad (c > 0) .$$

Then, for given $f \in L_2(\Omega)$, the extremal problem

$$J(u) = a(u,u) - 2(f,u)_o \rightarrow \min \quad , \qquad u \in K$$

has a unique solution $\bar{u}$. From $u \in K$ and variational inequality (7) we have (see [4], p.510)

$$L(u) = f \text{ in } \quad , \quad \frac{\partial u}{\partial n} \geq 0 \text{ and } u\frac{\partial u}{\partial n} = 0 \text{ on } \Gamma_2 ,$$

at least in the sense of weak solutions and traces. n denotes the conormal with respect to L(u).

(ii) The set K_c consists of all elements $v \in H^1$ satisfying the complementary variational inequality

$$a(v,u-v) - (f,u-v)_o \geq 0 \qquad \forall u \in K .$$

The inequality is satisfied if

$$L(v) = f \quad \text{in } \Omega \quad , \quad \frac{\partial v}{\partial n} \geq 0 \quad \text{on } \Gamma_2 \, , \quad \int_\Gamma v \frac{\partial v}{\partial n} d\Gamma \leq 0 \, .$$

Example 2.

(i) Consider the following boundary value problem for the stationary temperature distribution (absolute temperature) in a bounded region $\Omega \subset R^3$ with smooth boundary Γ consisting of three parts Γ_1 , Γ_2 and Γ_3 :

$$\Delta u = 0 \qquad \Omega \text{ in}$$

$$u = \alpha \quad \text{on } \Gamma_1 \, , \quad \frac{\partial u}{\partial n} = 0 \quad \text{on } \Gamma_2 \quad , \quad \frac{\partial u}{\partial n} + \beta u^4 = 0 \quad \text{on } \Gamma_3$$

($\alpha > 0$, $\beta > 0$) , i.e. given constant temperature on Γ_1 , no heat flow across Γ_2 and energy loss by radiation on Γ_3 according the law of Stefan - Boltzmann. The absolute temperature is always non negative.

Let $E = H^1(\Omega)$ as in Example 1 , and let

$$V = \{ u \in H^1 \ / \ u = 0 \text{ on } \Gamma_1 \}$$

$$K = \{ u \in H^1 \ / \ u = \alpha \text{ on } \Gamma_1 \, , \ u \geq 0 \quad \text{on } \Gamma_2 \cup \Gamma_3 \}$$

$$a(u,v) = \int_\Omega \operatorname{grad} u \operatorname{grad} v \, dx \qquad .$$

Assume

$$a(u,u) \geq c \, \|u\|_1^2 \qquad \forall \ u \in V \quad (c > 0) \, .$$

Then the extremal problem

$$J(u) = a(u,u) + 2 \int_{\Gamma_3} G(u) \, d\Gamma \quad \rightarrow \ \min \quad , \qquad u \in K$$

with

$$G(u) = \begin{cases} \frac{1}{5} \beta u^5 & \text{for } u \geq 0 \\ 0 & \text{for } u < 0 \end{cases}$$

has a unique solution $\bar{u} \in K$. (Note, that the functions in H^1 have boundary values in $L_2(\Gamma)$.) . The corresponding variational inequality is given by

$$a(u,v-u) + \int_{\Gamma_3} g(u)(v-u)\,d\Gamma \geq 0 \qquad \forall v \in K$$

where

$$g(u) = G'(u) = \begin{cases} \beta u^4 & \text{for} \quad u \geq 0 \\ 0 & \text{for} \quad u < 0 \end{cases} .$$

From $u \in K$ and the variational inequality one has $\Delta u = 0$ in Ω. Hence, by the maximum principle for harmonic functions, $u(x) > 0$ in Ω. Therefore from the variational inequality

$$\Delta u = 0 \text{ in } \Omega, \quad \frac{\partial u}{\partial n} = 0 \text{ on } \Gamma_2, \quad \frac{\partial u}{\partial n} + \beta u^4 = 0 \text{ on } \Gamma_3 .$$

(ii) The set K_c consists of all functions $v \in H^1$ satisfying

$$a(v,u-v) + \int_{\Gamma_3} g(v)(u-v)\,d\Gamma \geq 0 \qquad \forall u \in K .$$

The inequality is satisfied if

$$\Delta v = 0 \text{ in } \Omega, \quad \frac{\partial v}{\partial n} \geq 0 \text{ on } \Gamma_2, \quad \frac{\partial v}{\partial n} + g(v) \geq 0 \text{ on } \Gamma_3 ,$$

$$\alpha \int_{\Gamma_1} \frac{\partial v}{\partial n}\,d\Gamma \geq a(v,v) + \int_{\Gamma_3} g(v)v\,d\Gamma .$$

(iii) The set $\tilde{K}_c$ consists of all pairs $(v,w) \in H^1 \times H^1$ satisfying the inequality

$$a(v,u-v) + \int_{\Gamma_3} g(w)(u-w)\,d\Gamma \geq 0 \qquad \forall u \in K .$$

The inequality is satisfied if

$$\Delta v = 0 \text{ in } \Omega, \quad \frac{\partial v}{\partial n} \geq 0 \text{ on } \Gamma_2, \quad \frac{\partial v}{\partial n} + g(w) \geq 0 \text{ on } \Gamma_3 ,$$

$$\alpha \int_{\Gamma_1} \frac{\partial v}{\partial n}\,d\Gamma \geq a(v,v) + \int_{\Gamma_3} g(w)w\,d\Gamma .$$

Literature

[1] Courant und Hilbert , Methoden der Mathematischen Physik, vol.I, Springer-Verlag Berlin, Heidelberg, New York 1968 .(3.Auflage)

[2] Fichera,G., Problemi elastostatici con vincoli unilateri: il problema di Signorini con ambigue condizioni al contorno. Mem. Accad.Naz. Lincei, Ser.8, vol.7 (1964),pp. 91 - 140 .

[3] Friedrichs,K., Ein Verfahren der Variationsrechnung, das Minimum eines Integrals als das Maximum eines anderen Ausdruckes darzustellen. Nachrichten Ges.Wiss. Göttingen (1929),pp.13-20.

[4] Lions,J.L.and Stampacchia,G., Variational inequalities. Comm.Pure Appl.Math. vol.20 (1967),pp.493-519.

[5] Michlin,S.G., Variationsmethoden der Mathematischen Physik. Akademie-Verlag Berlin 1962 .

[6] Robinson,P.D., Complementary variational principles. In: Nonlinear functional analysis an applications. Ed. L.B.Rall. Academic Press New York, London 1971 . pp.507-576.

[7] Sewell,M.J.,Dual approximation principles. Phil.Trans. Roy.Soc. (London) vol.265 (1969),pp.319-351.

[8] Trefftz,E., Ein Gegenstück zum Ritzschen Verfahren. Verhandl.d.2.Internationalen Kongreß für Technische Mechanik (1926),pp.131-138.

Stampa: Editoriale Grafica - Roma - Tel. 5890154